Advances in Welding Metal Alloys, Dissimilar Metals and Additively Manufactured Parts

Advances in Welding Metal Alloys, Dissimilar Metals and Additively Manufactured Parts

Special Issue Editor

Giuseppe Casalino

MDPI • Basel • Beijing • Wuhan • Barcelona • Belgrade

MDPI

Special Issue Editor
Giuseppe Casalino
DMMM—politecnico di Bari
Land

Editorial Office
MDPI
St. Alban-Anlage 66
4052 Basel, Switzerland

This is a reprint of articles from the Special Issue published online in the open access journal *Metals* (ISSN 2075-4701) from 2016 to 2017 (available at: https://www.mdpi.com/journal/metals/special_issues/manufactured_parts)

For citation purposes, cite each article independently as indicated on the article page online and as indicated below:

LastName, A.A.; LastName, B.B.; LastName, C.C. Article Title. *Journal Name* **Year**, *Article Number*, Page Range.

ISBN 978-3-03897-372-0 (Pbk)
ISBN 978-3-03897-373-7 (PDF)

Cover image courtesy of Giuseppe Casalino.

Contents

About the Special Issue Editor

Giuseppe Casalino, Master in Mechanical Engineering and PhD in Advanced Manufactured Systems, Full Professor of Manufacturing Technologies and Systems. His research interests include welding, additive manufacturing, the quality evaluation of manufactured parts, and the numerical simulation of manufacturing processes and systems. Since 1998, he has published more than 130 papers in journals, congresses, and book chapters. He has been a visiting researcher and professor at the University of Michigan (2000), Dublin City University (2004–2009), Université de Cergy Pontoise (2007–2009), and Ecole Nationale Arts et Métiers Paris (2013). He has served as a Guest Editor for the journals Metals and Materials. He is also an Editor for the Section of Inventions and Innovation in Advanced Manufacturing of the journal Inventions. From 2015 to 2018, he was the Coordinator of Bachelor's Degrees in Industrial Engineering at Politecnico di Bari. He currently works at Politecnico di Bari, Bari, Italy.

Preface to "Advances in Welding Metal Alloys, Dissimilar Metals and Additively Manufactured Parts"

This book is dedicated to recent developments in welding technologies. The aim of this edition is to make you aware of the main subjects discussed in the scientific community and the procedures employed to address those subjects by specialists and scientists of welding. In 17 chapters, important aspects of recent welding technologies and materials weldability are addressed, including modeling and simulation; the evolution of microstructure and properties in welds; the prediction of residual stress, distortion, fatigue, and corrosion; weld quality and qualification; and destructive and non-destructive control.

Several important metal alloy weldments made of Ni-Base, titanium, HSLA, MART, Q&P and TRIP steels, Inconel, cast iron, and aluminum alloys are studied. The welding technologies include laser, arc, friction, resistance, and hybrid source. All the contributions outline problems and give solutions to achieve further progress in welding. Before I let you into the book, I would like to take this opportunity to thank all the authors for their contributions, and the reviewers for their expert review comments. I would also like to commend the Managing Editor, Natalie Sun, and the entire staff of the Metals Editorial Office on their advice and support during the preparation of this book.

Finally, I hope you enjoy reading "Advances in Welding Metal Alloys, Dissimilar Metals and Additively Manufactured Parts", and I hope you, too, will become a contributor to the development of welding as a leading technology in manufacturing.

Giuseppe Casalino
Special Issue Editor

metals

MDPI

Editorial

Advances in Welding Metal Alloys, Dissimilar Metals and Additively Manufactured Parts

Giuseppe Casalino

Dipartimento di Meccanica, Management, Matematica (DMMM), Politecnico di Bari, viale Japigia, 182, Bari 70126, Italy; giuseppe.casalino@poliba.it; Tel.: +39-080-596-2753; Fax: +39-080-596-2788

Received: 10 January 2017; Accepted: 23 January 2017; Published: 25 January 2017

Nowadays, strong, light-weight, multi-functional, high performing products are key for achieving success in the worldwide markets. Meeting those requirements calls for enabling technologies that lead to innovative and sustainable manufacturing [1].

A joint technique is one or a combination of the available mechanical, chemical, thermal processes to create a bond between materials with a number of combinations and geometries. Welding processes of metal alloys use a thermal energy source that can either melt materials of similar compositions and melting points or produce coalescence at temperatures essentially below the melting point of the base materials being joined. Such well-known welding processes include thermal fusion joining processes and solid-state joining processes; the latter are gaining renewed interest.

Among the thermal fusion joining processes, the most common process is electric arc welding. Several methods use the electric arc approach for fusion welding of steel [2], aluminum [3], titanium [4] and magnesium alloys [5]. The selection of filler material is critical for the quality of the dissimilar metal welds [6].

Laser beam and electron beam welding are high-energy beam welding methods that can operate in either melt-in/conduction or keyhole mode. In the latter mode, the laser beam is highly effective at welding metals. Similar and dissimilar weld can be produced for appliance, automotive, and aerospace applications [7–9].

Brazing and soldering involve a filler material, heated to its melting temperature, and applied between the mating parts, which do not melt. Recently, laser autogenous brazing has enabled the selective use of the unique properties exhibited by biocompatible materials such as stainless steel and shape memory materials, such as NiTi, to tailor the properties of implantable medical devices [10]. Important mid-temperature thermoelectric materials such as Pb Te-based alloys can be successfully brazed to form a thermoelectric module [11].

Resistance spot welding, which dominates the steel body-in-white production, involves a strong current through the metal combination that heats up and finally melts the metals at localized points. A force is applied before, during and after the application of the current to confine the contact area at the weld interfaces and, in some applications, to forge the workpieces. Similar [12] and dissimilar [13] weld can be easily produced.

Within the solid-state joining process, friction stir welding has gained a prominent position. Invented in 1991, it uses a non-consumable tool to join two facing workpieces without melting the workpiece material. Heat is generated by friction between the rotating tool and the workpiece material. This welding method has great capability at welding lightweight [14] and dissimilar weld [15]. Otherwise, friction welding is a process where the two pieces are moved relatively by means of an upsetting force. The relative motion heats the two pieces to a plastic-state. Two friction welding processes are available: linear friction welding and rotary friction welding [16].

Ultrasonic welding uses high frequency ultrasonic vibration for joining materials, such as in lithium-ion battery manufacturing, carbon fiber reinforced polymer–aluminum weld and dissimilar joining of aluminum to copper [17].

In solid-state welding by plastic deformation, a metallurgical bond can be obtained by large plastic deformation. The use of plastic deformation for joining parts potentially offers improved accuracy, reliability and environmental safety as well as creating opportunities to design new products through joining dissimilar materials [18].

Finally, hybrid welding is a joining process that simultaneously combines two welding processes in the same weld pool. The hybrid process has the individual advantages of both welding processes. The laser can be coupled with the arc for the well-established laser–arc hybrid welding of similar and dissimilar metals [19–21]. Moreover, the laser beam can assist the friction stir welding process via pre- and post-welding heating [22,23].

The Present Issue

This Special Issue is dedicated to welding technologies: modeling and simulation; evolution of microstructure and properties in welds; prediction of residual stress, distortion, fatigue, and corrosion; weld quality and qualification; and destructive and non-destructive control. Various important aspects have been addressed in the 17 papers that were published.

Park et al. have evaluated the fracture toughness of several different weldments for 7% nickel alloy steels. The weldment of 7% nickel alloy steel was fabricated by tungsten inert gas (TIG), flux cored arc welding (FCAW), and gas metal arc welding (GMAW) [24].

Cárcel-Carrasco et al. have examined the weldability of ductile cast iron when the root weld is applied with a tungsten inert gas (TIG) welding process employing an Inconel 625 source rod, and when the filler welds are applied with electrodes coated with 97.6% Ni [25].

Fall et al. have investigated the tool wear and its wear rate during friction stir welding (FSW) of Ti-6Al-4V alloy. A conical tungsten carbide tool was used to produce butt-type friction stir welded joints in two-millimeter thick Ti-6Al-4V sheets [26].

Russo Spena et al. have studied the spot weldability of a new advanced Quenching and Partitioning (Q & P) steel and a Transformation Induced Plasticity (TRIP) steel for automotive applications by evaluating the effects of the main welding parameters on the mechanical performance of dissimilar spot welds [27].

Song et al. have characterized the interfacial microstructures of 316L stainless steel (Fe–18Cr–11Ni) and a Kovar (Fe–29Ni–17Co or 4J29) diffusion, bonded via vacuum hot-pressing in a temperature range of 850–950 °C with an interval at 50 °C for 120 min and at 900 °C for 180 and 240 min, under a pressure of 34.66 MPa [28].

Gangwar et al. have presented the microstructural and mechanical properties of the joints for ATI-425 and TIMET-54M friction stir welding. The evolution of microstructure and concomitant mechanical properties were characterized by optical microscopy, microhardness, and tensile properties [29].

Sun et al. have presented a metallurgical and mechanical characterization of 2 mm thick 6061-T6 Al alloy plates [30].

Yuce et al. have achieved the process parameters' optimization procedure of fiber laser welding of dissimilar high strength low alloy (HSLA) and martensitic steel (MART) using a Taguchi approach. The influence of laser power, welding speed and focal position on the mechanical and microstructural properties of the joints was determined [31].

Yi et al. have demonstrated the effectiveness of using tungsten inert gas (TIG) dressing to remove weld pores, and changes in the mechanical properties due to the TIG dressing of Ti-3Al-2.5V weldments [32].

Ahmad et al. have performed multi-pass dissimilar material welding between Alloy 617 and 12Cr steel, performed under optimum welding conditions. The mechanical properties indicated that the yield strength and tensile strength of the dissimilar metal welded joint were higher than those of the Alloy 617 base metal [33].

Chai et al. have realized Hastelloy C-276 thin sheet—0.5 mm in thickness—weld, with filler wire using a pulsed laser. The results indicated that the weld pool geometry and microstructure were significantly affected by the duty ratio, which was determined by the pulse duration and repetition rate under a certain heat input [34].

Chang et al. have studied the effects of laser power on the remanence (Br), experimentally investigated in laser spot welding of a NdFeB magnet (N48H). Results show that the Br decreased with the increase of laser power. For the same welding parameters, the Br of magnets, that were magnetized before welding, were much lower than that of magnets that were magnetized after welding [35].

Oh et al., assuming non-uniform part-to-part gaps, have examined the effects of welding direction on the quality of the joint of galvanized steel sheets SGARC440 (lower part) and SGAFC590DP (upper part), examined using 2-kW fiber and 6.6-kW disk laser welding systems [36].

Kang et al. have studied CaO-added Mg alloy weld. Mechanical and metallurgical aspects of the weldments were analyzed after welding, and welding behaviors such as fume generation and droplet transfer were observed during welding [37].

Chen et al. have examined the effects of reflow time on the interfacial microstructure and shear strength of the SAC/FeNi-Cu connections. It was found that the amount of Cu6Sn5 within the solder did not have a noticeable increase after a long time period of reflowing [38].

Ji et al. have studied dissimilar joints with Ti-6Al-4V and Ti-5Al-2Sn-2Zr-4Mo-4Cr alloys via optical microscopy and scanning electron microscopy (SEM). The welds were obtained by linear friction welding [39].

Dewa et al. have comparatively investigated the low cycle fatigue behavior of Alloy 617 (INCONEL 617) weldments by the gas tungsten arc welding process at room temperature and 800 °C in the air to support the qualification, in high temperature applications, of the Next Generation-IV Nuclear Plant [40].

All the contributions have outlined problems and given solutions in order to achieve further progress in welding. The overall information provides a good foundation for future developments in welding processes and materials in manufacturing industries.

Finally, I would like to take this opportunity to thank all the authors for their contributions to this Special Issue, and the reviewers for their expert review comments. I would also like to thank the managing editor Natalie Sun and the entire staff of the *Metals* Editorial Office for their advice and support during the preparation of this Special Issue.

References

1. Martinsen, K.; Hu, S.J.; Carlson, B.E. Joining of dissimilar materials. *CIRP Ann. Manuf. Technol.* **2015**, *64*, 679–699. [CrossRef]

2. Vashishtha, H.; Taiwade, R.V.; Sharma, S.; Patil, A.P. Effect of welding processes on microstructural and mechanical properties of dissimilar weldments between conventional austenitic and high nitrogen austenitic stainless steels. *J. Manuf. Process.* **2017**, *25*, 49–59. [CrossRef]

3. Bonazzi, E.; Colombini, E.; Panari, D.; Vergnano, A.; Leali, F.; Veronesi, P. Numerical Simulation and Experimental Validation of MIG Welding of T-Joints of Thin Aluminum Plates for Top Class Vehicles. *Metall. Mater. Trans. A Phys. Metall. Mater. Sci.* **2017**, *48*, 379–388. [CrossRef]

4. Yang, M.; Zheng, H.; Qi, B.; Yang, Z. Effect of arc behavior on Ti-6Al-4V welds during high frequency pulsed arc welding. *J. Mater. Process. Technol.* **2017**, *243*, 9–15. [CrossRef]

5. Guo, J.; Zhou, Y.; Liu, C.; Wu, Q.; Chen, X.; Lu, J. Wire arc additive manufacturing of AZ31 magnesium alloy: Grain refinement by adjusting pulse frequency. *Materials* **2016**, *9*. [CrossRef]

6. Liu, K.; Li, Y.; Wang, J. Improving the Interfacial Microstructure Evolution of Ti/Stainless Steel GTA Welding Joint by Employing Cu Filler Metal. *Mater. Manuf. Process.* **2016**, *31*, 2165–2173. [CrossRef]

7. Casalino, G.; Guglielmi, P.; Lorusso, V.D.; Mortello, M.; Peyre, P.; Sorgente, D. Laser offset welding of AZ31B magnesium alloy to 316 stainless steel. *J. Mater. Process. Technol.* **2017**, *242*, 49–59. [CrossRef]

8. Yang, G.; Ma, J.; Wang, H.-P.; Carlson, B.; Kovacevic, R. Studying the effect of lubricant on laser joining of AA 6111 panels with the addition of AA 4047 filler wire. *Mater. Des.* **2017**, *116*, 176–187. [CrossRef]
9. Casalino, G.; Mortello, M.; Peyre, P. Yb-YAG laser offset welding of AA5754 and T40 butt joint. *J. Mater. Process. Technol.* **2015**, *223*, 139–149. [CrossRef]
10. Satoh, G.; Brandal, G.; Naveed, S.; Yao, Y.L. Laser Autogenous Brazing of Biocompatible, Dissimilar Metals in Tubular Geometries. *J. Manuf. Sci. Eng. Trans. ASME* **2017**, *139*, 041016. [CrossRef]
11. Chen, S.-W.; Wang, J.-C.; Chen, L.-C. Interfacial reactions at the joints of PbTe thermoelectric modules using Ag-Ge braze. *Intermetallics* **2017**, *83*, 55–63. [CrossRef]
12. Yuan, X.; Li, C.; Chen, J.; Liang, X.; Pan, X. Resistance spot welding of dissimilar DP600 and DC54D steels. *J. Mater. Process. Technol.* **2017**, *239*, 31–41. [CrossRef]
13. Ling, Z.; Li, Y.; Luo, Z.; Feng, Y.; Wang, Z. Resistance Element Welding of 6061 Aluminum Alloy to Uncoated 22MnMoB Boron Steel. *Mater. Manuf. Process.* **2016**, *31*, 2174–2180. [CrossRef]
14. Ahmed, M.M.Z.; Ataya, S.; El-Sayed Seleman, M.M.; Ammar, H.R.; Ahmed, E. Friction stir welding of similar and dissimilar AA7075 and AA5083. *J. Mater. Process. Technol.* **2017**, *242*, 77–91. [CrossRef]
15. Liu, X.; Chen, G.; Ni, J.; Feng, Z. Computational Fluid Dynamics Modeling on Steady-State Friction Stir Welding of Aluminum Alloy 6061 to TRIP Steel. *J. Manuf. Sci. Eng. Trans. ASME* **2017**, *139*, 5. [CrossRef]
16. Vairis, A.; Papazafeiropoulos, G.; Tsainis, A.-M. A comparison between friction stir welding, linear friction welding and rotary friction welding. *Adv. Manuf.* **2016**, *4*, 296–304. [CrossRef]
17. Ni, Z.L.; Ye, F.X. Dissimilar Joining of Aluminum to Copper Using Ultrasonic Welding. *Mater. Manuf. Process.* **2016**, *31*, 2091–2100. [CrossRef]
18. Mori, K.-I.; Bay, N.; Fratini, L.; Micari, F.; Tekkaya, A.E. Joining by plastic deformation. *CIRP Ann. Manuf. Technol.* **2013**, *62*, 673–694. [CrossRef]
19. Leo, P.; Renna, G.; Casalino, G.; Olabi, A.G. Effect of power distribution on the weld quality during hybrid laser welding of an Al-Mg alloy. *Opt. Laser Technol.* **2015**, *73*, 118–126. [CrossRef]
20. Casalino, G.; Campanelli, S.L.; Ludovico, A.D. Laser-arc hybrid welding of wrought to selective laser molten stainless steel. *Int. J. Adv. Manuf. Technol.* **2013**, *68*, 209–216. [CrossRef]
21. Subashini, L.; Phani Prabhakar, K.V.; Gundakaram, R.C.; Ghosh, S.; Padmanabham, G. Single Pass Laser-Arc Hybrid Welding of Maraging Steel Thick Sections. *Mater. Manuf. Process.* **2016**, *31*, 2186–2198. [CrossRef]
22. Baker, B.; McNelley, T.; Matthews, M.; Rotter, M.; Rubenchik, A.; Wu, S. Use of High-Power Diode Laser Arrays for Pre-and Post-Weld Heating during Friction Stir Welding of Steels. In Proceedings of the TMS Annual Meeting, Orlando, FL, USA, 15–19 March 2015; Volume 2015, pp. 21–36.
23. Campanelli, S.L.; Casalino, G.; Casavola, C.; Moramarco, V. Analysis and comparison of friction stir welding and laser assisted friction stir welding of aluminum alloy. *Materials* **2013**, *6*, 5923–5941. [CrossRef]
24. Park, J.Y.; Lee, J.M.; Kim, M.H. An Investigation of the Mechanical Properties of a Weldment of 7% Nickel Alloy Steels. *Metals* **2016**, *6*, 285. [CrossRef]
25. Cárcel-Carrasco, F.; Pérez-Puig, M.; Pascual-Guillamón, M.; Pascual-Martínez, R. An Analysis of the Weldability of Ductile Cast Iron Using Inconel 625 for the Root Weld and Electrodes Coated in 97.6% Nickel for the Filler Welds. *Metals* **2016**, *6*. [CrossRef]
26. Fall, A.; Fesharaki, M.H.; Khodabandeh, A.R.; Jahazi, M. Tool Wear Characteristics and Effect on Microstructure in Ti-6Al-4V Friction Stir Welded Joints. *Metals* **2016**, *6*. [CrossRef]
27. Spena, P.R.; de Maddis, M.; D'Antonio, G.; Lombardi, F. Weldability and Monitoring of Resistance Spot Welding of Q&P and TRIP Steels. *Metals* **2016**, *6*. [CrossRef]
28. Song, T.; Jiang, X.; Shao, Z.; Mo, D.; Zhu, D.; Zhu, M. The Interfacial Microstructure and Mechanical Properties of Diffusion-Bonded Joints of 316L Stainless Steel and the 4J29 Kovar Alloy Using Nickel as an Interlayer. *Metals* **2016**, *6*. [CrossRef]
29. Gangwar, K.; Ramulu, M.; Cantrell, A.; Sanders, D.G. Microstructure and Mechanical Properties of Friction Stir Welded Dissimilar Titanium Alloys: TIMET-54M and ATI-425. *Metals* **2016**, *6*. [CrossRef]
30. Sun, Y.; Tsuji, N.; Fujii, H. Microstructure and Mechanical Properties of Dissimilar Friction Stir Welding between Ultrafine Grained 1050 and 6061-T6 Aluminum Alloys. *Metals* **2016**, *6*. [CrossRef]
31. Yuce, C.; Tutar, M.; Karpat, F.; Yavuz, N. The Optimization of Process Parameters and Microstructural Characterization of Fiber Laser Welded Dissimilar HSLA and MART Steel Joints. *Metals* **2016**, *6*. [CrossRef]
32. Yi, H.; Lee, Y.; Lee, K. TIG Dressing Effects on Weld Pores and Pore Cracking of Titanium Weldments. *Metals* **2016**, *6*. [CrossRef]

33. Ahmad, H.W.; Hwang, J.H.; Lee, J.H.; Bae, D.H. An Assessment of the Mechanical Properties and Microstructural Analysis of Dissimilar Material Welded Joint between Alloy 617 and 12Cr Steel. *Metals* **2016**, *6*. [CrossRef]

34. Chai, D.; Wu, D.; Ma, G.; Zhou, S.; Jin, Z.; Wu, D. The Effects of Pulse Parameters on Weld Geometry and Microstructure of a Pulsed Laser Welding Ni-Base Alloy Thin Sheet with Filler Wire. *Metals* **2016**, *6*. [CrossRef]

35. Chang, B.; Du, D.; Yi, C.; Xing, B.; Li, Y. Influences of Laser Spot Welding on Magnetic Property of a Sintered NdFeB Magnet. *Metals* **2016**, *6*. [CrossRef]

36. Oh, R.; Kim, D.Y.; Ceglarek, D. The Effects of Laser Welding Direction on Joint Quality for Non-Uniform Part-to-Part Gaps. *Metals* **2016**, *6*. [CrossRef]

37. Kang, M.; Ahn, Y.; Kim, C. Gas Metal Arc Welding Using Novel CaO-Added Mg Alloy Filler Wire. *Metals* **2016**, *6*. [CrossRef]

38. Chen, Y.; Wu, X.; Wang, X.; Huang, H. Effects of Reflow Time on the Interfacial Microstructure and Shear Behavior of the SAC/FeNi-Cu Joint. *Metals* **2016**, *6*. [CrossRef]

39. Ji, Y.; Wu, S.; Zhao, D. Microstructure and Mechanical Properties of Friction Welding Joints with Dissimilar Titanium Alloys. *Metals* **2016**, *6*. [CrossRef]

40. Dewa, R.T.; Kim, S.J.; Kim, W.G.; Kim, E.S. Low Cycle Fatigue Behaviors of Alloy 617 (INCONEL 617) Weldments for High Temperature Applications. *Metals* **2016**, *6*. [CrossRef]

![metals logo] *metals*

MDPI

Article

An Investigation of the Mechanical Properties of a Weldment of 7% Nickel Alloy Steels

Jeong Yeol Park, Jae Myung Lee and Myung Hyun Kim *

Department of Naval Architecture and Ocean Engineering, Pusan National University, Busan 46241, Korea; cpu2565@pusan.ac.kr (J.Y.P.); jaemlee@pusan.ac.kr (J.M.L.)
* Correspondence: kimm@pusan.ac.kr; Tel.: +82-51-510-2486

Academic Editor: Giuseppe Casalino
Received: 31 August 2016; Accepted: 14 November 2016; Published: 19 November 2016

Abstract: During the last decade, the demand for natural gas has steadily increased for the prevention of environmental pollution. For this reason, many liquefied natural gas (LNG) carriers have been manufactured. Since one of the most important issues in the design of LNG carriers is to guarantee structural safety, the use of low-temperature materials is increasing. Among commonly employed low-temperature materials, nickel steel has many benefits such as good strength and outstanding corrosion resistance. Accordingly, nickel steels are one of the most commonly used low-temperature steels for LNG storage tanks. However, the study of fracture toughness with various welding consumables of 7% nickel alloy steel is insufficient for ensuring the structural safety of LNG storage tanks. Therefore, the aim of this study was to evaluate fracture toughness of several different weldments for 7% nickel alloy steels. The weldment of 7% nickel alloy steel was fabricated by tungsten inert gas (TIG), flux cored arc welding (FCAW), and gas metal arc welding (GMAW). In order to assess the material performance of the weldments at low temperature, fracture toughness such as crack tip opening displacement (CTOD) and the absorbed impact energy of weldments were compared with those of 9% nickel steel weldments.

Keywords: liquefied natural gas; nickel steel; weldments; crack tip opening displacement

---

1. Introduction

Recently, the demand for liquefied natural gas (LNG) is constantly increasing for the prevention of environmental pollution. In this trend, various types of LNG carriers have been developed and are operating worldwide. As shown in Figure 1, LNG carriers are divided into two categories—membrane and independent types, according to the classification of International Maritime Organization (IMO). The membrane-type tanks have a very thin primary barrier of Invar alloy or SUS 304L inside the cargo tank. The most common types of membrane-type tank are GTT Mark III and No. 96 types. Mark III consists of two layers of R-PUF (reinforced polyurethane foam) separated by triplex in order to configure an insulation system [1]. On the other hand, the No. 96 type composes a grillage structure made of plywood and filled with perlite in order to maintain tightness and insulation [2]. Independent-type tanks are perfectly self-supporting and do not come under a vessel's hull structure. The most common types of independent-type tank are Moss and SPB types. The Moss-type tank system is formed in such a way that a self-supporting spherical tank of an aluminum alloy is fixed to the hull by a cylindrical support structure (skirt) [3]. The SPB type, which is an IMO independent Type B tank, is designed using an aluminum alloy or 9% nickel steel [4]. One of the most important issues in the design of various types of vessel is the structural integrity of storage tanks under cryogenic temperature. Therefore, LNG storage tanks are typically manufactured using low-temperature materials considering the operation temperature of LNG.

Type	Membrane		Independent	
	GTT MARK III	GTT NO96-2	MOSS	IHI-SPB
Figure	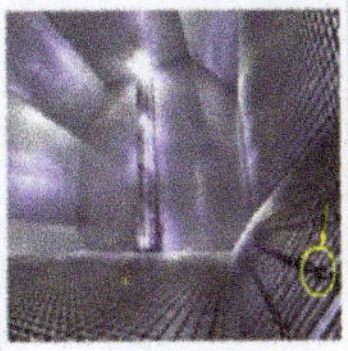		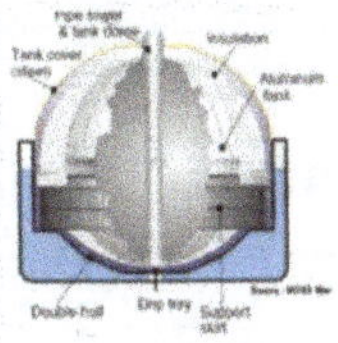	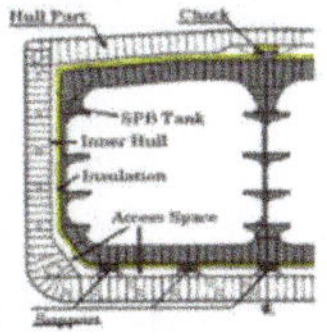
Material	SUS 304L	Invar alloy	Al alloy (Al 5083)	Al alloy or 9% Ni
Thickness	1.2 mm	0.7 ~ 1.5 mm	50 mm	30 mm

Figure 1. Detailed information of liquefied natural gas (LNG) storage tanks [4].

In particular, 9% nickel steel has been the most common material for LNG storage tanks over the last 50 years. Fatigue and fracture characteristic studies of 9% nickel steel have been consistently conducted by many researchers. Saitho et al. studied the mechanical properties of 9% nickel steel considering various welding consumables and processes [5]. Yoon et al. estimated fatigue crack growth characteristics of a 9% nickel steel welded joint [6]. Khourshid et al. investigated the influence of welding parameters on the brittle fracture of 9% nickel steel [7]. Despite the advantages at low temperature, the price of 9% nickel steel fluctuates substantially according to the price of nickel. In particular, the price of welding consumables is also greatly influenced by the price of nickel. Therefore, shipyards have a considerable interest in employing 7% nickel alloy steel instead of 9% nickel steel in order to reduce the cost of the material. However, research on the weldment of 7% nickel alloy steel is not sufficient compared with the weldment of 9% nickel steel.

In this regard, the major objective of this study was to evaluate the mechanical properties of 7% nickel alloy steel weldments considering several different welding consumables. Based on the results of tensile, Charpy-V impact, and crack tip opening displacement (CTOD) tests performed in this study, the most suitable weldment for 7% nickel alloy steel was determined. In addition, fracture performances of a weldment for 7% nickel alloy steel were compared with those of 9% nickel steel for applications at cryogenic temperature, e.g., $-163\,°C$.

2. Materials and Methods

Nickel steel has many benefits, such as good strength, outstanding corrosion resistance, and applicability in a wide range of temperatures. As shown in Table 1, the International Gas Code (IGC) provides the proper amount of nickel according to low-temperature applications [8].

Table 1. Various temperature application of nickel steels [8].

Minimum Design Temperature (°C)	Chemical Composition and Heat Treatment	Application
−60 −65	1.5% nickel steel—normalizded 2.25% nickel steel—normalized or normalized and tempered	Liquefied Propane Gas
−90 −105	3.5% nickel steel—normalized or normalized and tempered 5% nickel steel—normalized or normalized and tempered	Liquefied Ethane Gas
−165	9% nickel steel—double nomalized and tempered or quenched and temperd	Liquefied Natural Gas

In this study, 7% nickel alloy steel was considered for replacing the conventional 9% nickel steel for LNG storage tanks. The chemical composition of 7% nickel alloy steel, considered in this study, is summarized in Table 2. The 7% nickel alloy steel in this study was treated with a thermo-mechanical control process (TMCP), as shown in Figure 2. The TMCP applied to direct quenching, lamerallizing, and tempering (DQ-L-T), in which lamerallizing was inserted between direct quenching and tempering, was utilized to form a more stable retained austenite, making it finely dispersed and increasing the content [9].

Table 2. The chemical composition of 7% nickel alloy steel.

Material	C	Si	Mn	P	S	Cr	Ni	Mo
7% nickel alloy steel	0.04	0.06	0.78	0.002	0.004	0.46	7.13	0.09

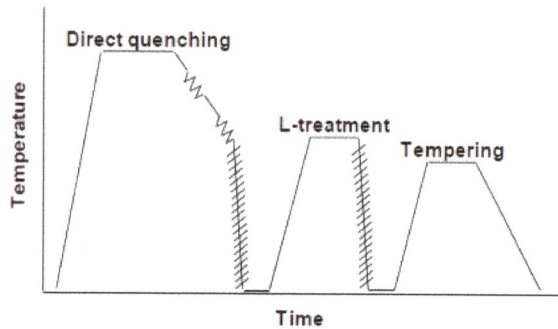

Figure 2. Schematic illustration of the manufacturing processes [9].

In Figure 3, the mixture of tempered-martensite, bainite, and retained austenite was observed in the microstructure of 7% nickel alloy steel. The weldments of 7% nickel alloy steel were produced by using three different welding procedures—flux cored arc welding (FCAW), gas metal arc welding (GMAW), and tungsten inert gas (TIG). The welding consumable employed for FCAW and GMAW was ERNiCrMo-3 in the AWS 5.14 code. On the other hand, a high manganese welding consumable was used for the TIG welding. The chemical compositions of the welding consumables are summarized in Table 3.

Figure 3. Microstructure of 7% nickel alloy steel.

Table 3. The chemical composition of welding consumables [13].

Welding Consumable	C	Si	S	Mn	Ni	Cr	Mo	Co	Nb	N
High manganese	0.01	-	0.41	7.25	15.77	20.59	2.9	0.13	0.06	0.16
ERNiCrMo-3	0.018	0.47	0.001	0.36	62.91	22.51	9.01	-	3.65	-

The tensile test was conducted in accordance with ASTM E8 [10]. In addition, the fracture toughness tests, i.e., the CTOD and Charpy-V impact tests, were carried out according to BS 7448 Part 2 and ASTM E23, respectively [11,12]. As shown in Figure 4, compact tension specimens 12 mm in thickness, 48 mm in width, and 57.6 mm in height were used for fracture toughness tests. CTOD test specimens for the heat affected zone (HAZ) were fabricated 1 mm from the weld fusion line (F.L.).

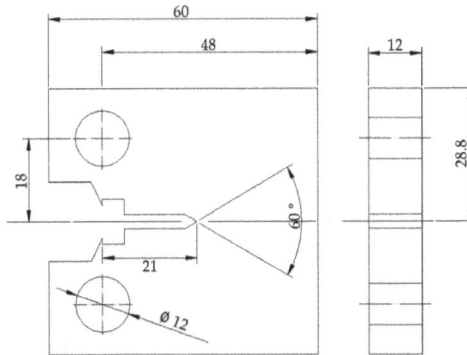

Figure 4. Compact tension specimen of crack tip opening displacement (CTOD) test.

Before the CTOD test, a fatigue pre-crack of 3 mm was inserted into the notch tip of each specimen. The fatigue pre-crack can simulate a natural crack well enough to provide a satisfactory CTOD test result [11]. The conditions of the fatigue pre-crack were a sinusoidal wave, a stress ratio of 0.1, and a frequency at 10 Hz. The CTOD test was conducted at room (25 °C), low (−100 °C), and cryogenic (−163 °C) temperatures to assess the applicability of each material for LNG storage tanks.

The test equipment used for the CTOD test was a servo hydraulic testing machine (Instron 8800, INSTRON, High Wycombe, UK) with a maximum load capacity of ±500 kN. In addition, low and cryogenic temperatures were controlled by the cryogenic chamber (ILWON FREEZER, Namyangju-si, Korea). Test temperatures were maintained by a liquid nitrogen gas inlet-outlet control system. The total equipment for the CTOD test are shown in Figure 5.

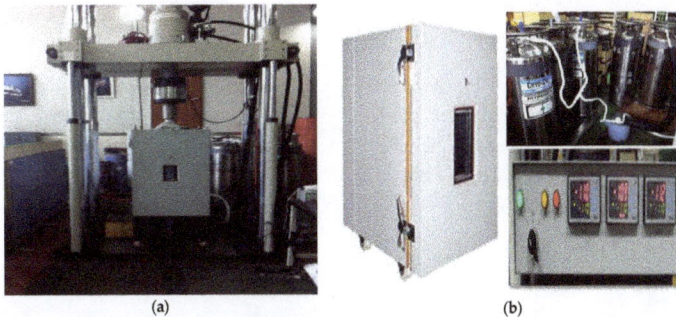

Figure 5. Test equipment: (**a**) servo hydraulic testing machine; (**b**) temperature control system.

3. Test Results

3.1. Mechanical Properties

Table 4 summarizes the tensile test results of the three weldments. The weldments of ERNiCrMo-3 with FCAW and GMAW exhibit a similar tendency in yield and tensile strengths as expected.

Table 4. The mechanical properties of the weldments [13–15].

Classification	YS (Mpa)	TS (Mpa)	E.L. (%)
High manganese & TIG	473	614	36
ERNiCrMo-3 & FCAW	509	781	68.7
ERNiCrMo-3 & GMAW	506	753	75.0
BV requirement [14]	480	670	22
DNV requirement [15]	490	640	25

On the other hand, the weldment of high manganese with TIG has the lowest mechanical properties. In addition, it does not satisfy the requirements of mechanical properties from Bureau Veritas (BV) and Det Norske Veritas (DNV) [14,15]. It has been well known since the 1980s that high manganese welding consumables have several problems, such as weldability and strength [16]. Therefore, the weldment of high manganese with TIG was determined to be unsuitable for 7% nickel alloy steel. Accordingly, ERNiCrMo-3 with FCAW and GMAW are considered in the latter part of this study.

3.2. Charpy-V Impact Test

Table 5 shows the Charpy-V impact test results of the three weldments. The Charpy-V impact tests were carried out at −196 °C. As a result, all weldments satisfied the requirements of BV and DNV [14,15]. However, the weldments of ERNiCrMo-3 and FCAW were about 55% and 58% lower than that of the high manganese with TIG and ERNiCrMo-3 with GMAW, respectively. Therefore, ERNiCrMo-3 and GMAW were more applicable to the 7% nickel alloy steel weldment than were ERNiCrMo-3 and FCAW.

Table 5. Charpy-V impact test results of the weldments [13–15].

Classification	Test Temp. (°C)	Charpy-V Impact Test (J)			
		1	2	3	Average
High manganese & TIG		91	105	89	95
ERNiCrMo-3 & FCAW		47	39	41	42
ERNiCrMo-3 & GMAW	−196	104	100	96	100
BV requirement [14]		-	-	-	34
DNV requirement [15]		-	-	-	

Figure 6 presents the Charpy-V impact test results for various locations of the HAZ. In the case of FCAW, F.L. exhibited the lowest impact absorbed energy, while GMAW had the lowest value at F.L. + 1 mm. This result is attributed to the difference in heat input from each welding process. For all locations, absorbed energies of GMAW were about 12% (the maximum) higher than those of FCAW except F.L. + 1 mm. Therefore, it appears that GMAW is slightly better than FCAW in terms of impact absorbed energy.

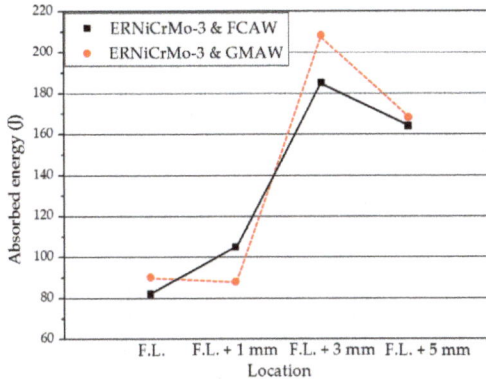

Figure 6. Charpy-V impact test results for various locations of the HAZ [13].

3.3. The CTOD Test

Based on the above results, the CTOD tests were performed on the weldment of ERNiCrMo-3 and GMAW as an optimal candidate for 7% nickel alloy steel. Figure 6 shows the CTOD values of weld metal. As shown in Figure 7, the CTOD values at cryogenic temperature satisfies the DNV requirement. It is estimated that the fracture resistance of the weldments of 7% nickel alloy steel decrease with decreasing temperature from room to cryogenic temperatures. Figure 8 shows the CTOD values of GMAW being compared with other welding processes such as shielded metal arc welding (SMAW) and gas tungsten arc welding (GTAW) [17]. In addition, chemical compositions of other welding consumables are summarized in Table 6. As the nickel content of the welding consumable increases, CTOD values also tend to increase. In addition, welding consumables of the 70% Ni type with GTAW has the highest CTOD value at cryogenic temperatures. As is well known, higher heat input results in a greater possibility of brittle fracture [18]. Therefore, it appears that SMAW has higher heat input than GTAW.

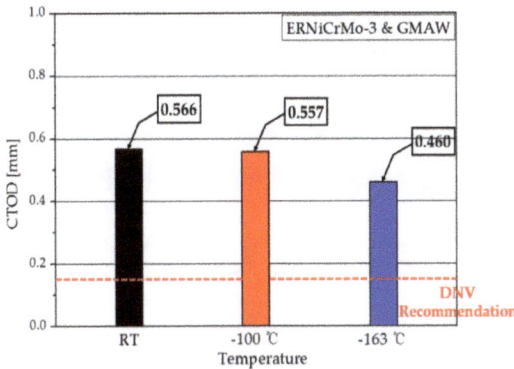

Figure 7. CTOD values of the weldment at various temperatures.

Table 6. The chemical composition of other welding consumables [13,17].

Welding Consumable	C	Si	S	Mn	Ni	Cr	Mo
ERNiCrMo-3	0.018	0.47	0.001	0.36	62.91	22.51	9.01
70% Ni type	-	-	-	-	70	-	-

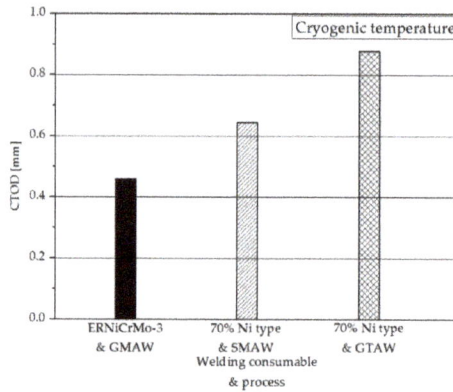

Figure 8. Comparison of the CTOD values of the weldments for other welding processes [17].

Figure 9 shows CTOD values at the HAZ. The CTOD values at the HAZ are more sensitive to the temperature decrease than those of the weldment. At a cryogenic temperature, the CTOD value at the HAZ satisfies the requirement of DNV. As shown in Figure 10, the CTOD values at the HAZ by SMAW and GTAW has a similar tendency. On the other hand, the CTOD value at the HAZ by GMAW is about 50% lower than those by GTAW [17].

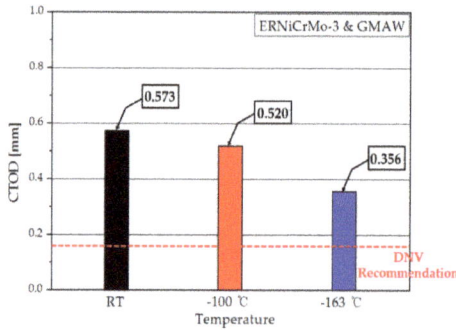

Figure 9. The CTOD values of the HAZ from room to cryogenic temperatures.

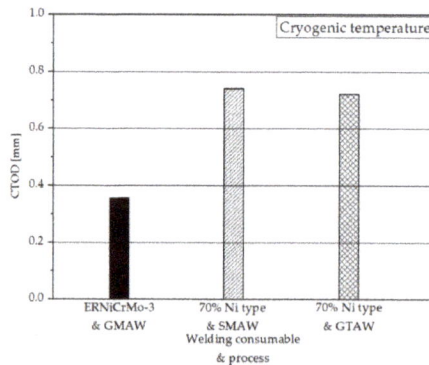

Figure 10. Comparison of the CTOD values of the HAZ for other welding process [17].

4. Discussion

The CTOD values of the weldment of 7% nickel alloy steel were compared with those for 9% nickel steel considering welding processes, as indicated in Figure 11 [5].

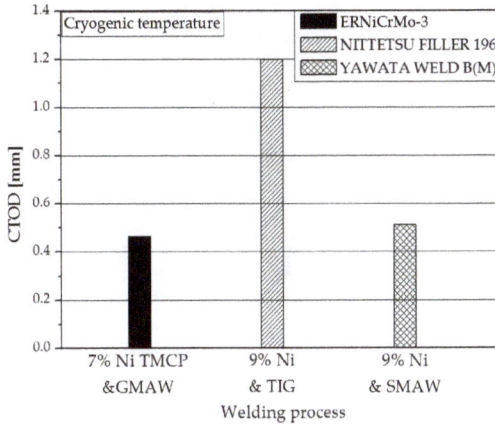

Figure 11. CTOD values of the weldment for 9% nickel steel and 7% TMCP nickel steel [5].

The chemical compositions of welding consumables for 9% nickel and 7% nickel alloy steel are summarized in Table 7 [5]. The welding consumable with TIG welding has the highest nickel content among welding consumables. It is known that a high nickel content has been used extensively to improve fracture toughness [19]. Therefore, the weldment of TIG has the highest CTOD value at cryogenic temperature. In addition, the weldment of 7% nickel alloy steel exhibits similar CTOD values compared to that of 9% nickel steel with SMAW. In this regard, the welding consumable affects the fracture toughness more than the welding process. Therefore, the weld metal of 7% nickel alloy steel is considered to replace 9% nickel steel in terms of fracture toughness.

Table 7. Chemical composition of welding consumables for 9% and 7% TMCP nickel steels [5].

Welding Consumable	C	Si	Mn	Ni	Cr	Mo	W	Nb	Fe
NITTETSU FILLER 196 (9% Ni & TIG)	0.04	0.40	0.45	72.5	-	19.0	2.91	-	3.5
YAWATA WELD B(M) (9% Ni & SMAW)	0.09	0.20	3.22	65.1	15.8	3.35	-	1.60	10.2
ERNiCrMo-3 (7% TMCP Ni & GMAW)	0.018	0.47	0.36	62.91	22.51	9.01	-	3.65	-

5. Conclusions

In this study, the mechanical characteristics of 7% nickel alloy steel weldments were evaluated based on tensile, Charpy-V impact, and CTOD test results. In addition, the fracture performances of 7% nickel alloy steel weldments were compared with those of 9% nickel steel for LNG storage tank applications. The conclusions from this study are summarized as follows:

- In the 7% nickel alloy steel, the weldment of ERNiCrMo-3 with FCAW had the highest yield and tensile strengths among other weldments. The mechanical properties of high manganese weldment with TIG did not satisfy the minimum requirements of BV and DNV. Therefore, the weldment of high manganese with TIG was determined to be unsuitable for 7% nickel alloy steel in terms of mechanical properties.

- The weldment of ERNiCrMo-3 with GMAW is about 2.3 and 1.05 times higher than that of ERNiCrMo-3 with FCAW and high manganese with TIG in terms of the absorbed energy at $-196\ °C$, respectively. Based on the tensile and Charpy-V impact test results, the weldment of ERNiCrMo-3 with GMAW is the most appropriate for 7% nickel alloy steel.
- The weldment of ERNiCrMo-3 with GMAW exhibited the lowest CTOD values compared with other conventional weldments of 7% nickel alloy steel. It is estimated that the CTOD value of the weldment was affected by the nickel content of the welding consumables.
- Compared with the CTOD value of 9% nickel steel weldment, 7% nickel alloy steel weldment exhibits comparable CTOD values except the TIG welding process. Therefore, the weld metal of 7% nickel alloy steel is considered to be a viable alternative to 9% nickel steel from a cost-effective perspective.

Acknowledgments: This work was supported by the National Research Foundation (NRF) of Korea, grant funded by the Ministry of Education, Science, and Technology (MEST) of the Korean Government through GCRC-SOP (No. 2011-0030013). This work was also supported by Human Resource Training Program for Regional Innovation and Creativity through the Ministry of Education and NRF of Korea (NRF-2014H1C1A1073088).

Author Contributions: Myung Hyun Kim formulated this research with cooperation from Jae Myung Lee. Jeong Yeol Park performed the experimental work and, with the help of Myung Hyun Kim, interpreted the results and prepared the manuscript. All co-authors contributed to the manuscript proof and submissions.

Conflicts of Interest: The authors declare no conflict of interest.

References

1. Chun, M.S.; Kim, M.H.; Kim, W.S.; Kim, S.H.; Lee, J.M. Experimental investigation on the impact behavior of membrane-type LNG carrier insulation system. *J. Loss Prev. Process Ind.* **2009**, *22*, 901–907. [CrossRef]
2. Kim, M.H.; Lee, S.M.; Lee, J.M.; Noh, B.J.; Kim, W.S. Fatigue strength assessment of MARK-III LNG cargo containment system. *Ocean Eng.* **2010**, *37*, 1243–1252. [CrossRef]
3. Ishimatu, J.; Kawabata, K.; Morita, H.; Ikkai, H.; Suetake, Y. *Building of Advanced Large Sized Membrane Type LNG Carrier*; Mitsubishi Heavy Industries, Ltd.: Tokyo, Japan, 2004.
4. Oh, D.J.; Lee, J.M.; Noh, B.J.; Kim, W.S.; Ando, R.; Matsumoto, T.; Kim, M.H. Investigation of fatigue performance of low temperature alloys for liquefied natural gas storage tanks. *J. Mech. Eng. Sci.* **2015**, *229*, 1300–1314. [CrossRef]
5. Saitoh, N.; Yamaba, R.; Muraoka, H.; Saeki, O. *Development of Heavy 9% Nickel Steel Plates with Superior Low-Temperature Toughness for LNG Storage Tanks*; Nippon Steel: Tokyo, Japan, 1993.
6. Yoon, Y.K.; Kim, J.H.; Shim, K.T. Mechanical characteristics of 9% Ni steel welded joint for LNG storage tank at cryogenic. *Int. J. Mod. Phys. Conf. Ser.* **2012**, *6*, 355–360. [CrossRef]
7. Khourshid, A.E.F.M.; Ghanem, M.A. The influence of welding parameters on brittle fracture of liquefied natural gas storage tank welded joint. *Mater. Sci. Appl.* **2013**, *4*, 198–204. [CrossRef]
8. IA104E. International code for the construction and equipment of ships carrying liquefied gas in bulk. In *ICG Code*; IGC: London, UK, 2014; p. 99.
9. Furuya, H.; Kawabata, T.; Takahashi, Y.; Kamo, T.; Inoue, T.; Okushima, M.; Ando, R.; Onishi, K. Development of Low-Nickel Steel for LNG Storage Tanks. In Proceedings of the 2013 International Conference & Exhibition on Liquefied Natural Gas, Houston, TX, USA, 16–19 April 2013.
10. ASTM E8. Standard test methods for tension testing of metallic materials. In *Annual Book of ASTM Standards*; ASTM International: Baltimore, MD, USA, 2009.
11. BS 7448-Part 2. Fracture mechanics toughness test. Method for determination of KIc, critical CTOD and critical J values of welds in metallic materials. In *British Standard*; British Standard Institution: London, UK, 1997.
12. ASTM International. Standard test methods for notched bar impact testing of metallic materials. In *Annual Book of ASTM Standards*; ASTM International: Baltimore, MD, USA, 2007.
13. Kim, M.H. *JDP for Assessment of 7% Nickel Steel for Type-B LNG Tanks*; Hyundai Heavy Industries: Busan, Korea, 2013.

14. NR 216 DT R07 E. Rules on materials and welding for the classification of marine units. In *Annual Book of BV Rule*; BV: Courbevoie, France, 2014; p. 211.

15. Rule for Ships Pt.2 Ch.3. Materials and welding. In *Annual Book of DNV Rule*; DNV: Oslo, Norway, 1996; p. 37.

16. Kim, K.S.; Park, C.Y.; Kang, J.K. Availability evaluation of high Mn steel by comparison with current materials available in cryogenic environment. In Proceedings of the 2013 International Offshore and Polar Engineering, Anchorage, AK, USA, 30 June–5 July 2013.

17. Nishigami, H. Development of 7% Ni-TMCP steel plate for LNG storage tanks. In Proceedings of the 2011 International Gas and Union Research Conference, Seoul, Korea, 19–21 October 2011.

18. Shin, Y.T.; Kang, S.W.; Kim, M.H. Evaluation of fracture toughness and microstructure on FCA weldment according to heat input. *J. Weld. Join.* **2008**, *26*, 51–60. [CrossRef]

19. Scheid, A.; Felix, L.M.; Martinazzi, D.; Renck, T.; Kwietniewski, C.E.F. The microstructure effect on the fracture toughness of ferritic Ni-alloyed steels. *Mater. Sci. Eng. A* **2016**, *661*, 96–104. [CrossRef]

metals

Article

An Analysis of the Weldability of Ductile Cast Iron Using Inconel 625 for the Root Weld and Electrodes Coated in 97.6% Nickel for the Filler Welds

Francisco-Javier Cárcel-Carrasco *, Miguel-Angel Pérez-Puig, Manuel Pascual-Guillamón and Rafael Pascual-Martínez

ITM, Universitat Politècnica de València, Valencia 46022, Spain; mipepui@mcm.upv.es (M.-A.P.-P.); mpascual@mcm.upv.es (M.P.-G.); rapasmar@hotmail.com (R.P.-M.)
* Correspondence: fracarc1@csa.upv.es; Tel.: +34-963-87-7000; Fax: +34-963-87-9459

Academic Editor: Giuseppe Casalino
Received: 26 July 2016; Accepted: 14 November 2016; Published: 18 November 2016

Abstract: This article examines the weldability of ductile cast iron when the root weld is applied with a tungsten inert gas (TIG) welding process employing an Inconel 625 source rod, and when the filler welds are applied with electrodes coated with 97.6% Ni. The welds were performed on ductile cast iron specimen test plates sized 300 mm × 90 mm × 10 mm with edges tapered at angles of 60°. The plates were subjected to two heat treatments. This article analyzes the influence on weldability of the various types of electrodes and the effect of preheat treatments. Finally, a microstructure analysis is made of the material next to the weld in the metal-weld interface and in the weld itself. The microstructure produced is correlated with the strength of the welds. We treat an alloy with 97.6% Ni, which prevents the formation of carbides. With a heat treatment at 900 °C and 97.6% Ni, there is a dissolution of all carbides, forming nodules in ferritic matrix graphite.

Keywords: weldability; pre-heating; ductile cast iron; microstructure; root pass

1. Introduction

Castings such as steel are basically alloys of iron, carbon, and silicon that favor the formation of graphite, and carbon content may vary between 2% and 6.67%. Carbon percentages of up to about 4% are often used commercially because a large percentage of carbon may affect brittleness. Alloys obtained in casting processes are generally neither ductile nor malleable. The EN 1563 normative defines ductile cast iron as a "casted material based on iron, carbon and silicon, and in which carbon is primarily in the form of spheroidal particles".

Classification depends on the metallographic structure and the percentages of carbon, alloy elements, and impurities. The poor mechanical properties of some castings are due to the presence of graphite flakes [1] that produce discontinuities in the matrix and so lead to the presence of stress concentrators.

These alloys have a high carbon content that makes them difficult to weld because of the formation of martensite and brittle iron carbides during the cooling of the joining process [1,2]. Due to this, welding on cast iron generally involves repair operations and not joining casting between them, although this last operation can be performed following some precautions [3]. These precautions are mainly oriented to the selection of the filler metal [3–5] and the reduction of the cooling rate of the weld in order to reduce martensitic transformations and carbide precipitations, which would with ease lead to the cracking of the joint.

Adding pure magnesium, magnesium ferrosilicon, or magnesium-nickel to an alloy encourages the formation of graphite nodules that improve the characteristics of the casting by combining the mechanical strength, toughness, and ductility of steel with themoldability of gray iron.

The process of welding the filler partially determines its mechanical properties. The result is also influenced by preheating treatments and cooling rates. A subsequent normalization process also improves the properties.

This study examined the weldability of ductile cast iron (in two circumstances: before and after an annealing post-treatment to dissolve any hard precipitates and restore the microstructure of the heat affected zone) when the root weld is applied with a tungsten inert gas (TIG) welding process employing an Inconel 625 source rod, and when the filler welds are applied with electrodes coated with 97.6% Ni.

2. Materials and Methods

2.1. Composition and Mechanical Characteristics of the Castings Used

Graphite appears as nodules in ductile cast iron because of the presence of small amounts of magnesium [1] retained by the iron and distributed evenly throughout the matrix. Magnesium eliminates the discontinuity caused by the graphite grain in gray cast iron and produces significant improvements in the mechanical characteristics in comparison with gray iron. The chemical composition of ductile cast iron expressed in percentagescan be seen in Table 1. The mechanical properties of the castings considered in this study are shown in Table 2.

Table 1. Alloying elements of ductile cast iron.

Cast	%C	%Si	%Mn	%S	%P	%Ni	%Cu	%Cr	%Mo	%Mg
Ductile	3.71	2.54	0.04	<0.01	0.02	0.02	0.026	0.03	<0.01	0.03

Table 2. Mechanical properties of ductile cast iron.

Mechanical Characteristics	Values
Tensile strength (MPa)	370
Yield strength (MPa)	320
Elongation (%)	6
Elastic modulus (MPa)	160,000
Brinell hardness	190
Fatigue limit (MPa)	280

EN 1563 specifies the characteristics for ductile iron, while EN 1561 specifies the characteristics for gray cast iron.The process of welding the specimens of ductile iron was made using the following techniques:tungsten inert gas (TIG) welding and manual arc welding with coated electrodes (SMAW).

TIG welding is one of the most common techniques for welding elements whose thickness is less than 8 mm. The process is recommended for root welding pieces of even greater thickness to ensure sufficient penetration in the bonding processes, and for welding ductile iron castings that require prior preparation. The welds were made on ductile cast iron plates described in Table 2 using ER Inconel 625 source rods after preheating the test specimens. Various largely irrelevant distortions may have been present in the test plates.

Arc welding is one of the techniques most used for welding ductile cast iron, and filler material was added with two SMAW passes over the TIG root weld [6]. The test plates may have revealed largely irrelevant distortions after being preheated.The filler weld was made using 97.6% Ni electrodes in order to minimize carbide formation as well as improving the service conditions.The weld was carried out on a series of test plates that were pre-heated to 350 °C. Once the welds were finished, some of the plates were air-cooled, while others were immediately annealed at 900 °C in order to observe the different metallurgical behavior produced by these treatments [7].

The specimen test plates were produced in a sand casting process and originally measured 300 mm × 95 mm × 11 mm. They were subsequently milled to 300 mm × 95 mm × 10 mm (coupon). The edges were prepared for electric welding with a covered electrode by cutting a 30° chamfer so that with proper spacing a perfect union could be achieved along the entire thickness of both plates. This preparation was made using an adjustable band saw [6].

The electrodes used for the welding were selected to obtain optimum results [8] in terms of the mechanical properties in the process of joining the castings. Compositions are shown below in Table 3.

Table 3. Electrodes used for ductile iron welds.

Electrode	Composition	
	Element	%
	Si	0.5
	Mn	0.5
	Ni	58
Inconel 625	Cr	20–25
	Nb	3.5
	Ta	4.5
	Mo	8–10
	Fe	Other
	C	<0.1
	Si	<0.4
Ni 97.6%	Mn	0.20
	Ni	97.61
	Fe	Other

2.2. Welding Processes

TIG root welding using Inconel 625 source material (1.5 mm in diameter) was carried out using a continuous welding current of between 120 and 130 A, straight polarity, and an argon flow of 12 L/min. The welding seam was performed in a single circular pass in a horizontal right to left motion and an inclination angle of the tungsten electrode of between 70° and 80° when moving forward, and an angle of about 20° above horizontal for the source rod. The plate was preheated to 350 °C before welding.Shielded metal arc welding (SMAW) was performed with direct current and reverse polarity (due to the basic character of the coating) and using 140 A and a 3.2 mm electrode diameter that was previously heated at 90° for 24 h at 100 °C (to improve fluidity and hydrogen inclusion and thereby prevent cracking). The weld was carried out with two filler passes from left to right horizontally with an electrode angle of inclination of approximately 60°.

Because of the difficulty in welding the castings and to avoid fractures due to the stresses generated during cooling, we preheated the castings to 350 °C before welding, lightened the bead in sections separated by more than 40 mm, and hammered the plates vigorously. Various problems arose due to the breakdown of the electrodes, excessive intensity as the weld progressed, difficulties in removing the coating, and a lack of fluidity.

To solve these problems, a suitable preparation was made that left a separation adequate to prevent sticker breaks [5] and ensure good penetration. It was also decided to change the electrodes when one-half was consumed and thus avoid deterioration by decomposition (thereby preventing the formation of porosities or internal inclusions). The electrodes were reused once cold.

Due to the heating of the material, a decreasing amount of intensity was required; however, if intensity was reduced too much, then sticker breaks may have formed. Changing the electrode before it was consumed provided sufficient cooling time for the material without having to reduce intensity. Moreover, it was possible to vary the speed of the welding movement to prevent filler source burning.

The test plates were initially root welded using Inconel and then welded with coated electrodes for the filler. The plates (once adequately prepared) were welded after a preheat at 350 °C in order to reduce the tensions, slow the cooling, increase the fluidity of the welds (preventing pores forming and cracks in the bead), and thus facilitate a reduction in intensity to about 120–130 A. Some of the test plates were annealed at 900 °C after welding and cooled in the furnace.

The corrosive liquid used to make the microstructure of ferrite, white iron, and martensite clearer and to improve the quality of photographs was called "Nital 3".

3. Results

We measured the Vickers micro hardness (HV), yield strength, tensile strength, and elongation as the average of tests using five different sizes.

The HV micro hardness was measured with 300 g loads for 10 s with a diamond point at 136°. Measurements were taken in the zone adjacent to the weld, in the metal-weld interface, and the weld zone (a total of nine measurements) in accordance with UNE-EN 876. The mechanical properties are shown in Table 4 (the micro hardness was made in areas close to Figures 3 and 6). The tensile tests to determine the mechanical properties were carried out in accordance with UNE-EN 10002-1for tensile tests at room temperature with a universal testing machine and a maximum force of 10 t. The results are shown in Table 4.

Table 4. Mechanical properties of the ductile iron welds.

Welding	Rod	S (MPa)	YS (MPa)	A%	Hv Interface	Hv Welding	Hv HAZ
Pre-heating 350 °C and air cooling	Inconel 625 root Ni 97.6% filler	320	310	5.5	730 490	510 191	295
Pre-heating 350 °C and annealed at 900 °C	Inconel 625 root Ni 97.6% filler	300	285	7	610 230	400 160	210

Figures 1–8 show the microstructures produced by the different types of welds. Figure 9 shows the different types of fractures that occurred during standard testing [9,10].

4. Analysis of Results and Discussion

Micrographs of TIG root welds (made with Inconel 625) of the ductile iron test plate following a preheat treatment at 350 °C (but without subsequent annealing treatment) (Figure 1) reveal in the interface structures formed by traces of white iron in which most of the graphite has combined to form cementite. Very small and uniformly distributed spherulites can also be seen. The rate of cooling also produced martensite precipitates, a partial consequence of the effect of Ni on the dissolution of the carbides being counterbalanced by the effect of Cr as a stabilizer. The Vickers hardness was 460, mostly in areas where the carbides precipitated. The consequence was the beginnings and expansion of a brittle fracture in the zone of the interface. The structure can be defined as a martensite white iron matrix.

The micrograph shows the manual weld of ductile cast iron using an electrode coated with 97.6% Ni (Figure 2) made on top of an Inconel root weld. The image reveals a structure in which traces of white iron have significantly diminished, resulting in a higher concentration of nodules of uniformly distributed graphite that are larger than those shown in the previous micrograph. This is largely due to the Ni content of the coated filler electrode, largely preventing the formation of carbides in the dilution zone; however, it is also possible to see martensite precipitate in the austenitic matrix (in which the Vickers hardness is 700 as a result of the rapid cooling of the weld). The fracture has extended in the interface zone, which has become brittle because of the cooling rate.

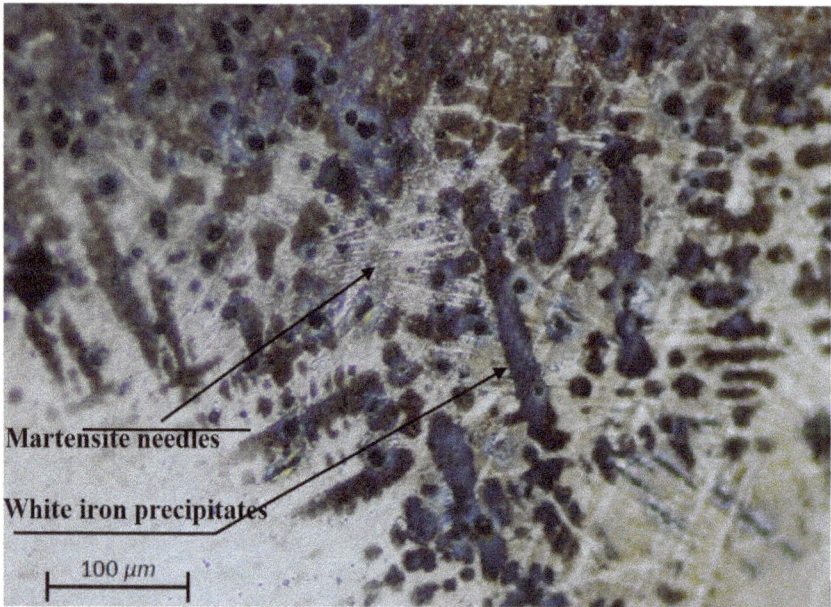

Figure 1. Micrograph of the ductile iron tungsten inert gas (TIG) root weld using Inconel 625 and without subsequent annealing.

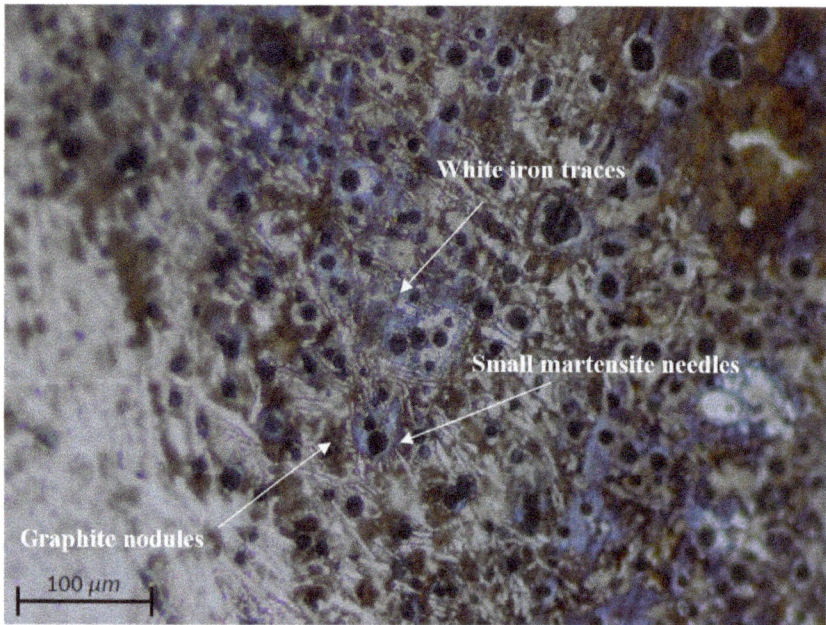

Figure 2. Micrograph of ductile iron weld using 97.6% Ni coated electrode on an Inconel root weld without heat treatment.

In a micrograph of a filler weld made with a 97.6% Ni electrode with a maximum dilution of the base material and preheating to 350 °C followed by air cooling (Figure 3), three zones can be distinguished: weld, base material, and interface [11,12].

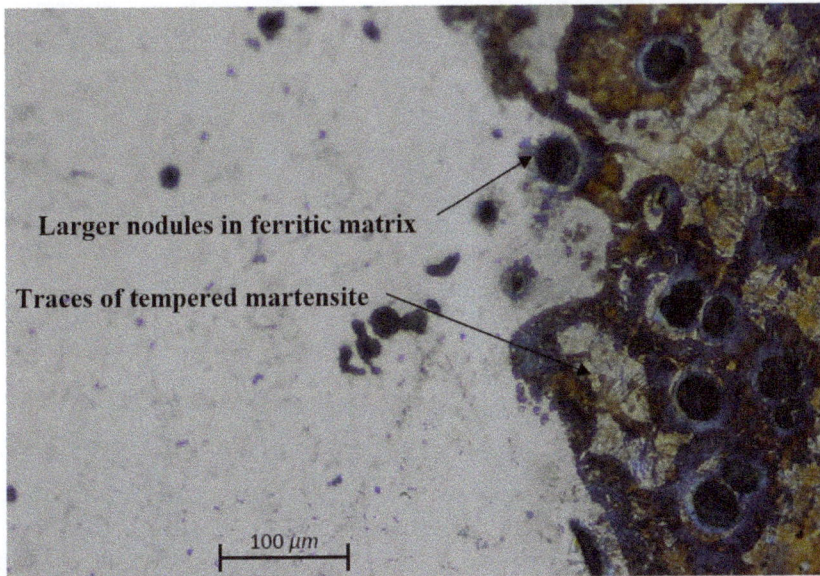

Figure 3. Micrograph of a weld performed with a 97.6% Ni electrode on a ductile iron casting with heat treatment at 350 °C and air cooling.

Ferritic-pearlitic ductile iron can be seen in the vicinity of the base material near the weld. This material does not provide high levels of hardness (360 HV) and becomes more similar to the base material in hardness as we move away from the area affected by heat. At the interface of the weld, we can see martensite precipitates tempered in a ferrite matrix (as a result of the Ni content in the filler metal) with slight traces of cementite and graphite nodules (smaller than those in the casting) that are uniformly distributed over the entire zone. The hardness produced at this stage is greater than the base metal. In the region of the weld, the graphite appears as nodules that are smaller than those in the interface and evenly distributed across the weld. The hardness is lesser than that obtained in the interface, and fractures occur in the weld zone near the brittle interface.

In the micrograph of the base material (Figure 4), structural formations of material near the weld and at the boundary of the heat affected zone can be seen. These ferriticpearlitic structures contain uniform nodules that are larger than those visible in the micrographs of the interface between the filler material and the ductile iron. These structures are uniformly distributed and similar to the structures of the material from which the test plates are made albeit generally harder.

The micrograph (Figure 5) shows the formation of nodules in the weld made with filler material applied with a 97.6% Ni electrode where a uniform distribution of ferritic material and traces of retained austenite can be found with sizes smaller than those found on the base material. A distribution similar to that shown above can be found in the structure on the Inconel 625 filler material used in the root weld.

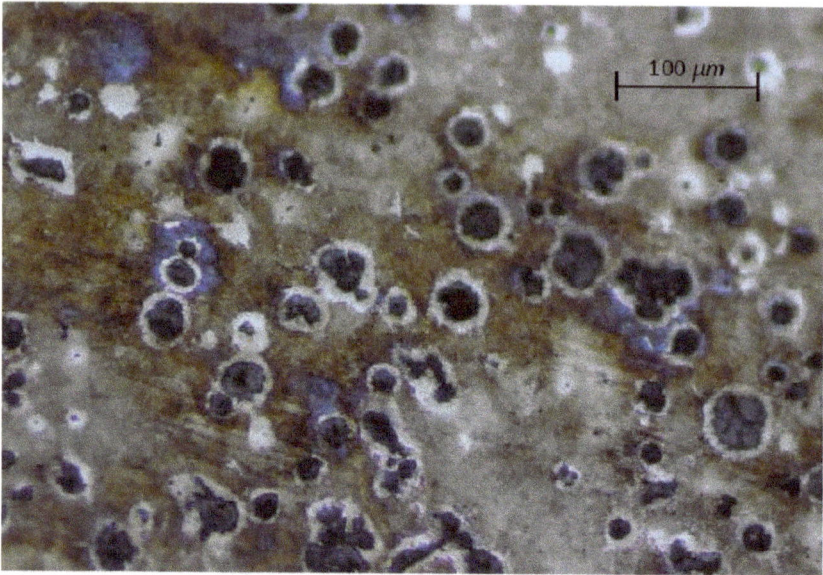

Figure 4. Micrograph of the structure produced on the base material at the edge of the heat affected zone.

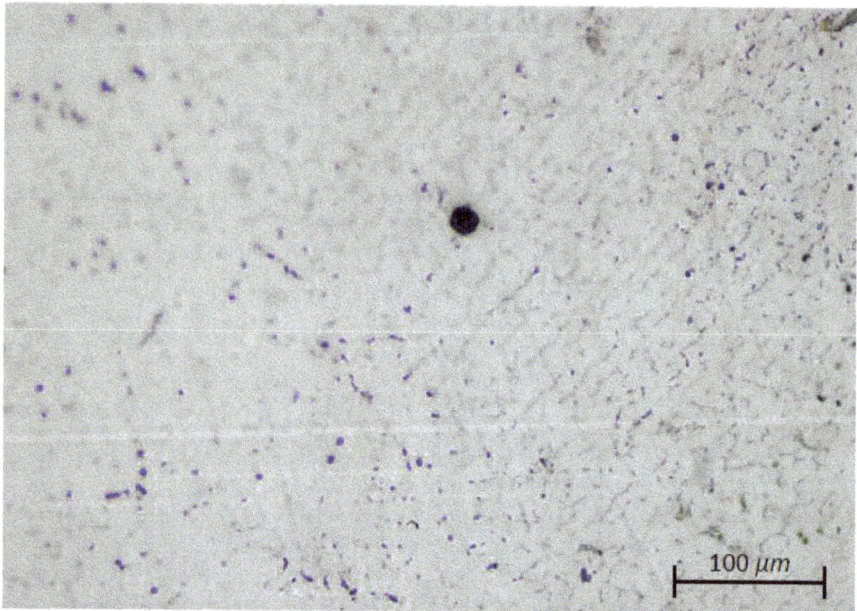

Figure 5. Micrograph of the material deposited on the ductile iron weld.

A transformation of the ferrite structure in the interface near the metal base can be seen in the micrograph of the weld obtained with a 97.6% Ni coated electrode on ductile iron in the filler zone on the second pass—and with subsequent heat treatment at 900 °C (Figure 6). The transformation is a consequence of thermal annealing and the increased Ni in the filler. This implies a considerable reduction in the strength of the material and an increase in the elongation at fracture with respect to the filler metal (in which nodules that are smaller than those in the interface are precipitated) [13,14] and the base metal where the hardness is lesser than samples made without preheating. The fracture is more ductile than in the previous cases and usually occurs in the cast zone near the interface [15].

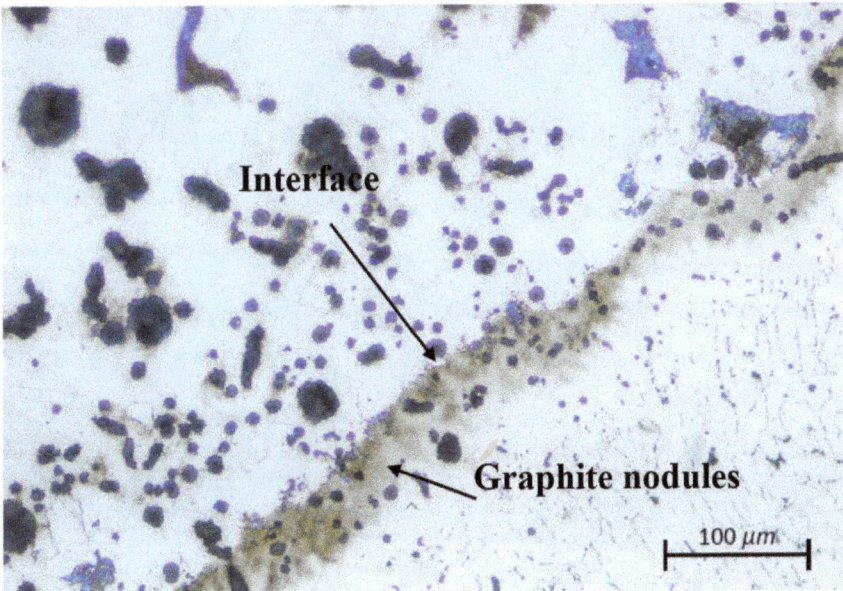

Figure 6. Micrograph of the filler zone in a ductile iron weld using a 97.6% Ni coated electrode with subsequent thermal annealing at 900 °C.

In the micrograph of the interface of the ductile iron TIG root weld made with an Inconel 625 electrode and preheating to 350 °C and with subsequent annealing treatment at 900 °C (Figure 7), it can be seen that, despite the heat treatment, small traces of white iron in a martensitic matrix remain undissolved. Vickers hardness levels of 600 are produced—this being below the levels obtained using air cooling because the Cr content in the Inconel has retained part of the carbides formed in the weld. The illustration shows a higher amount of graphite nodules that are smaller than those found in the air-cooled base material. The fracture was brittle in this zone and started at the interface.

The emergence of iron carbide (cementite) is influenced by the speed of cooling metastable, while Ni (nickel) is an gammageno element, which prevents the formation of carbides, given the amount of carbon present in the composition smelting, and added to the fast cooling of the welding speed, resulting in the precipitation of carbides of iron that can later be dissolved with annealing heat treatments.

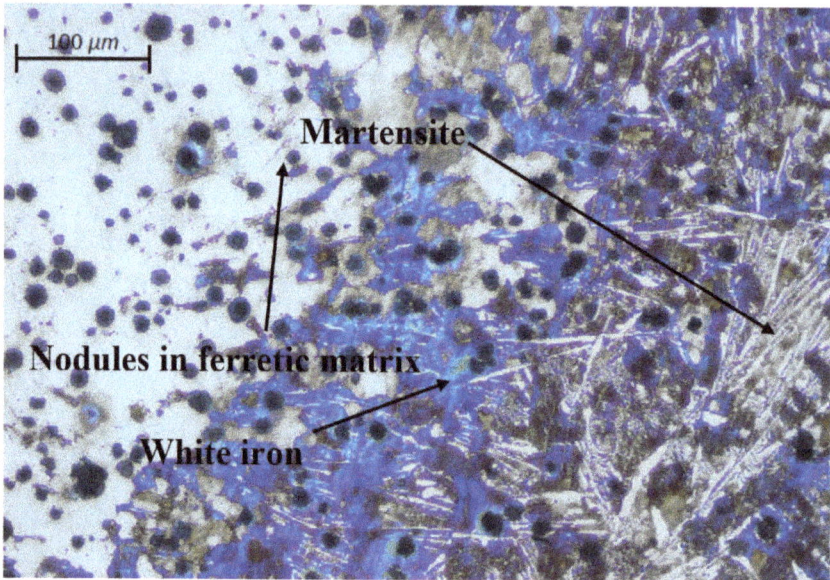

Figure 7. Micrograph of a root weld in ductile iron made with Inconel 625 and subsequent heat treatment at 900 °C.

The left part of the micrograph (Figure 8) shows that, after the annealing step, the ductile iron passed from ferritic-pearlitic ductile iron to ferritic iron with slightly larger and uniformly distributed nodules. The hardness is less than the original casting but with greater elongation at fracture. The micrograph on the right shows a uniform distribution of spherulites in the weld filler metal, and these are smaller in the ferrite matrix as a consequence of the Ni content and the furnace cooling rate.

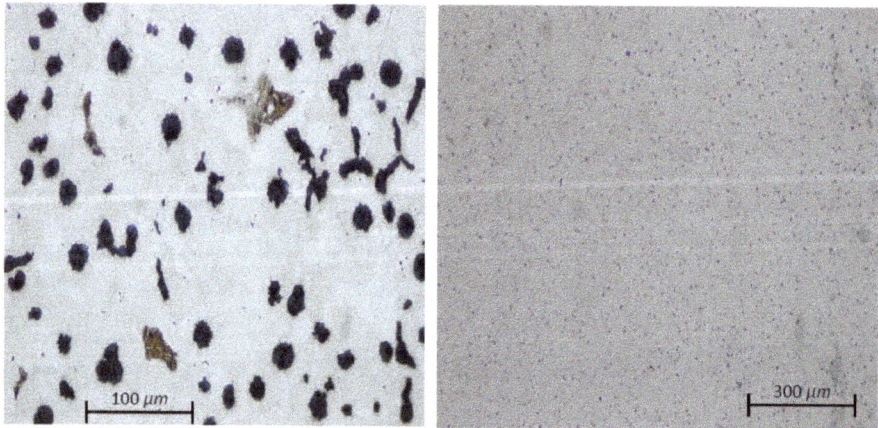

Figure 8. Micrographs of ductile iron showing the distribution of the nodules on the weld after annealing at 900 °C.

Manual arc welding with an electrode of 97.6% Ni is straightforward, the weld not being excessively fluid and high levels of carbide formation not occurring. Hardness is considerably less than that obtained in the root weld made with an Inconel source rod, although mechanical properties are acceptable and there is greater elongation at fracture. Micrographs show that an annealing treatment caused the martensitic structures to disappear and dissolved the traces of white iron producing a structure of nodular graphite on top of the ferrite matrix with an increased level of deformability [16–18].

Fractures according to the type of weld are shown in Figure 9.

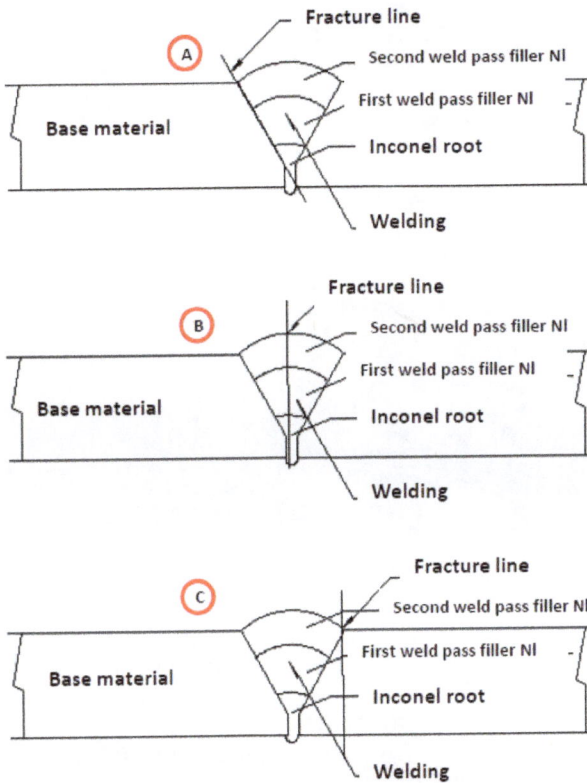

Figure 9. Zones of metallographic observation: (**A**) weld base metal base interface, (**B**) weld bead, (**C**) base metal.

The scanning electron microscopy (SEM) in Figure 10 shows the interface zone in the root weld applied with Inconel 625 but without heat treatment. Figure 11 shows the filler weld made with two passes of a 97.6% Ni coated electrode without treatment and the corresponding spectra.The spectrum in Figure 10 shows a high amount of chromium in the weld interface and the root weld where an Inconel 625 electrode was used—which influenced the formation of carbide precipitates that give rise to traces of white iron. The effect was to increase hardness and counteract the effect of Ni, as can be observed in the micrograph obtained for this zone in Figure 1. Another effect is the production of the beginnings of brittle fractures and a progressive growth of fractures in response to traction and bending.

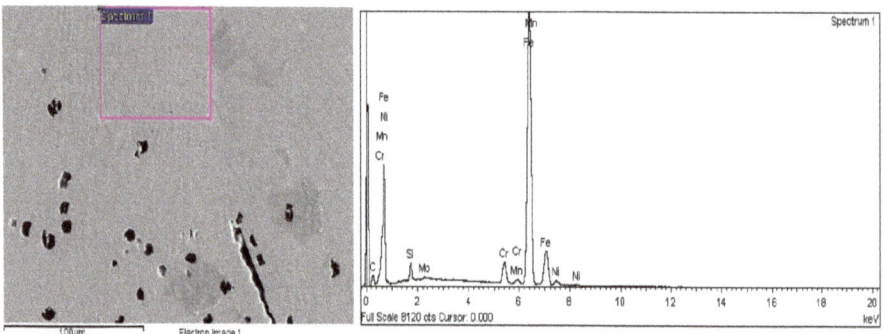

Figure 10. A SEM image and micrograph of the interface zone in a root weld without heat treatment and where a high amount of Cr can be observed in zones welded with Inconel.

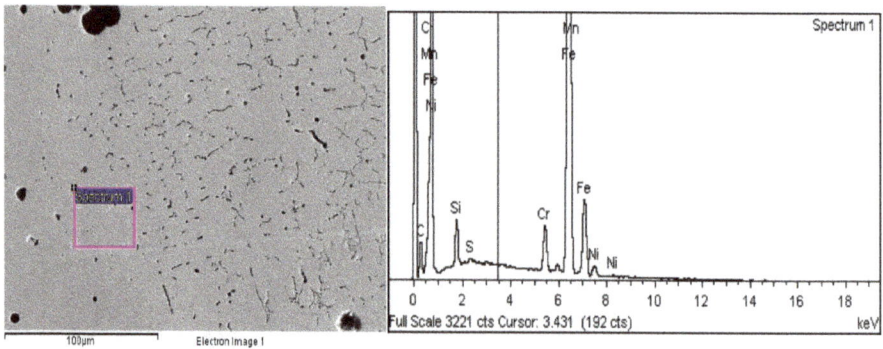

Figure 11. SEM and micrograph of the weld interface with filler applied using a 97% Ni electrode without subsequent heat treatment. Note the low Cr content.

The spectrum shown in Figure 11 corresponds to the interface of the first filler pass (97.6% Ni electrode) on the root pass weld (made using an Inconel 625 electrode). A remarkable decrease in the Cr content can be observed in the spectrum, causing a substantial decrease in traces of white iron and a greater influence of Ni on the structure (causing a decrease in hardness). A fracture initiated by bending and traction expands through the interface zone—as can be seen in the corresponding optical micrograph shown in Figure 2.

The SEM images in Figure 12 basically show the interface zone of the weld applied in the second filler pass with a 97.6% Ni electrode over the weld applied in the first pass (97.6% Ni) with subsequent annealing. Figure 13 shows the root weld on the same plate applied with Iconel 625 and annealed at 900 °C and the corresponding spectra.

The spectrum shown in Figure 12 corresponds to the second pass of the filler electrode (97.6% Ni) and annealing treatment at 900 °C with furnace cooling. We can observe a virtual absence of Cr, which directly influences the structure of the welding interface: dissolving the carbides and producing an absence of white cast iron and a substantial decrease in hardness. The result is a greater elongation at fracture. The graphite is precipitated in a fully nodular form, and the ferriticmatrix can be seen in the optic micrograph in Figure 6. Tensile and bending fracture occurs in this zone near to the base material.

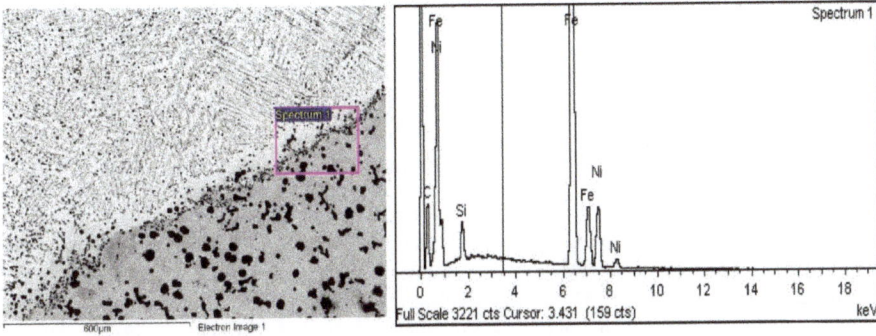

Figure 12. This figure shows micrographic and spectrum images of the filler weld interface of the top pass (electrode 97.6% Ni) and with a heat treatment at 900 °C. A higher content of free carbon is observed due to the dissolution of the white iron traces, while Cr is barely detectable (meaning greater ductility and less hardness).

The spectrum shown in Figure 13 corresponds to the interface of the root weld applied with an Inconel 625 rod and an annealing treatment at 900 °C and furnace cooling. The spectrum shows a decrease in Cr that is less pronounced than in the specimen with no heat treatment. The result is a partial reduction in carbides as a consequence of incomplete dissolution through heating and the speed of furnace cooling. As a consequence, there is a reduction in hardness at the interface although it remains very hard (as can be seen in the optical micrograph in Figure 7). The result is a substantial increase in strain capacity, although the fracture is brittle and starts in the root zone.

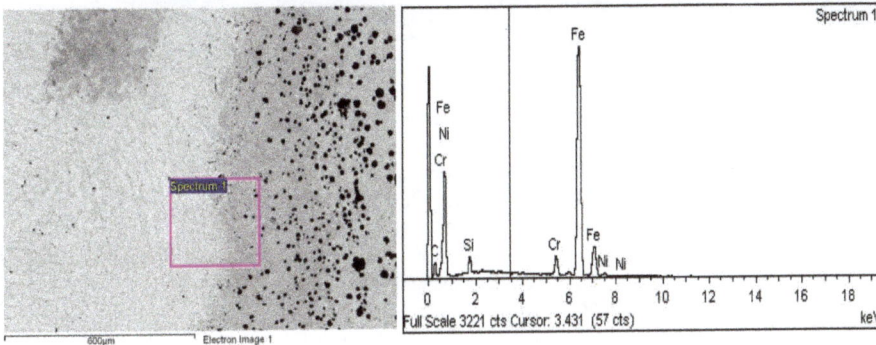

Figure 13. Micrograph and SEM taken at the root weld interface (Inconel 625) with annealing at 900 °C that shows a continuing high Cr level. The hardness level is high but lower than without heat treatment although the structure remains brittle.

5. Conclusions

- Before welding ductile cast iron test plates using an Inconel 625 electrode for the root weld in a single pass and a 97.6% Ni-based electrode for two subsequent filler passes, we preheated the test plates to improve weld fluidity and discourage the formation of brittle structures by favoring penetration and thus preventing fractures from starting. We welded short sections that had been strongly beaten to release residual stress (a process that proved beneficial) and obtained significant mechanical values for resistance and hardness. However, because of the cooling rate, significant values were not obtained for elongation at fracture. By applying a 900 °C annealing treatment and cooling some of the welded test pieces in the furnace, a considerable improvement was obtained in terms of elongation at fracture, the mechanical values being smaller for breaking strength and elasticity.

- Manual arc welding with an electrode of 97.6% Ni is straightforward, the weld not being excessively fluid and high levels of carbide formation not occurring. Hardness is considerably less than that obtained in the root weld made with an Inconel source rod, although mechanical properties are acceptable and there is greater elongation at fracture.

In the root weld made with an Inconel electrode, the subsequent application of an annealing treatment did not significantly change the initial conditions, and part of the white iron structures and traces of martensitic matrix remained with only small variations in the decrease in hardness and elongation. When the preheat treatment at 350 °C was employed, the fluidity in the pool increased and the mechanical characteristics were higher, and elongation at fracture increased.

Scanning electron microscopy verified the chemical compositions of the structural elements produced and confirmed that annealing treatment after both types of welding produced a greater working capacity in which hardness and tension diminished to enable more ductile fracturing and larger strains. The result is an improvement in the characteristics of the welded unions.

We treated an alloy with 97.6% Ni, which prevented the formation of carbides.

With heat treatment at 900 °C and 97.6% Ni, there is a dissolution of all carbides, forming nodules in ferritic matrix graphite.

Acknowledgments: We would like to give thanks to Language Translators Service of the Languishes Department of UniversitatPolitècnica de València, Spain.

Author Contributions: In this investigation, F.J.C.-C. and M.P.-G. conceived and designed the experiments; F.J.C.-C., R.P.-M., and M.A.P.-P. performed the experiments; F.J.C.-C., R.P.-M., and M.P.-G. analyzed the data; M.A.P.-P. contributed materials/analysis tools; F.J.C.-C. and M.P.-G. wrote the paper.

Conflicts of Interest: The authors declare no conflict of interest.

References

1. Apraiz, J. *Fundiciones*; Editorial Dossat: Madrid, Spain, 1981; pp. 144–145.
2. *Eutectic Castolin: Practice Handbook (Spanish Edition)*; Castolin: Madrid, Spain, 1993.
3. Huke, E.E.; Udin, H. Welding Metallurgy of Nodular Cast Iron. *Weld. J.* **1953**, *32*, 378s–385s.
4. Zhang, X.Y.; Zhou, Z.F.; Zhang, Y.M.; Wu, S.L.; Guan, L.Y. Influence of nickel-iron electrode properties and joint shapes on welded joint strength of pearlitic nodular iron. *Weld. J. Incl. Weld. Res. Suppl.* **1996**, *75*, 280s.
5. Pease, G.R. The Welding of Ductile Iron. *Weld. J.* **1960**, *39*, 1–9.
6. Nonast, R. *SoldeoElectrico Manual al Arco Metálico*; Gráficas Summa: Madrid, Spain, 1973; p. 84.
7. Fatahalla, N.; Bahi, S.; Hussein, O. Metallurgical parameters, mechanical properties and machinability of ductile cast iron. *J. Mater. Sci.* **1996**, *31*, 5765–5772. [CrossRef]
8. *Manual de Utilización*; CastolinEspaña: Madrid, Spain, 1993; p. 64.
9. Jeshvaghani, R.A.; Jaberzadeh, M.; Zohdi, H.; Shamanian, M. Microstructural study and wear behavior of ductile iron surface alloyed by Inconel 617. *Mater. Des.* **2014**, *54*, 491–497. [CrossRef]
10. Steglich, D.; Brocks, W. Micromechanical modelling of the behaviour of ductile materials including particles. *Comput. Mater. Sci.* **1997**, *9*, 7–17. [CrossRef]
11. Pascual, M.; Cembrero, J.; Salas, F.; Martínez, M.P. Analysis of the weldability of ductile iron. *Mater. Lett.* **2008**, *62*, 1359–1362. [CrossRef]
12. Cembrero, J.; Pascual, M. Soldabilidad de las fundiciones de grafitoesferoidal. *Rev. Metal.* **1999**, *35*, 392–401. [CrossRef]
13. El-Banna, E.M.; Nageda, M.S.; El-Saadat, M.M.A. Study of restoration by welding of pearlitic ductile cast iron. *Mater. Lett.* **2000**, *42*, 311–320. [CrossRef]
14. El-Banna, E.M. Effect of preheat on welding of ductile cast iron. *Mater. Lett.* **1999**, *41*, 20–26. [CrossRef]
15. Bayati, H.; Elliott, R. Influence of matrix structure on physical properties of an alloyed ductile cast iron. *Mater. Sci. Technol.* **1999**, *15*, 265–277. [CrossRef]

16. Seferian, D. *Las Soldaduras: Técnica y Control*; Editorial Urmo.: Bilbao, Spain, 1965; pp. 295–299.
17. Kiser, S.D.; Faws, P.E.; Northey, M. Welding Cast Iron: Straightforward. *Can. Weld. Assoc. J.* **2005**, 1–4. Available online: http://selector.specialmetalswelding.com/publica/cwafall2005.pdf (accessed on 18 November 2016).
18. Pouranvari, M. On the Weldability of Grey Cast Iron Using Nickel Based Filler Metal. *Mater. Des.* **2010**, *31*, 3253–3258. [CrossRef]

![metals logo] *metals*

MDPI

Article

Tool Wear Characteristics and Effect on Microstructure in Ti-6Al-4V Friction Stir Welded Joints

Ameth Fall [1],*, Mostafa Hashemi Fesharaki [2], Ali Reza Khodabandeh [2] and Mohammad Jahazi [1],*

[1]   Ecole de Technologie Supérieure (E.T.S), Département de Génie Mécanique, 1100 Rue Notre-Dame Ouest, Montréal, QC H3C 1K3, Canada
[2]   Science and Research Branch, Islamic Azad University, Hesarak, Teheran 982100, Iran; sdma.hf@gmail.com (M.H.F.); ar_khd@yahoo.com (A.R.K.)
*   Correspondence: ameth-maloum.fall.1@ens.etsmtl.ca (A.F.); mohammad.jahazi@etsmtl.ca (M.J.); Tel.: +1-438-490-3643 (A.F.); +1-514-396-8974 (M.J.)

Academic Editor: Giuseppe Casalino
Received: 24 September 2016; Accepted: 4 November 2016; Published: 10 November 2016

Abstract: In the present paper, tool wear and the rate of wear during friction stir welding (FSW) of Ti-6Al-4V alloy are investigated. A conical tungsten carbide tool was used to produce butt-type friction stir welded joints in two-millimeter thick Ti-6Al-4V sheets. An original design of a movable pin allowed for the examination of the tool damage for each process condition. The influence of tool degradation on the quality of the welded joints and the damage brought to the microstructure are examined and discussed. For this purpose, optical and scanning electron microscopies as well as EDX analyses were used to examine the tool wear and the resulting macrostructures and microstructures. The type and nature of the defects are also analyzed as a function of FSW processing parameters. Important geometry and weight variations were observed on the pin and shoulder for all welding conditions, in particular when low tool rotation and travel speeds were used. Experimental results also show that the radial wear of the pin is not uniform, indicating the presence of important frictional temperature gradients through the thickness of the joint. The maximum wear was measured at a location of about one millimeter from the pin root center. Finally, tool rotation was determined as the most significant process parameter influencing both tool wear and microstructure of the joints.

Keywords: friction stir welding; titanium; tool wear; microstructure

1. Introduction

Manufacturing of structural components made of welded Ti-alloy sheets or thin plates is continuously increasing in the transportation and energy industries mainly because these alloys possess an excellent combination of low-density, superior mechanical properties along with high corrosion and erosion resistance. Fusion welding techniques such as tungsten inert gas (TIG) or laser are extensively used to join titanium alloys. However, the formation of a brittle cast structure, residual stresses, and undesirable deformation limits their applications to critical structural components [1,2]. Friction stir welding (FSW) as a solid state welding process is one of the most promising techniques for joining sheets and thin plates made of titanium alloys, avoiding a large number of difficulties arising from the use of fusion welding processes, and could be used for the manufacturing of large size components for aerospace applications [3–6]. Moreover, FSW could also be used as an alternative manufacturing technology for the fabrication of hollow components made of Ti alloys which are traditionally produced by a combination of diffusion bonding and superplastic forming [7,8]. Details regarding the principles of the FSW process are well discussed in other publications [6,7].

While FSW has been extensively used for joining light alloys such as Al and Mg, very few reports are available on the application of the technique to titanium alloys. The main difficulty arises from the challenges related to FSW tool material. The high mechanical strength and low thermal conductivity of titanium results in very high frictional forces and heat generation at the tool-workpiece interface, thereby limiting possible tool materials compared with Al or Mg alloys [8]. In the case of Al and Mg alloys, due to their low strength and melting points, the welding tool is commonly made of tool steel [9–11]. In contrast, the situation is different for high strength materials such as titanium alloys as their high strength and high melting point do not allow use of tool steels for FSW. Specifically, the tool material should maintain enough strength, to deform and stir the alloy, be resistant to fatigue, fracture and mechanical wear, as well as inert to chemical reactions with both the weld material and the atmosphere at temperatures higher than 1000 °C, which is not the case for any tool steel. Tungsten-based alloys have been used as tool and pin materials for FSW of nickel-aluminum alloys and titanium alloys. Sanders et al. [5] and Farias et al. [12] recommended four tungsten alloys as tool materials: WC, W-25%Re, Densimet and W-1%LaO$_2$, while others have recommended TiC or polycrystalline boron nitride (pcBN) [13,14].

Indeed, even with such refractory materials, tool wear is still a major issue and its influence on microstructure damage, and hence, weld quality, is of prime importance. However, few data is available on tool wear during FSW with most of the publications being focused on steel and aluminum alloys [15,16]. For example, Park et al. and Yutaka et al. [17,18] conducted a detailed study on the tool wear in FSW stainless and ferritic steel welds using pcBN and Co tools, respectively. They reported that Cr-rich borides were formed at the interface between the workpiece and the pcBN tool, resulting in significant mechanical/chemical wear. Also, Weinberger et al. [16] reported that tools made of Rhenium (Re) and W-based Rhenium alloys (e.g., 25% Re + W) show very similar performance characteristics to that of pcBN. Zhang et al. [19] reported wear of pcBN tool during FSW of pure Ti and indicated that pcBN might have reacted with Ti. However, no details were provided on the influence of the FSW parameters on tool wear and the nature of the wear products.

Among the above tool materials, WC is the most cost effective with relatively good machinability and chemical stability. However, the influence of processing parameters during FSW of Ti-alloys on WC tool damage and its impact on microstructural changes and defect generation have not been quantified. In the present work, two-millimeter thick Ti-6Al-4V sheets were friction stir welded using a WC tool. The tool wear characteristics and its evolution are analyzed and discussed in relation to FSW process parameters (tool rotational and travel speeds), temperature gradients and different frictional conditions in the joint. Microstructural damages and changes in relation to the evolution of tool wear are also analyzed and discussed. The impact of each process parameter on wear rate is quantified and the most influential process parameters are identified. On the basis of the obtained results, the process conditions leading to optimum weld quality are determined.

2. Experimental Materials and Methods

Commercially Ti-6Al-4V alloy was used in this investigation with the following chemical composition: (wt. %) of Al 6.09, V 4.02, C 0.011, Fe 0.14, N 0.008, H 0.0023 and balance Ti. The material was received in annealed state and its mechanical properties are provided in Table 1.

Table 1. Mechanical properties of the as received base material Ti-6Al-4V.

Property	T.S. (MPa)	Y.S. (MPa)	El. (%)	Hardness (VHN)
Value	994	910	17.2	344

The dimensions of the samples were 100 mm in length, 50 mm in width and 2 mm in thickness and they were FSWed in butt configuration with the different investigated processing conditions listed in Table 2.

Table 2. Friction stir welding (FSW) parameter used on the samples.

Weld Number	Rotational Speed (rpm)	Travel Speed (mm/min)
1	500	100
2	600	100
3	700	100
4	1000	100
5	1250	100
6	1500	100

Edwards and Ramulu [20] reported that a cylindrical pin tool is not indicated for FSW of Ti because the heat generated in the shoulder is not able to flow to the root of the joint. On this basis, a conical shape pin with a flat shoulder made of commercial grade WC tool, with the characteristics shown in Table 3, was used. Details regarding tool design and pin geometry are provided in Figure 1a,b.

Table 3. Original size of the welding tool (mm) Shoulder height is considered to be the same as the pin length.

Tool Material	Shoulder Diameter	Shoulder Height	Pin Diameter	Pin Length
WC	15	8	6	1.8

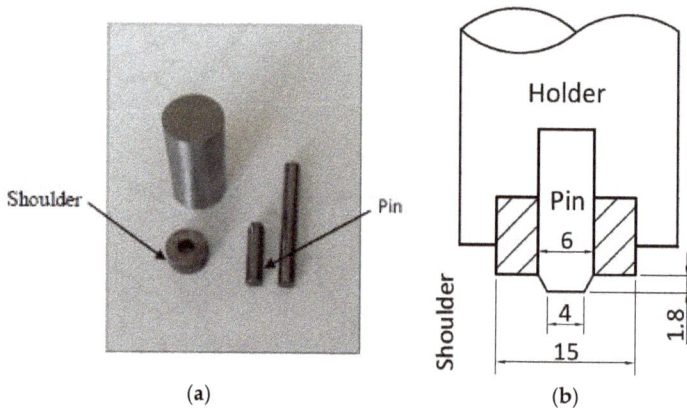

(a)   (b)

Figure 1. Geometry and dimensions of the shoulder and pin tool (a) shoulder and pin tool schematics in (mm) (b).

The original design of a "movable" pin allowed for the removal and examination of pin's wear at different stages of the process and therefore, better quantification of the impact of pin damage on the microstructure. Continuous protection of the joint area from oxidation was achieved by argon gas protection over the entire joint area. Further details regarding the experimental setup have been published elsewhere [21] and, to avoid repetition, will not be provided here.

In order to examine the effect of process parameters on tool wear and its rate, the weight of the device (pin and tool) were measured before and after FSW using a micro balance with an accuracy of 1 mgr. The wear is measured for all process conditions after 10 cm weld length. Also, specimens for optical microscopy (OM, Olympus DSX-500, Montreal, QC, Canada) and scanning electron microscopy (FEG-SEM, Hitachi SU8230, Toronto, ON, Canada) analyses were cut perpendicular to the welding direction in the mid weld location© for all processing conditions and mechanically polished with 6, 3 and 1 μm diamond paste. The final polishing was accomplished using colloidal silica of about 40 nm in diameter, followed by etching in Kroll's reagent (2 vol. % HF and 4 vol. % HNO_3 in water).

Finally, grain size measurements and particle size were made according to ASTM E112 and ASTM D422 standards, respectively using Image analysis software MIP4 (Nahamin Pardaz, Tehran, Iran) [22].

3. Results and Discussion

3.1. Effect of Wear on Microstructures

Figure 2 illustrates several micrographs of FSWed cross-sections at essentially the same initial travel speed of 100 mm/min, for several rotational speeds. Examination of weld cross-sections shows the presence of three distinct zones for all process conditions: Base Metal (BM), Heat Affected Zone (HAZ), and Stirred Zone (SZ). The advancing and retreating sides of the joint are identified in Figure 2a–g. Also, the HAZ, represented here by the dark area, and the SZ, revealed in white color, are also visible. It is worth noting that, the presence of a Thermomechanically Affected Zone (TMAZ), which has been reported in linear friction welded [23] as well as friction stir processed Ti-6Al-4V alloy [24] was not observed in the present investigation.

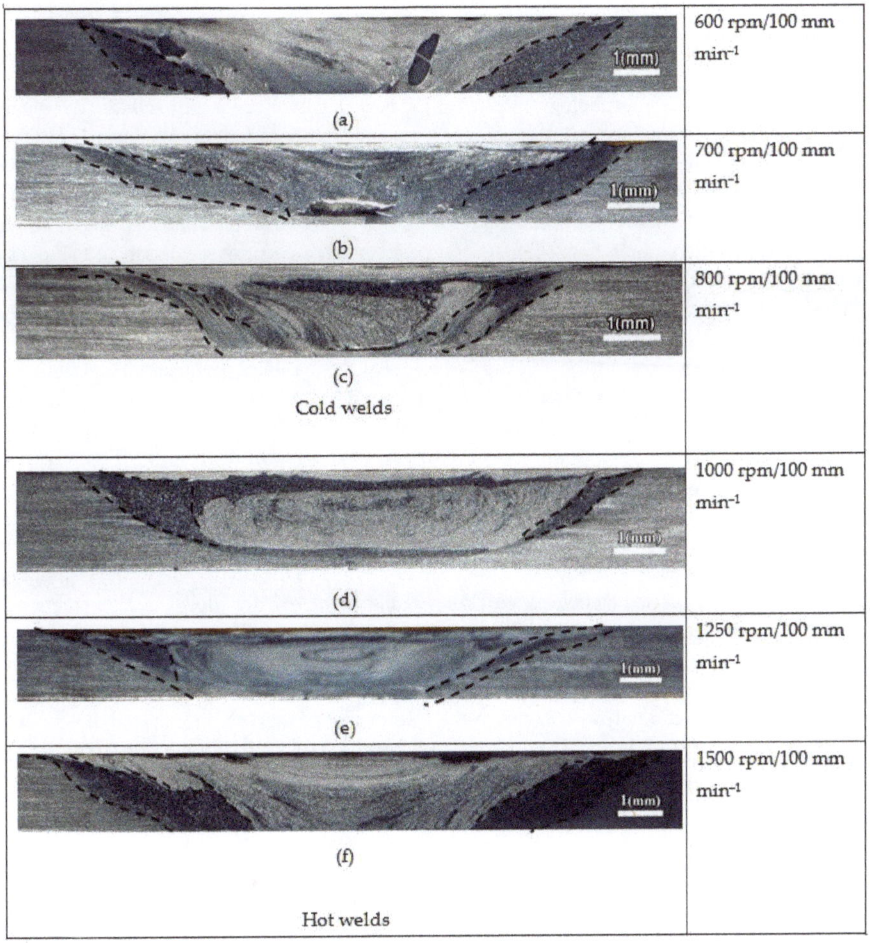

Figure 2. Macrostructures of the weld joint under cold weld parameters ((**a**) 600, (**b**) 700 and (**c**) 800 rpm) and hot weld condition ((**d**) 100, (**e**) 1250 and (**f**) 1500 rpm) for 100 mm·min^{-1} tool travel speed.

Measurement of the SZ volume for the cold and hot weld conditions revealed a two percent increase from one 600 rpm to 1500 rpm rotational speed. A similar trend was also observed for the size of the HAZ, which increased by about one percent when passing from the lowest rotational speed to the highest one. The error in volume calculation is equal to 0.15%. The observed difference has been related to the heat generated, due to a combination of frictional heating and plastic deformation heating, during the FSW process. Using a computational model for FSW of 7075 Al alloy, Bastier et al. [25] reported that the overall magnitude of the heat generated due to plastic deformation accounts only for 4.4% of the total heat generation whereas frictional heat generation provides 95.6%. Therefore, the low or high tool rotational speeds influence the frictional conditions and hence the heat distribution in the joint, resulting in different sizes for the SZ and the HAZ.

The influence of the tool rotational speed on the weld quality and the macrostructure of the joint is also reported in Figure 2. Two main processing conditions can be observed: from 600 to 800 rpm, called hereafter cold weld, and from 1000 to 1500 rpm, called hereafter hot weld. Volumetric defects were mainly observed in the cold weld conditions, with the presence of cavity in the joint visible at 600, 700 and 800 rpm, respectively as shown in Figure 3.

Detailed examination of the joints revealed also the presence of particles of various sizes, which were analyzed by OM, SEM, and EDX and were determined to be WC, as will be discussed in more detail in the following paragraphs.

Figure 3 illustrates the morphology and size of the particles. It can be seen that under hot welding conditions the particles are finer and more homogeneously distributed in the top surface of the joint, just below the pin. In contrast, for cold weld conditions, they are much larger and located close to the SZ-HAZ interface. For instance, for the coldest welding conditions (i.e., 600 rpm) their size is in the range of 500 μm to 800 μm; while at 1000 and 1500 rpm they are between 2 and 10 μm, as shown in Figure 3c. The SEM micrographs shown in Figure 4 show that WC particles are distributed all over the titanium matrix. The EDX analyses in locations far from the top surface and at the top surface of the weld joint, presented in Figure 4, confirm the presence of tool material in the microstructure of the weld joint.

The above observations may be analyzed in terms of the intensity of the vertical material flow during FSW. Prado et al. [26] used threaded and unthreaded tool to study the influence of processing conditions on material flow during FSW of metal matrix aluminum composites. While they did not identify explicitly cold or hot weld conditions; however, an analysis of their results (Figure 2 in [26]) indicate that under hot weld conditions, material flow in the vertical direction is significantly higher than under cold weld conditions. The high temperatures generated under hot weld conditions provide a vigorous material stirring in the SZ, including stronger vertical flow, which break down the WC particles into smaller ones and distribute them more uniformly in the matrix. In contrast, under cold weld conditions, lower stirring combined with lower temperatures do not allow easy movement of the WC particles resulting in limited vertical flow and their agglomeration near the SZ-HAZ boundary.

A comprehensive thermomechanical analysis is required to provide an accurate evaluation and prediction of the size, location, and distribution of the particles and their relation to wear conditions. The development of such analysis was not in the framework of this study; however, based on the published literature [27,28], it could be said that under cold weld conditions, due to the stronger thermal gradient and weaker vertical flow, more intense wear is expected.

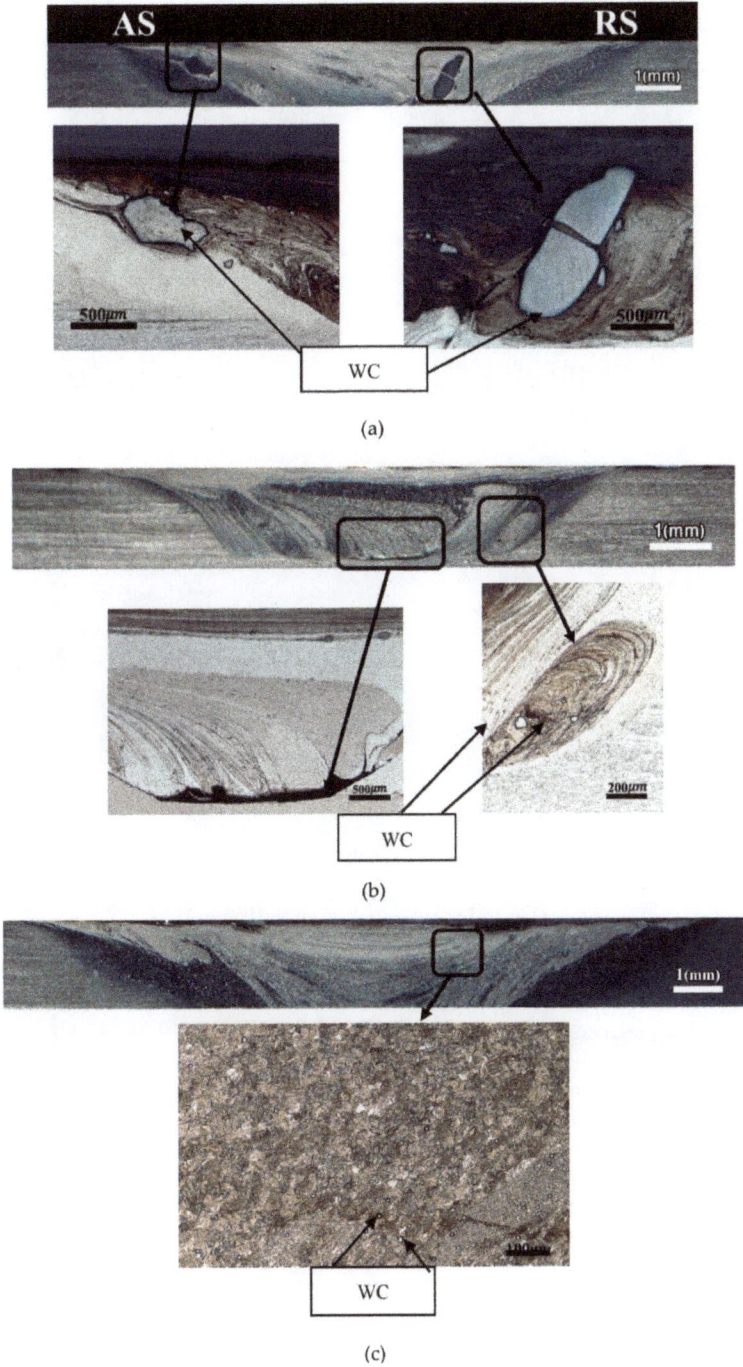

Figure 3. Effect of tool wear on the microstructure of the cold weld joints: (a) 600 rpm, (b) 800 rpm and (c) 1500 rpm.

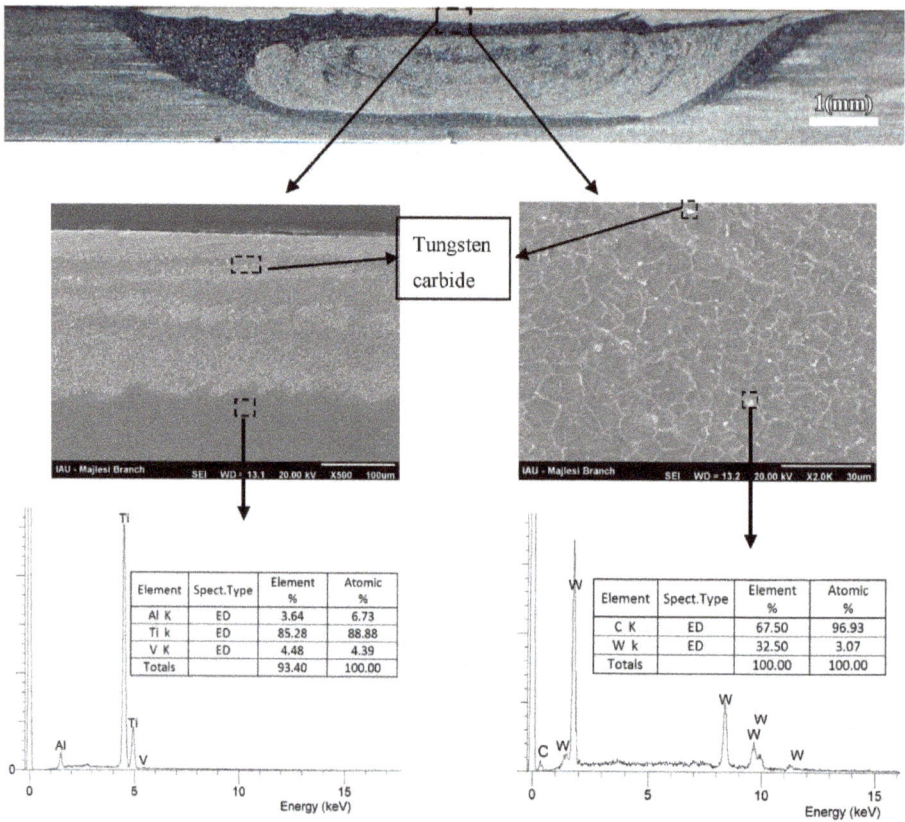

Figure 4. Tool wear effect on the microstructure of the hot weld joint (1000 rpm) and EDX analysis.

The evolution in the grain size and morphology from the BM to HAZ and then SZ is shown in Figure 5, where backscattered SEM images of the weld joint microstructure are reported. The grain size measurements were made in the in upper SZ and closest SZ in the HAZ. The results indicate that, for all the investigated processing conditions, a fully lamellar structure was developed in the BM and HAZ. The presence of a fully transformed β grain structure indicates that the temperature in the FSW was above the β-transus temperature of the alloy, which is estimated to be 980 °C. Grain size analysis revealed that the BM is characterized by elongated primary α and transformed β grains with an average grain size of 20 μm (Figure 5a,d). For instance, the average grain size changes from 20 μm in the BM to about 15 μm at the start of the HAZ and then further reduces to 5 μm in the SZ. The influence of tool rotational speed on the evolution of the microstructure from BM to HAZ is shown in Figure 5b,e. Also, a comparison between Figure 5a,c shows that the grain size in the BM (Figure 5a) is much larger than in the SZ (Figure 5c).

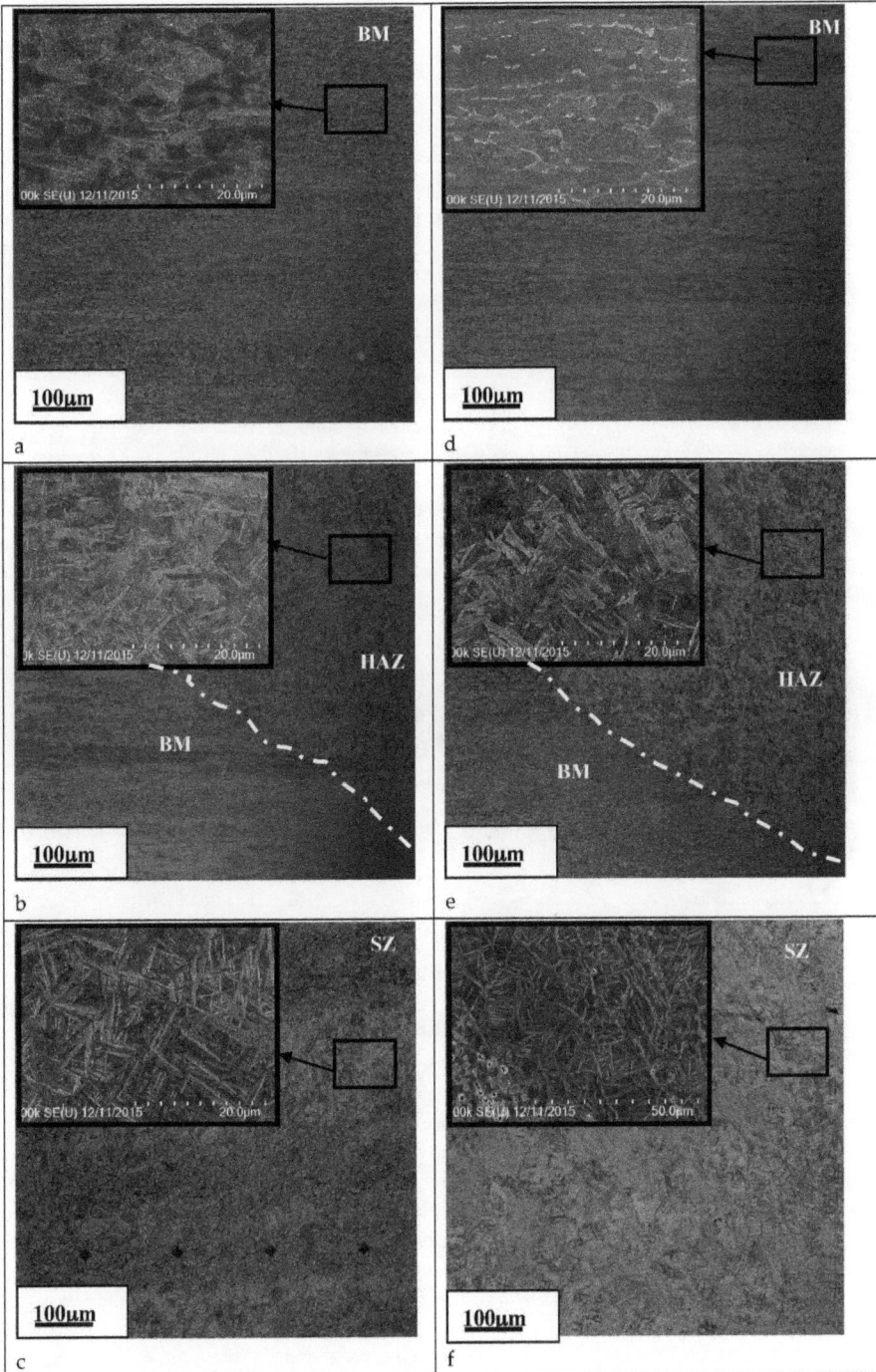

Figure 5. Optical and SEM images showing the microstructure flow through forwarding sides of the different weld region identified in Figure 5 at 1000 rpm (**a–c**) and 1500 rpm (**d–f**).

The above results indicate that the high temperatures and strains reached in the SZ during FSW of the Ti-6Al-4V alloy have been above the critical levels for the initiation of dynamic recrystallization. The occurrence of dynamic recrystallization during FSW of aluminum has been reported by Mishra et al. and Sanders et al. [5,7]. Several mechanisms, based on dynamic recrystallization, have been proposed to explain grain refinement in the SZ of various aluminum alloys. Dynamic recrystallization in Ti alloys has been reported by other authors at much lower strain and strain rates than the ones encountered in FSW [29,30]. It is therefore reasonable to assume that this phenomenon also takes place during FSW of the Ti-6Al-4V investigated alloy and is responsible for the observed significant grain refinement in the SZ. Specifically, the average grain size measured in the SZ was about one micrometer for the highest tool rotational speed and about seven micrometers for the lowest one. Finally, it must be noted that, considering the size of the WC particles (between 3 and 200 µm), it is not expected that they would affect the kinetics of dynamic recrystallization of the matrix during the FSW process.

3.2. Quantification of the Tool Wear

Figure 6 illustrates some typical sequences of tool/pin photographs illustrating apparent wear under selected processing conditions. It can be seen that all the tools suffered different degrees of wear. Some oxidations were also observed around the edge of some pins, which is attributed to the presence of tungsten oxide, as also reported by other authors [31]. The most severe wear was observed in the area between the pin center and pin edge, with the pin center being the least worn region. This is an interesting result as it could be used for optimum selection of tool materials. Analysis of the worn tool shape also showed that maximum wear took place under cold weld conditions (Figure 2).

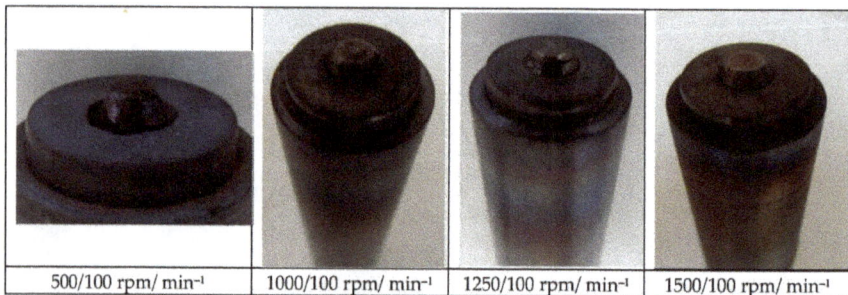

| 500/100 rpm/ min⁻¹ | 1000/100 rpm/ min⁻¹ | 1250/100 rpm/ min⁻¹ | 1500/100 rpm/ min⁻¹ |

Figure 6. Photographs of tungsten-based alloy tool after welding at different process conditions.

The difference in the tool weight measured with the micro balance before and after processing is illustrated in Table 4. Using the data in Table 4, the percentage of weight variation was calculated and correlated with the tool rotational speed as illustrated in Figure 7a. It can be seen in this figure that the effective tool weight is considerably affected by the rotational speed during the process. A decrease is noticed for all processing conditions; however, it is significantly higher for the lowest tool rotational speed. The pin length variation was also evaluated for the different processing conditions and is represented in Figure 7b,c. It can be seen that the length is also affected and some geometrical changes can be seen in the pin under certain processing conditions and mainly for low rotational speeds. This implies that a consumption of tool material, as indicated by some authors in aluminum alloys [32,33] also occurs during FSW of titanium. However, as mentioned by Casalino et al. [33], tool consumption could be reduced during FSW of aluminum alloys by using coated tools. Finally, tool wear (measured as percentage of weight change) decreases for increased rotational speeds. The above data were used to evaluate tool height variation and wear rate as shown in Figure 7c,d.

Table 4. Tungsten based alloy tool weight variation after welding at different process conditions.

rpm/min^{-1}	W1 (gr)	W2 (gr)	ΔW (gr)	F (%)
500/100	23.760	23.55	0.210	0.99
1000/100	23.742	23.661	0.081	0.34
1250/100	23.967	23.898	0.069	0.29
1500/100	23.752	23.693	0.059	0.25

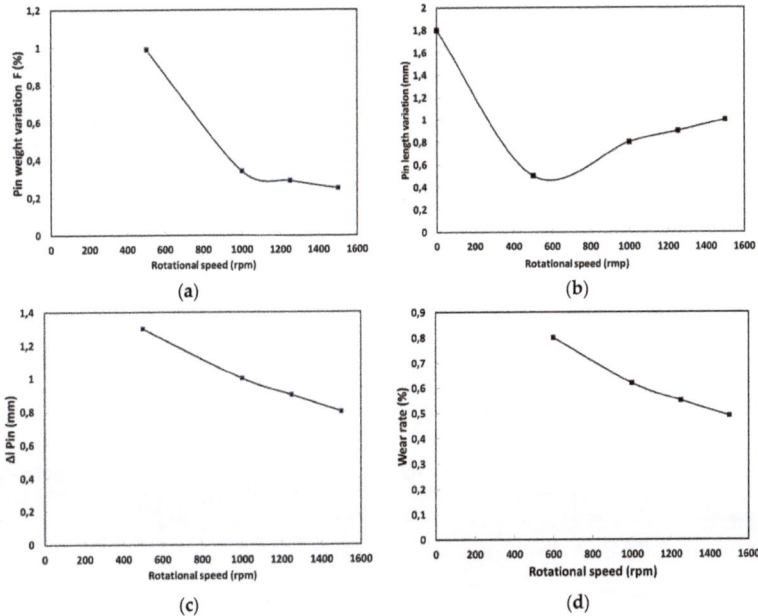

Figure 7. Wear measurement after 10 cm welding: (**a**) Weight loss data as function of rational speed; (**b**) Pin length evolution as function of processing parameter; (**c**) Pin length variation with rotational speed; (**d**) Wear rate as function of rotational speed.

Figure 7d shows the wear rate evolution with rotational speed. The initial wear rate (in % wear) is represented by the initial slope of the corresponding curves. It can be seen that the wear rate decreases with increasing the rotational speed in the range of process parameter studied in the present work. These findings are in agreement with those reported by Prado et al. [26] and Alidokht et al. [34] in FSW of aluminum alloys, where a decrease of wear rate was noted with sliding distance and rotational speed. The results suggest that a high wear rate is expected during the initial stages of welding, after which it decreases significantly. This finding is also important and could be used in the optimum material selection for FSW of Ti alloys or similar high strength and low thermal conductivity materials. Furthermore, pins used under high rotational speeds display slightly higher wear resistance compared with those used at lower rotational speeds. This may be related to the higher temperatures and therefore better material workability reached under high rotational speed welding conditions.

SEM analyses of the surface microstructure of the worn tools suggested two possible tool wear mechanisms during FSW. In presumed order of importance, these are oxidative wear of WC (Figures 4 and 6), and the increased brittleness due to the transformation of the WC binder. Lofaj et al. [35] and Casas et al. [36] reported that when WC reacts with oxygen, its volume expands by about 300% via an oxidation process that generates CO gas in the solid. The pressure of this gas is greater than the fracture strength of the WC and therefore leads to crack formation. Furthermore, oxide layers are easily fractured, due to residual tensile stresses associated with different coefficients of thermal expansion between the substrate and the carbide particle. The second mechanism is the increased brittleness due to the transformation of the WC binder [5,20]. The WC binder phase is continuously exposed to severe repetitive conditions due to its contact with the workpiece during FSW. It has been shown that WC exposed to fatigue loading exhibits reduced fracture toughness, erosion resistance and thermal shock resistance [35]. The repetitive cycles of contacts between the tool and material creates fatigue like conditions, resulting in particle fracture in the affected regions. However, it was difficult to determine the contribution of each mechanism as both processes occur almost at the same time.

4. Conclusions

Tool wear during friction stir welding of titanium alloy Ti-6Al-4V was investigated under several processing conditions. The following main conclusions can be drawn from the present study:

1. Tool wear is strongly affected by the tool rotational speed. The highest tool wear was obtained under low rotational speeds.
2. The radial wear of the pin is very different at different locations of the pin, and the maximum wear is produced at about two millimeters from the pin root center under cold weld conditions. The welding speed has a decisive effect on the radial wear rate of the pin, and the maximum wear rate was measured for the lowest rotational speed.
3. Microscopic analysis of the welded joints showed that sound FSW joints of Ti-6Al4V were obtained under the following processing conditions: conical WC pin, WC shoulder, tool rotational speed between 1000 and 1500 rpm.

Author Contributions: M. H. Fesharaki, M. Jahazi and A. Khodabandeh conceived and designed the experiments; M. H. Fesharaki performed the experiments; A. Fall and M. H. Fesharaki carried out metallography analyses and sample preparation; A. Fall and M. Jahazi analyzed the data and wrote the paper.

Conflicts of Interest: The authors declare no conflict of interest.

References

1. Thomas, W.M.; Nicholas, E.D. Friction stir welding for the transportation industries. *Mater. Des.* **1997**, *18*, 269–273. [CrossRef]
2. Lütjering, G.; Williams, J.C. *Titanium*; Springer: Berlin, Germany, 2003; Volume 2.
3. Zwicker, U. *Titanium and Titanium Alloys*; Springer: Berlin, Germany, 2007.
4. Saresh, N.; Pillai, M.G.; Mathew, J. Investigations into the effects of electron beam welding on thick Ti-6Al-4V titanium alloy. *J. Mater. Process. Technol.* **2007**, *192*, 83–88. [CrossRef]
5. Sanders, D.G.; Ramulu, M.; McCook, E.J.K.; Edwards, P.D.; Reynolds, A.P.; Trapp, T. Characterization of Superplastically Formed Friction Stir Weld in Titanium 6Al-4V: Preliminary Results. *J. Mater. Eng. Perform.* **2008**, *17*, 187–192. [CrossRef]
6. Thomas, W.M.; Threadgill, P.L.; Nicholas, E.D. Feasibility of friction stir welding steel. *Sci. Technol. Weld. Join.* **1999**, *4*, 365–372. [CrossRef]
7. Mishra, R.S.; Mahoney, M.W. *Friction Stir Welding and Processing*; ASM International: Geauga County, OH, USA, 2007.
8. Lee, W.-B.; Lee, C.-Y.; Chang, W.-S.; Yeon, Y.-M.; Jung, S.-B. Microstructural investigation of friction stir welded pure titanium. *Mater. Lett.* **2005**, *59*, 3315–3318. [CrossRef]
9. Badarinarayan, H.; Yang, Q.; Zhu, S. Effect of tool geometry on static strength of friction stir spot-welded aluminum alloy. *Int. J. Mach. Tools Manuf.* **2009**, *49*, 142–148. [CrossRef]
10. Rai, R.; De, A.; Bhadeshia, H.K.D.H.; DebRoy, T. Review: Friction stir welding tools. *Sci. Technol. Weld. Join.* **2011**, *16*, 325–342. [CrossRef]
11. Tongne, A.; Jahazi, M.; Feulvarch, E.; Desrayaud, C. Banded structures in friction stir welded Al alloys. *J. Mater. Process. Technol.* **2015**, *221*, 269–278. [CrossRef]
12. Farias, A.; Batalha, G.F.; Prados, E.F.; Magnabosco, R.; Delijaicov, S. Tool wear evaluations in friction stir processing of commercial titanium Ti–6Al–4V. *Wear* **2013**, *302*, 1327–1333. [CrossRef]

13. Sato, Y.S.; Harayama, N.; Kokawa, H.; Inoue, H.; Tadokoro, Y.; Tsuge, S. Evaluation of microstructure and properties in friction stir welded superaustenitic stainless steel. *Sci. Technol. Weld. Join.* **2013**, *14*, 3. [CrossRef]
14. Park, S.H.C.; Sato, Y.S.; Kokawa, H.; Okamoto, K.; Hirano, S.; Inagaki, M. Rapid formation of the sigma phase in 304 stainless steel during friction stir welding. *Scr. Mater.* **2003**, *49*, 1175–1180. [CrossRef]
15. Hovanski, Y.; Santella, M.L.; Grant, G.J. Friction stir spot welding of hot-stamped boron steel. *Scr. Mater.* **2007**, *57*, 873–876. [CrossRef]
16. Weinberger, T.; Enzinger, N.; Cerjak, H. Microstructural and mechanical characterization of friction stir welded 15–5PH steel. *Sci. Technol. Weld. Join.* **2013**, *14*, 210–215. [CrossRef]
17. Park, S.H.C.; Sato, Y.S.; Kokawa, H.; Okamoto, K.; Hirano, S.; Inagaki, M. Boride formation induced by pcBN tool wear in friction-stir-welded stainless steels. *Metall. Mater. Trans. A* **2009**, *40*, 625–636. [CrossRef]
18. Yutaka, S.S.; Masahiro, M.; Shinichi, S.; Hiroyuki, K.; Toshihiro, O.; Koyohito, I.; Shinya, I.; Seung, H.C.P.; Itto, S.; Satoshi, H. Performance Enhancement of Co-Based Alloy Tool for Friction Stir Welding of Ferritic Steel. In *Friction Stir Welding and Processing VIII*; TMS: San Diego, CA, USA, 2015; pp. 39–46.
19. Zhang, Y.; Sato, Y.S.; Kokawa, H.; Park, S.H.C.; Hirano, S. Stir zone microstructure of commercial purity titanium friction stir welded using pcBN tool. *Mater. Sci. Eng. A* **2008**, *488*, 25–30. [CrossRef]
20. Edwards, P.; Ramulu, M. Effect of process conditions on superplastic forming behaviour in Ti-6Al-4V friction stir welds. *Sci. Technol. Weld. Join.* **2009**, *14*, 669–680. [CrossRef]
21. Fall, A.; Jahazia, M.; Khodabandehb, A.R.; Fesharakib, M.H. Effect of process parameters on microstructure and mechanical properties of friction stir welded Ti-6Al-4V joints. *Int. J. Adv. Manuf. Technol.* **2016**, *87*, 1–13. [CrossRef]
22. Nahamin Pardazan Asia. Iran, 2014. Available online: http://en.metsofts.ir (accessed on 6 August 2016).
23. Dalgaard, E.; Frederik, C.L.; Rabet, M.J.; Priti, W.; John, J.J. Texture Evolution in Linear Friction Welded Ti-6Al-4V. *Adv. Mater. Res.* **2010**, *89–91*, 124–129. [CrossRef]
24. Ma, Z.Y.; Pilchak, A.L.; Juhas, M.C.; Williams, J.C. Microstructural refinement and property enhancement of cast light alloys via friction stir processing. *Scr. Mater.* **2008**, *58*, 361–366. [CrossRef]
25. Bastier, A.; Maitournam, M.H.; Van, K.D.; Roger, F. Steady state thermomechanical modelling of friction stir welding. *Sci. Technol. Weld. Join.* **2006**, *11*, 278–288. [CrossRef]
26. Prado, R.A.; Murr, L.E.; Soto, K.F.; McClure, J.C. Self-optimization in tool wear for friction-stir welding of Al6061 + 20%Al$_2$O$_3$ MMC. *Mater. Sci. Eng. A* **2003**, *349*, 156–165. [CrossRef]
27. Tongne, A.; Desrayaud, C.; Jahazi, M.; Feulvarch, E. On material flow in Friction Stir Welded Al alloys. *J. Mater. Process. Technol.* **2017**, *239*, 284–296. [CrossRef]
28. Krishnan, K.N. On the formation of onion rings in friction stir welds. *Mater. Sci. Eng. A* **2002**, *327*, 246–251. [CrossRef]
29. Vo, P.; Jahazi, M.; Yue, S. Recrystallization During Beta Working of IMI834. *Adv. Mater. Res.* **2007**, *15–17*, 965–969. [CrossRef]
30. Yoon, S.; Rintaro, U.; Hidetoshi, F. Effect of initial microstructure on Ti-6Al-4V joint by friction stir welding. *Mater. Des.* **2015**, *88*, 1269–1276. [CrossRef]
31. Zackrisson, J.; Jansson, B.; Uphadyaya, G.S.; Andr'en, H.-O. WC-Co based cemented carbides with large Cr$_3$C$_2$ additions. *Int. J. Refract. Met. Hard Mater.* **1998**, *16*, 417–422. [CrossRef]
32. Gerlich, A.; Su, P.; North, T.H. Tool penetration during friction stirs spot welding of Al and Mg alloys. *J. Mater. Sci.* **2005**, *40*, 6473–6481. [CrossRef]
33. Casalino, G.; Sabina, C.; Michelangelo, M. Influence of shoulder geometry and coating of the tool on the friction stir welding of aluminium alloy plates. *Proced. Eng.* **2014**, *69*, 1541–1548. [CrossRef]
34. Alidokht, S.A.; Zadeh, A.A.; Soleymani, S.; Saeid, T.; Assadi, H. Evaluation of microstructure and wear behavior of friction stir processed cast aluminum alloy. *Mater. Charact.* **2012**, *63*, 90–97. [CrossRef]
35. Lofaj, F.; Yu, S.K. Kinetics of WC-Co oxidation accompanied by swelling. *J. Mater. Sci.* **1995**, *30*, 1811–1817. [CrossRef]
36. Casas, B.; Ramis, X.; Anglada, M.; Salla, J.M.; Llanes, L. Oxidation-induced strength degradation of WC-Co hardmetals. *Int. J. Refract. Met. Hard Mater.* **2001**, *19*, 303–309. [CrossRef]

metals

Article

Weldability and Monitoring of Resistance Spot Welding of Q&P and TRIP Steels

Pasquale Russo Spena [1],*, Manuela De Maddis [2], Gianluca D'Antonio [2] and Franco Lombardi [2]

[1]   Faculty of Science and Technology, Free University of Bozen-Bolzano, 39100 Bolzano, Italy
[2]   Department of Management and Production Engineering, Politecnico di Torino, 10129 Torino, Italy;
      manuela.demaddis@polito.it (M.D.M.); gianluca.dantonio@polito.it (G.D.A.);
      franco.lombardi@polito.it (F.L.)
*   Correspondence: pasquale.russospena@unibz.it; Tel.: +39-0471-017-112

Academic Editor: Giuseppe Casalino
Received: 12 October 2016; Accepted: 4 November 2016; Published: 8 November 2016

Abstract: This work aims at investigating the spot weldability of a new advanced Quenching and Partitioning (Q&P) steel and a Transformation Induced Plasticity (TRIP) steel for automotive applications by evaluating the effects of the main welding parameters on the mechanical performance of their dissimilar spot welds. The welding current, the electrode tip voltage and the electrical resistance of sheet stack were monitored in order to detect any metal expulsion and to evaluate its severity, as well as to clarify its effect on spot strength. The joint strength was assessed by means of shear and cross tension tests. The corresponding fracture modes were determined through optical microscopy. The welding current is the main process parameter that affects the weld strength, followed by the clamping force and welding time. Metal expulsion can occur through a single large expulsion or multiple expulsions, whose effects on the shear and cross tension strength have been assessed. Longer welding times can limit the negative effect of an expulsion if it occurs in the first part of the joining process. The spot welds exhibit different fracture modes according to their strengths. Overall, a proper weldability window for the selected process parameters has been determined to obtain sound joints.

Keywords: dissimilar resistance spot welding; quenching and partitioning steel; transformation induced plasticity steel; welding parameters; welding monitoring; mechanical strength; microstructures; fracture modes

1. Introduction

Advanced high strength steels (AHSSs) are used extensively in the automotive industry for the fabrication of more resistant and lighter components with the main aim of reducing fuel consumption and gas emissions, and of improving passenger safety. Dual Phase (DP), Transformation Induced Plasticity (TRIP), martensitic, complex phase and hot stamping boron steels are the most commonly used AHSS grades for such applications. New AHSSs are currently under research and development to achieve better combinations of ductility (e.g., crashworthiness and sheet formability) and mechanical strength (e.g., impact resistance). In this context, Quenching and Partitioning (Q&P) steels appear to be one of the most innovative and promising solutions. They are characterized by a microstructure that consists of retained austenite and martensite: the former phase provides ductility and toughness, the latter mechanical strength. On the basis of carbon content and volume fraction of austenite and martensite, tensile strength can usually vary from 700 to 1300 MPa, while elongation at fracture can vary from 10% to 25% [1,2]. At present, only car body prototypes are made of Q&P steels.

The AHSSs used in the automotive industry are often welded together in dissimilar (different steel grades joined together) configurations in order to assemble car body parts and body frames. Dissimilar welding usually requires more precautions than conventional similar welding since the steels to be joined may have different melting points, thermal conductivities, thicknesses, or need different filler metals or pre-heats. Therefore, optimal welding parameters are often a compromise based on the properties of steels. Generally, the reduction in heat input can limit some harmful effects that occur during a joining process, such as cracks, thermal distortions, chemical segregations in the fusion zone, wide heat affected zones, thereby promoting more sound dissimilar joints [3]. In other cases, hybrid welding technologies (i.e., different welding techniques used simultaneously) are helpful solutions to join steels that have very different chemical and physical properties [4]. Resistance spot welding (RSW) is known to be the leading joining technique in the automotive industry due to its suitability for automation and high operating speeds. Thousands of spot welds are usually performed to join doors, body-in-white and other components in a vehicle. The quality and the mechanical performances of RSW joints are crucial for the safety and durability design of a vehicle. In fact, many of these joints are used in structural assemblies that are involved in transferring loads through the body frame during a crash event, and may act as fold initiation sites to manage impact energy [5]. Moreover, the integrity and mechanical properties of spot welds also affect their fatigue and fracture resistance and, in turn, the overall performance of a car body frame in terms of vibrations, noise and harshness [6]. The main issues of an RSW process are ascribable to the complexity of the chemical (e.g., composition in the fusion zone) and physical phenomena (e.g., heat input) that are involved during sheet joining. AHSS spot welds generally exhibit lower mechanical performances than those of the base materials: the peculiar complex microstructures of AHSSs, which are obtained by means of strictly controlled industrial thermo-mechanical processes, are destroyed completely in the fusion zone and altered in the heat affected zone, where they are replaced by more brittle metallurgical constituents.

Therefore, the weldability of Q&P steels is one of the most important key factors in controlling the possible usage of this steel grade in the automotive industry. Only a few preliminary studies about the weldability of Q&P steels in similar and dissimilar configurations have been carried out so far. Wang et al. [7] found that the fatigue performances of RSW joints of Q&P980 steels are similar to those of DP steels with the same tensile strength during cross and shear tension tests. Russo Spena et al. studied dissimilar RSW between a Q&P980 steel, a Twinning Induced Plasticity (TWIP1000) [8] and a TRIP800 steel [9], and mainly assessed the welding parameters effects on the shear tension strength of spot welds. In both cases, the welding parameters had to be carefully controlled in order to obtain spot welds with an adequate shear tension strength for the automotive industry (with reference to American Welding Society (AWS) and International Organization for Standardization (ISO) standards), as well as to limit defects, such as metal expulsion and voids.

The spot welding of Q&P steels requires a rigorous control of the joining parameters in order to achieve a suitable mechanical strength, fatigue resistance and energy absorption capability for the automotive industry. For all of these reasons, the monitoring of RSW can be considered a useful tool to collect information about the joining process and to help in determining proper weldability windows. It is common practice to monitor the welding current and electrode tip voltage to detect metal expulsion. The loss of molten metal from the nugget normally induces the formation of cracks, voids and metal splashes. These defects reduce the mechanical performance of spot welds, in particular under dynamic loads [10]. Moreover, metal expulsion has a harmful effect on weld bonding (spot welding in conjunction with adhesive bonding) as it damages the adhesive layer [11]. As a result, it is necessary to limit metal expulsion as much as possible since it may lead to nonconforming spot welds, in terms of strength and defects, for the assembling of car body parts.

The aim of this study was to investigate the weldability of dissimilar Q&P/TRIP spot welds by evaluating the effects of the welding current, clamping force and welding time on their microstructure and mechanical strength. The current and electrode tip voltage were monitored during the welding process in order to detect any metal expulsion and to evaluate its severity and effect on the load-carrying and absorption capability of the Q&P/TRIP joints. A design of experiment was adopted to realize dissimilar welded lap- and cross-joint samples with different welding parameters. These specimens were then subjected to shear and cross tension tests, respectively, in order to measure the maximum load and the corresponding absorbed energy and displacement (an index of ductility) on the load-crosshead displacement curves. The fracture modes of the welded shear- and cross-joint samples were assessed from a macroscopic and microscopic standpoint. Optical and SEM microscopy, as well as Vickers measurements, were used to determine the microstructures and hardness throughout the dissimilar spot welds.

2. Materials and Methods

2.1. Q&P and TRIP Steels

A Q&P (grade Q&P980) sheet steel and a TRIP (ISO 1.0948) sheet steel were selected for the dissimilar spot welding tests. These sheets were fabricated by means of industrial thermo-mechanical rolling processes and provided in an annealed condition. The Q&P steel had a thickness of 1.1 ± 0.05 mm and was uncoated, while the TRIP steel was 1.5 ± 0.05 mm thick and was hot-dipped zinc coated. The Q&P steel microstructure consists of a mixture of retained austenite and martensite, whose volume fractions are about 60% and 40%, respectively (Figure 1a). TRIP steel exhibits a more complex microstructure that is made up of a mixture of ferrite, retained austenite and bainite (although the presence of a small amount of martensite cannot be excluded) (Figure 1b). In these steels, ferrite and austenite contribute to formability and toughness, whereas martensite and bainite ensure mechanical strength.

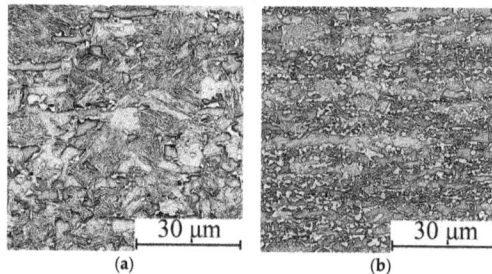

(a)   (b)

Figure 1. SEM micrographs of the (a) Quenching and Partitioning (Q&P) and (b) Transformation Induced Plasticity (TRIP) steels in as-received conditions.

The chemical compositions and the main mechanical properties of Q&P and TRIP steels are listed in Tables 1 and 2, respectively.

Table 1. Chemical composition (wt. %) of the Quenching and Partitioning (Q&P) and Transformation Induced Plasticity (TRIP) steels as measured by means of optical emission spectroscopy.

Steel	C (%)	Si (%)	Mn (%)	P + S (%)	Al (%)	Nb (%)
Q&P	0.22	1.41	1.88	<0.02	0.04	<0.001
TRIP	0.20	0.31	2.23	<0.02	1.05	0.022

Table 2. Main mechanical properties of the Q&P and TRIP steels. UTS: ultimate tensile strength; YS: yield strength; e_f: elongation at fracture.

Steel	YS (MPa)	UTS (MPa)	e_f (%)	Hardness (HV0.5)
Q&P	655	1000	22	260
TRIP	525	890	27	225

2.2. RSW Welding and Monitoring

Q&P and TRIP sheets were cut from large industrial sheets into coupons of 105 mm × 45 mm and 150 mm × 50 mm, which were then tested by means of shear and cross tension tests, as discussed hereafter. Both types of coupons were cut with the largest length along the rolling direction. Spot welding tests were performed using an industrial medium frequency direct current machine (MFDC) (Matuschek Messtechnik GmbH, Alsdorf, Germany) with an alternating current of 1 kHz. The smaller coupons were welded together in a lap-joint configuration, Figure 2a, whereas the larger coupons where welded in a cross configuration (Figure 2b). In both cases, spot welds were realized in the middle of the overlapped area by means of two copper-chromium electrodes, each with a face diameter of 6 mm, according to the AWS D8.9M standard [12].

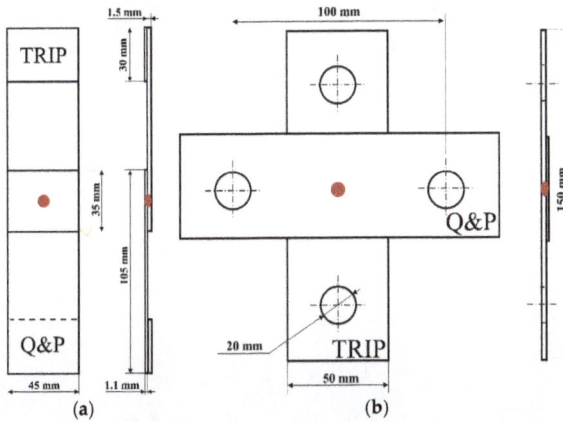

Figure 2. Geometrical size of the welded (**a**) shear tension and (**b**) cross tension specimens. Drawings not to scale.

The sheet coupons were welded at varying currents, clamping forces and welding times, based on the design of experiment, L-9(3^3) orthogonal array, shown in Table 3. The welding parameters range was defined after a preliminary pilot experimentation, following the recommendation of the ISO 18278-2 standard [13] and the industrial practice: the minimum current intensity was set to ensure the occurrence of a minimal nugget size that allowed the formation of a complete button pull fracture during a peel test (e.g., an incomplete pull out could be obtained for lower currents than 6 kA), whereas the maximum current intensity was determined as a significant metal expulsion occurred. In order to achieve consistent results, at least three samples were welded in a lap-joint configuration and two samples in a cross-joint configuration for each welding combination.

Table 3. L-9(3^3) orthogonal array for the welding tests, both for the shear and cross tension configurations. I_{weld}: welding current; F_{clamp}: clamping force; t_{weld}: welding time.

Run	I_{weld} (kA)	F_{clamp} (kN)	t_{weld} (ms)
1	6	2	200
2	6	3	325
3	6	4	450
4	7.5	2	325
5	7.5	3	450
6	7.5	4	200
7	9	2	450
8	9	3	200
9	9	4	325

The electrode tip voltage was measured with two electrical wires directly clipped to the electrode tips. Twist pairs were used to reduce the induced voltage noise of the alternating welding current. The current was measured with an electrical transducer that consisted of a Rogowski coil and an integrator circuit (Figure 3). The transducer provided an output voltage that was proportional to the welding current, in mV/A. The Rogowski coil was looped around the bottom electrode arm as close as possible to the electrode tips to obtain signals related directly to the joining process.

Figure 3. The electrical transducer (Rogowski coil and integrator device) used in the welding tests [14].

The high alternating current (I_{weld}) through the electrode arms induced a variable magnetic field in the surrounding environment and, in turn, an induced voltage (V_{coil}) in the Rogowski coil, as follows:

$$V_{coil} = H\,(dI_{weld}/dt) \tag{1}$$

where H is the coil sensitivity (in Vs/A), which is a characteristic feature of the coil itself. V_{coil} had to be integrated to reproduce the welding current waveform, so that

$$V_{out} = S \int V_{coil} dt = S_H I_{weld} \tag{2}$$

where S is a characteristic factor of the integrator circuit and S_H is the overall transducer sensitivity (in mV/A). As a result, it was possible to calculate the welding current by measuring the output voltage of the integrator and multiplying it by the transducer sensitivity (0.22 mV/A). The electrode tip voltage and welding current signals were collected simultaneously, each at a rate of 100,000 samples per second, by means of an NI USB-6216 acquisition system (National Instruments, Austin, TX, USA).

2.3. Microstructural and Mechanical Characterization of Spot Welds

The lap- and the cross-joint samples were tested by using an axial testing machine (Easydur Italiana, Induno Olona, Italy), according to the AWS D8.9M standard [12]. All the tests were conducted with a crosshead speed set at 10 mm/min. Based on the AWS standard, the maximum load reached during each test, the energy absorbed and the displacement up to the peak load were collected from each test. The fracture modes of all the welded samples, an index of the fracture resistance capability, were visually assessed and classified in accordance with the AWS D8.1M standard [15].

Some welded specimens were cut along the middle section, polished, etched with chemical solutions (2% nital and/or picral reagents) and then examined by means of optical (Optika srl, Ponteranica, Italy) and SEM (Phenom-World, Eindhoven, The Netherlands) microscopy to detect the microstructures of the parent metals, the heat affected zones (HAZs), and the weld nugget. The spot weld geometry was characterized for each welding condition by evaluating the nugget diameter and spot thickness. A metallographic examination was also performed on the fractured shear and cross tension specimens to determine the regions where cracks propagated.

Vickers microhardness measurements were performed to characterize the hardness of the microstructures throughout the spot welds from the Q&P side, through the fusion zone and HAZ, to the TRIP side. Hardness indentations were made using a 200 g load (HV0.2). The distance between two successive indentations was either 0.2 or 0.4 mm.

2.4. Statistical Analysis

A multifactor analysis of variance (ANOVA) of the shear and cross strength values was carried out to determine which welding parameters had a statistically significant effect on the spot weld strength and their contributions. A multiple range test was used to determine how each factor affected the mechanical strength of the Q&P/TRIP spot welds.

3. Results

3.1. Spot Weld Microstructural and Hardness Characterization

Dissimilar spot welds are usually characterized by chemical heterogeneity in the fusion zone, due to the different chemical compositions of joined steels, and by nugget asymmetry, because of their different thicknesses, melting points, thermal and electrical conductivities [8]. Figure 4 displays a typical nugget obtained from the Q&P/TRIP welds, as well as the average nugget diameter and spot thickness for the different welding runs. It can be seen that the nugget is slightly asymmetric with respect to the faying surface (Figure 4a). This can mainly be attributed to the different thicknesses of the welded sheets. As already known from the literature, the Q&P/TRIP nuggets widen as the welding current is increased. The clamping force exhibits the opposite behavior instead: a larger force reduces the shear strength of spot welds. This is coherent with the effect of the clamping force on a microscopic scale: the area and the number of regions where the sheets and the electrode tips are in direct contact increase as the clamping force increases. As a result, the total electrical contact resistance and, hence, the current density decrease, and, in turn, the heat input also decreases. Therefore, the spot welds tend to exhibit the smallest nuggets for the same current level when the highest clamping force is used. The joint thickness is mainly reduced for increased clamping forces (Figure 4b).

Figure 5 shows the microstructural changes of the Q&P and TRIP steels in the HAZ regions (from near the fusion zone to near the base metal), whereas two representative spot weld microhardness profiles (welding configurations No. 3 and 7) are displayed in Figure 6. Overall, the microstructures of the fusion zone and the HAZ of the Q&P and TRIP steels are similar in all the welding configurations; however, their size changes at varying welding parameters. The heat input involved in the joining process has a notable effect on the microstructures and, in turn, on the hardness of the joints and steels. The high cooling rate in the fusion zone (between the two water-cooled electrode tips) induces the formation of a full martensite microstructure, as also pointed out by the corresponding high hardness values. The scattering in the hardness of the fusion zone can mainly be attributed to a local inhomogeneous chemical composition (e.g., segregations). The regions close to the fusion zone on both steel sides fully austenitized, thereby inducing the formation of a full martensitic structure. The hardness of the martensitic structures in the HAZs of the Q&P and TRIP steels is quite similar, with the former steel having a slightly higher hardness in some cases (consistently with the carbon content of the two steels) (Figure 6a). Moving toward the parent steels, the thermal cycle promoted intercritical transformations or tempering of the microstructures. On the Q&P side, martensite formed by cooling from the intercritical regions. Far away, the previous martensite tempered, showing a decrease in hardness with the presence of local minimum values, whereas the retained austenite transformed into bainite, with a corresponding increase in hardness close to the base metal. On the TRIP side, the previous bainite tempered, the retained austenite transformed into bainite, whereas the ferrite grains slightly enlarged. Overall, a continuous decrease in hardness occurred from the martensitic region, close to the nugget, to the base metal.

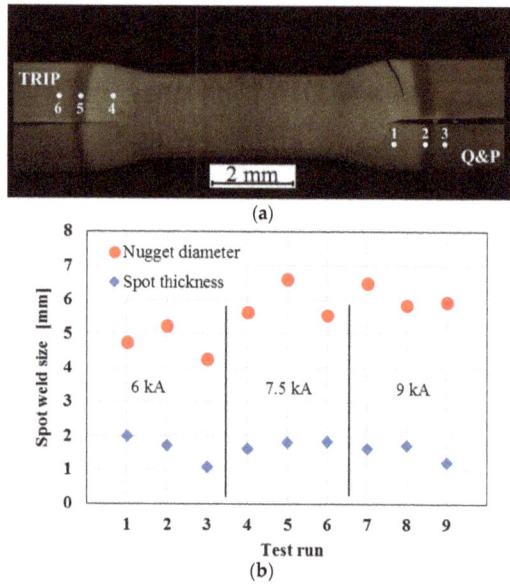

(a)

(b)

Figure 4. Q&P/TRIP spot weld: (**a**) typical cross section appearance and (**b**) average nugget diameter and joint thickness obtained for the different welding conditions. The numbers from 1 to 6 define the locations of the microstructures displayed in Figure 5.

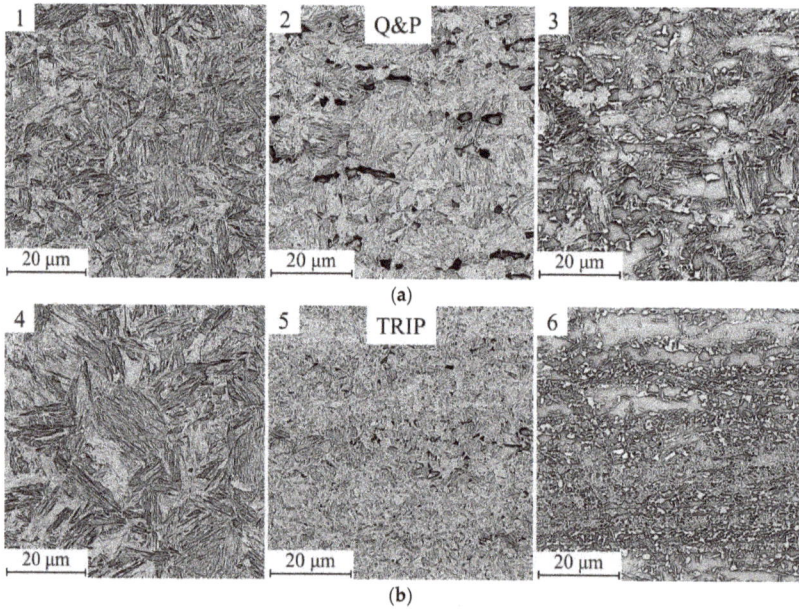

(a)

(b)

Figure 5. SEM micrographs of the microstructure in the heat affected zones (HAZs) on the (**a**) Q&P and (**b**) TRIP sides: from close to the fusion zone (left-hand images) to the base metals (right-hand images). The numbers from 1 to 6 refer to the locations pointed out in Figure 4.

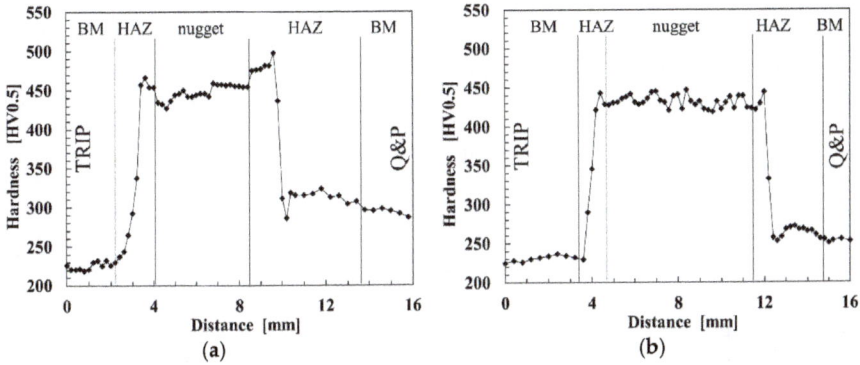

Figure 6. Typical microhardness profiles of the Q&P/TRIP welds. Welding configurations: (**a**) run No. 3 (6 kA, 4 kN, 450 ms); and (**b**) run No. 7 (9 kA, 2 kN, 450 ms).

3.2. Monitoring of the Welding Parameters

Figure 7 shows the types of signals that were obtained from the monitoring of the welding current and electrode tip voltage. As metal expulsion did not occur, Figure 7a, the current increased at the beginning of the joining process until it reached the set value (root mean square, RMS). Then, the current was held constant by the electric inverter of the welding machine. Consequently, the electrode tip voltage and electrical resistance modified based on the changes in the growth of the nugget.

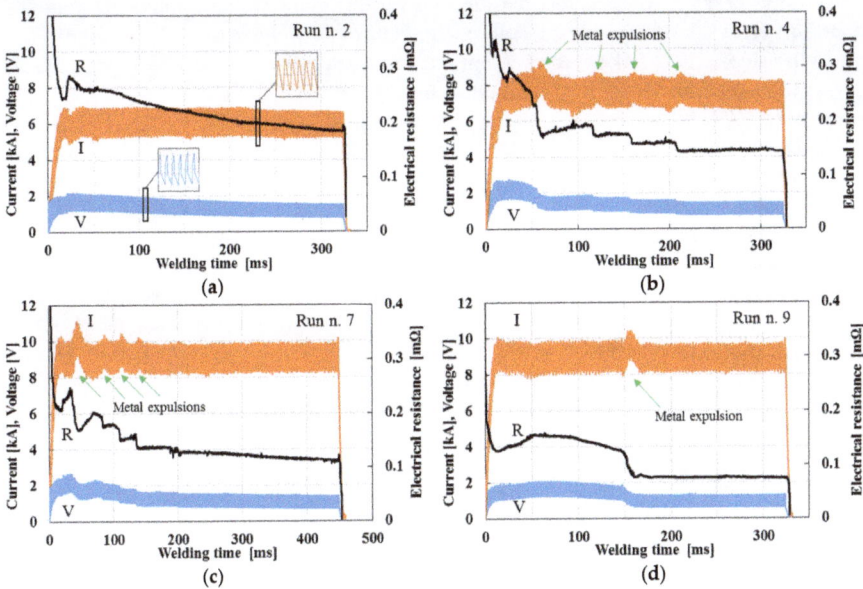

Figure 7. Typical results of the monitoring of the welding current and electrode tip voltage, as well as of the calculated electrical resistance: (**a**) run No. 2 without metal expulsion; (**b**) run No. 4; (**c**) No. 7; and (**d**) No. 9 with metal expulsion. I: welding current; R: electrical resistance; V: electrode tip voltage.

The presence of a metal expulsion abruptly reduced the electrical resistance of the sheet stack, inducing a local drop in the welding current and electrode tip voltage (Figure 7b–d). This drop in electrical resistance is amenable to the collapse around the nugget that reduces the sheet stack thickness through which any additional current must flow along with the widened effective contact area due to the expelled molten metal entrapped on the faying surface [16]. The resistance drop can be used to quantify the severity of a metal expulsion: the larger the loss of material in an expulsion event, the larger the reduction in dynamic resistance (in mΩ) [17]. Metal expulsion could happen through a single large expulsion, Figure 7d, or through multiple expulsions (Figure 7b,c). The former type occurred for run No. 9 and for some samples No. 8, whereas the latter occurred for runs No. 4 and 7 and for some samples No. 8. The appearance of a single expulsion is probably due to the largest clamping force that was used (4 kN) in the welding configuration No. 9. Metal expulsion occurred as soon as molten metal temporarily exceeded the compressive force of the surrounding solid, but the high clamping force was then able to prevent further metal losses. In all the samples of No. 9, the resistance drops occurred in the 140–170 ms range along with a reduction of 0.04–0.05 mΩ. The number of expulsions increased as the clamping force was reduced since the molten metal was able to reach a critical pressure several times during the joining process. Two or even fewer expulsions were observed in the samples No. 8 (3 kN), but they were more numerous in the samples No. 4 and 7, which were welded with the lowest clamping force (2 kN). It can be observed from Figure 7b,c that a larger metal expulsion occurred at the beginning of the joining process, in the 40–60 ms range, and this was then followed by other less severe expulsions. These expulsions took place over shorter times for the samples No. 7, whereas they were distributed over longer times for the samples No. 4. The largest expulsions were similar to those detected for the samples No. 9, which underwent a resistance drop of 0.04–0.05 mΩ. The successive expulsions, instead, exhibited a decrease in resistance of 0.01–0.015 mΩ.

3.3. Shear and Cross Tension Tests

The results of the shear and cross tension tests for the nine welding configurations are listed in Table 4, whereas Table 5 shows the results obtained from the multifactor ANOVA carried out on the strength values. The ANOVA results point out that all the welding parameters are statistically significant, at the 95% confidence level, in affecting the shear strength of welded samples. The welding current is the most important parameter influencing strength (55.9%), followed by the clamping force (22.4%) and the welding time (5.6%). Owing to the lack of more experimental combinations, the possible interactions among the welding parameters cannot be calculated, although their contribution is below 16.1% (residual contribution). The ANOVA performed on the cross tension values shows that the welding parameters are not statistically significant at the 95% confidence level. This is due to the slight changes in the strength values at varying welding combinations. However, some considerations can be made if the cross strength is normalized to the spot weld size, as discussed hereafter.

Table 4. Results of the shear and cross tension tests for the different welding configurations. Displ.: displacement; α: cross tension strength normalized to the spot weld size.

Run	Shear Tension			Cross Tension			α
	Strength (kN)	Energy (J)	Displ. (mm)	Strength (kN)	Energy (J)	Displ. (mm)	(kN/mm^2)
1	15.8 ± 0.9	21.9 ± 3.1	2.7 ± 0.3	4.96	30.4	10.9	0.95
2	16.1 ± 0.5	23.5 ± 1.3	3.0 ± 0.3	4.68	39.4	12.3	0.81
3	14.6 ± 0.1	18.8 ± 2.0	2.5 ± 0.3	4.65	25.6	9.3	1.00
4	19.1 ± 1.0	30.8 ± 3.8	3.0 ± 0.2	4.83	50.7	14.5	0.78
5	19.9 ± 0.8	35.8 ± 2.9	3.5 ± 0.1	5.33	67.3	17.0	0.71
6	17.2 ± 0.7	25.3 ± 2.6	2.8 ± 0.1	4.49	44.8	14.2	0.74
7	21.4 ± 0.8	37.8 ± 5.8	3.4 ± 0.4	5.03	68.4	16.9	0.70
8	19.1 ± 0.6	33.1 ± 1.8	3.3 ± 0.1	5.05	49.8	14.0	0.78
9	17.2 ± 0.6	23.9 ± 2.7	2.7 ± 0.2	4.69	52.5	15.6	0.72

Table 5. ANOVA analysis of the shear tension strength values of the spot welds. SS: sum of squares; DoF: degree of freedom; MS: mean square; Fischer (F)-ratio; probability (*p*)-value.

Parameter	SS	DoF	MS	F-Ratio	*p*-Value	Contribution (%)
Current	75.06	2	37.53	46.1	$<10^{-5}$	55.9
Force	31.12	2	15.56	19.1	$<10^{-5}$	22.4
Time	9.01	2	4.50	5.5	0.012	5.6
Residual	16.28	20	0.81			16.1
Total	131.46	26				

The influences of the welding parameters on the shear strength are summarized in the interval plots of Figure 8. It can be seen that the welding current increases the shear strength of the spot welds from 6 to 9 kA. However, the increase in strength is limited when the current passes from 7.5 to 9 kA. This may be attributed to the occurrence of metal expulsions, which limit the improvement in strength that could be obtained with a higher welding current. The effect of the clamping force on the shear strength is similar at 2 and 3 kN, whereas it significantly reduces the shear strength at 4 kN. Two main factors are responsible for the reduction in the spot strength as the clamping force is increased: (i) higher clamping forces reduce the heat input and, in turn, the nugget size (as previously mentioned); (ii) the indentation of the electrode tips on the sheet surfaces induces high stress concentrations in the regions around the weld nugget [18,19]. The welding time has the same effect on the shear strength up to 325 ms, whereas it increases the strength for 450 ms. However, its contribution is not important, as can be deduced from the ANOVA table.

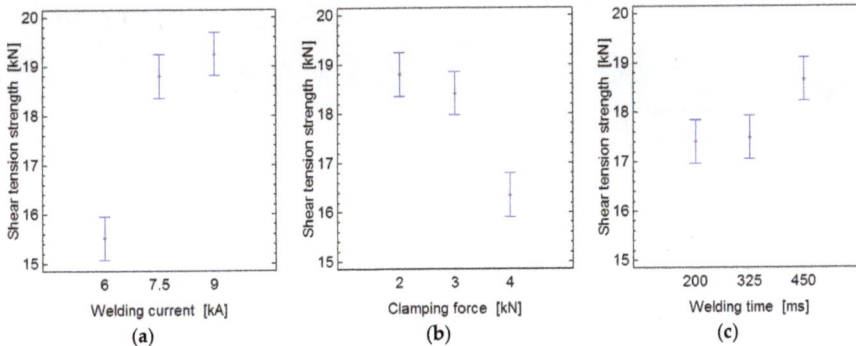

Figure 8. Mean shear tension strength and standard error (95% confidence intervals) for each level of the welding parameters: (**a**) welding current; (**b**) clamping force; and (**c**) welding time.

Figures 9 and 10 display the shear and cross tension strength as functions of the nugget diameter. Since the displacement values have the same trend as the absorbed energy values, they have not been plotted so as to avoid redundant data. As expected, the shear tension strength and absorbed energy increase as the nugget size increases. Test run No. 7 (9 kA, 2 kN, 450 ms) gives the maximum strength and absorbed energy coherently with the maximum current and welding time used for this configuration. In this regard, even though metal expulsion occurred during run No. 7, due to the high current, it only happened at the beginning of the joining process and not for most of the time (for about 300 ms), as shown in Figure 7c. This promoted the formation of large nuggets and, in turn, the highest strengths. The metal expulsion had a more negative effect on the samples No. 8 and 9 for the same current level. In fact, these samples exhibited similar shear strengths to those of the samples No. 4 and 6, which were welded with a lower current. Since they were welded with a shorter time, and large expulsions occurred in the middle of the joining process, the capability of the nugget to growth was probably reduced.

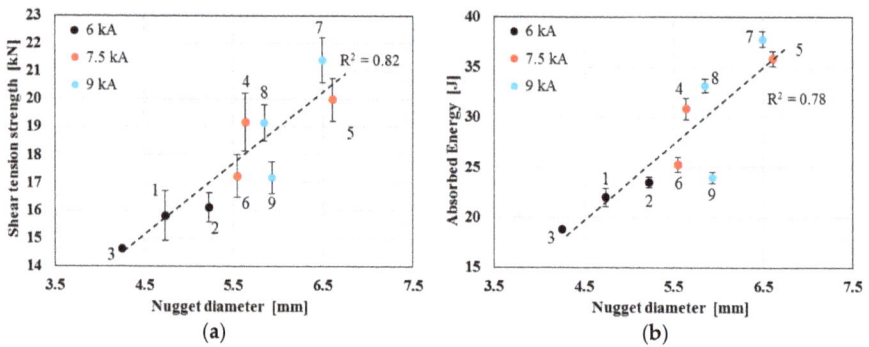

Figure 9. (a) Shear tension strength and (b) absorbed energy of the spot welds for the different welding conditions. The numbers in the graph refer to the experimental runs.

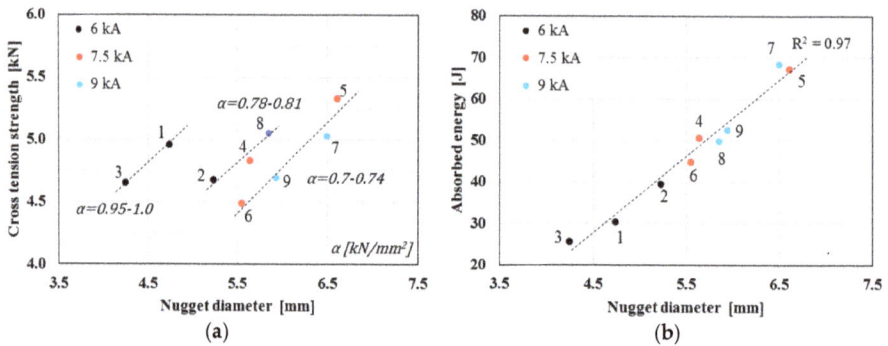

Figure 10. (a) cross tension strength and (b) absorbed energy of the spot welds for the different welding conditions. The numbers in the graph refer to the experimental runs.

As can be seen from Figure 9, although the samples No. 5 exhibit a slightly larger nugget and they did not experience any expulsion, they exhibit lower strength values than samples No. 7. Considering that the welding time has a low contribution to the shear strength (see the ANOVA table), this discrepancy may be attributed to the stronger electrode indentation caused by the higher clamping force. This would also justify the higher strength values of the samples No. 4 than those of the samples No. 6, which have a similar nugget but were subjected to a higher clamping force, as well as the lower strength values obtained for the samples (welded with the same current level) that were clamped with the highest force. Absorbed energy shows the same trend as the shear strength. Therefore, metal expulsion also reduces the energy absorbing attitude of the spot welds during mechanical loading, and also presumably their tendency to withstand impulsive loads in the case of vehicle accidents.

A common way of assessing the mechanical response of cross-welded samples is to make use of the α ratio normalized to the spot weld size, in kN/mm^2, as follows [20]:

$$\alpha = CTS/(d_n \times th_s), \tag{3}$$

where CTS is the cross tension strength, d_n is the nugget diameter and th_s is the sheet thickness. Since the experimental welding tests involved steels with different thicknesses, α has been computed considering the minimum thickness, based on the ISO standard [13], where the thinnest sheet also guides welding parameters. On the basis of the α values, three different groups can be defined from Figure 10a, where the typical linearity between stress and nugget size is held. The first group includes runs No. 1 and 3, where α is about 1. The second group consists of samples welded with different welding currents, with the cross tension strength increasing as the current is increased, where α is about 0.8. In the third group, α is about 0.7. The graph shows roughly that the capability of spot welds to withstand cross tension stresses, in terms of strength per unit of area, reduces as the nugget increases and metal expulsion occurs. Samples with large nuggets that experienced metal expulsions at high currents (runs No. 7–9) exhibit similar cross tension strengths to those of the samples welded with lower currents. These results are coherent with those of a previous work by Huin et al. about the dissimilar welding of DP and hot stamping boron steels [21].

3.4. Spot Weld Fracture

During the shear tension tests, the Q&P/TRIP spot welds could fail either by interfacial fracture or by button pull (Figure 11). Interfacial fractures occurred in all of the samples welded during the run No. 1 and in some samples obtained with runs No. 2 and 3. This fracture mode is due to the small size of the nuggets, which were not able to sustain large shear stresses. It can be noted that the tendency of interfacial to button pull fracture from low to high current is coherent with the increase in the size of the nugget.

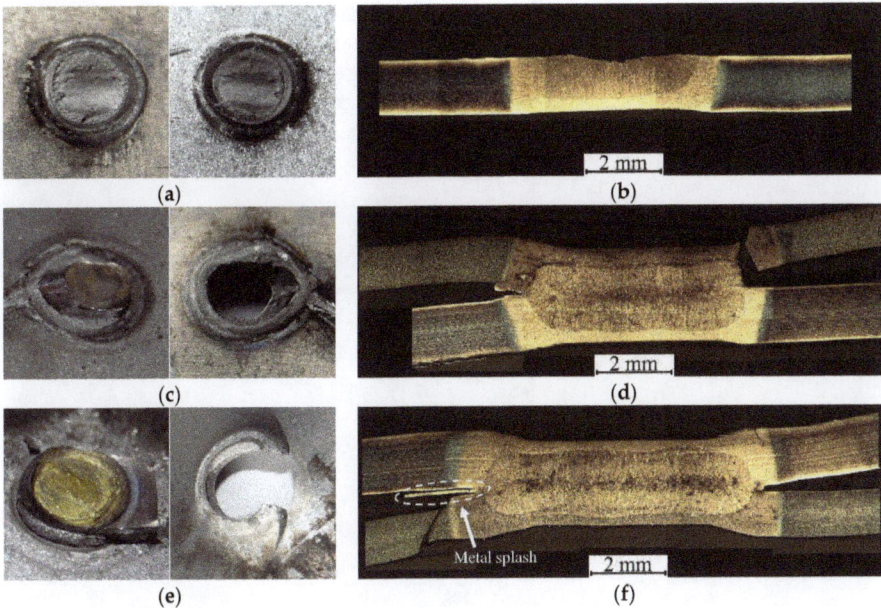

Figure 11. Modes of fractures of the spot welds obtained after the shear tension tests: (**a,b**) interfacial fracture; (**c,d**) button pull with the fracture path mainly along the nugget border; and (**e,f**) button pull with the fracture path through the steel sheet.

The button pull fractures could appear in two different modes, depending on the crack propagation path. In one case, a fracture grew along the lateral border of the nugget (with respect to the cross section in Figure 4a) and through the sheet in the samples with an intermediate shear tension strength, as occurred for the runs No. 6 and 9, and partially for runs No. 2–5. In other cases, cracks spread in the HAZ of the Q&P steels, particularly in the region where the original microstructures tempered due to the heat input, as occurred for runs No. 7 and 8 and partially for runs No. 4 and 5. In both of the button pull fractures, failure mainly occurred on the Q&P side, due to its reduced thickness compared to the TRIP steel. The transition from the two types of failures is induced by the bending moment caused by the rotation of the welded joint during the mechanical test. In fact, if the weld nugget is sufficiently large, it can rotate during the test. Therefore, the stress condition, which is initially shear, primarily becomes tensile along the sheets [19]. At this state of stress, the region with the lowest tensile strength breaks, thus leading to the final fracture. This explanation justifies the fracture in the HAZ of Q&P steels where the martensite tempered, this being the region with the minimum hardness (see Figure 6) and, in turn, the minimum strength.

The fracture modes that have been obtained from the cross tension tests are summarized in Figure 12. The spot welds exhibit two different modes of fracture: interfacial fracture with button pull and button pull by partial dome fracture. Coherently with the thinner thickness, the Q&P sheet deformed more than the TRIP sheet, and the fractures mainly propagated on the Q&P side. The former fracture only occurred for the samples welded during run No. 1, Figure 12a,b, due to the small size of the nuggets generated by the low heat input. The cracks nucleated at the notch tip of the faying surface, which is a site of stress concentration, and then propagated in an interfacial mode; at a given distance, the bending moment involved in the nugget changed the crack path from the faying surface to through the Q&P sheet, due to the cross load and the presence of the crack itself, up to the final failure. The partial dome fractures could be characterized either by a crack propagation along the border of the nugget and then through the Q&P sheet (Figure 12d) or by a fracture that could also involve the HAZ regions of the two steels (Figure 12f). The transition between the two types of button pull failures is attributable to the nugget size and to the angle between the faying surface and the border of the nugget close to the notch tip. The lower the angle, the greater the tendency of a fracture to spread along the nugget border [21]. Moreover, the presence of metal splashes at the notch tip, referring to the left-hand notch tip in Figure 7f, could promote crack propagation in the HAZ regions, but not along the nugget border. This fracture mode was in fact observed in almost all of the samples that underwent metal expulsion.

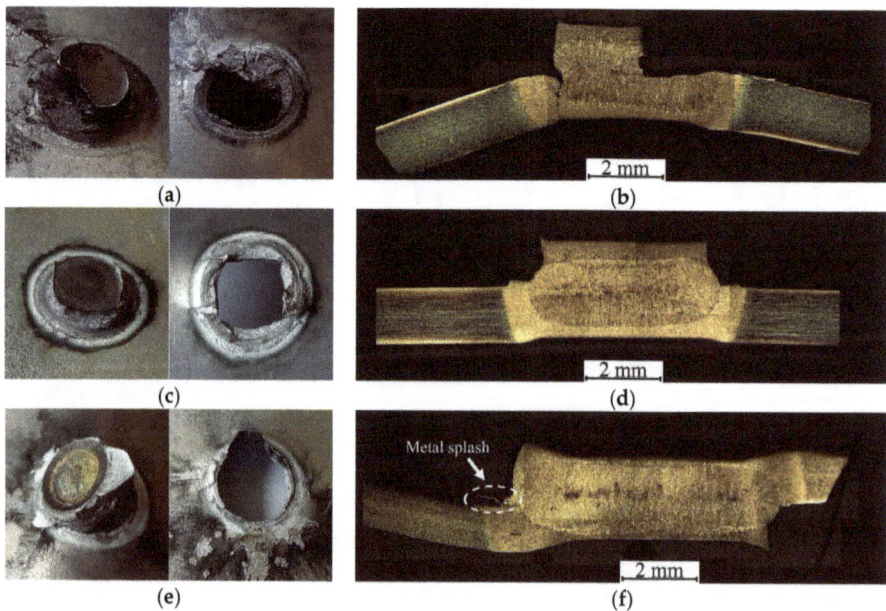

Figure 12. Modes of fractures of spot welds obtained after the cross tension tests, macroscopic and microscopic views. Typical fracture: (**a,b**) interfacial fracture with button pull; (**c,d**) partial dome fracture; and (**e,f**) button pull along nugget border and through HAZ steel sheet.

4. Conclusions

This work has investigated the spot weldability of a Q&P steel with a TRIP steel for assembly applications for the automotive industry and the severity and effect of metal expulsion on load-carrying capability of Q&P/TRIP joints. The main results are summarized as follows:

- Based on the monitored signals, expulsion can appear as a single event or multiple events during the joining process, its severity being assessed by the extent of the dynamic resistance drop.
- The welding current is the most important parameter that affects the shear strength, followed by the clamping force and welding time. No statistical significant parameters have been found for the cross tension strength; however, the cross tension strength normalized by to spot size (α) points out the detrimental effect of a metal expulsion.
- If the expulsion occurs at the beginning of the joining process and for a short time, its harmful effect on shear strength is more limited for the longest welding time (sample No. 7, 450 ms).
- The shear-welded samples failed by interfacial and button pull fractures. Button pull fractures could occur by crack propagation along the nugget border or in the Q&P HAZ where previous martensite tempered.
- Cross-welded samples failed by interfacial and button pull, and button pull by partial dome fracture. Metal splashes promoted the crack propagation through the HAZ regions in partial dome fractures.
- Run No. 5 (7.5 kA, 3 kN, 450 ms) represents the best welding combination since it ensures spot welds with high shear and cross tension strengths and the absence of metal expulsion.

Acknowledgments: The authors are grateful to the Free University of Bozen–Bolzano that supported this research work (Grant No. TN2001, scientific coordinator: Pasquale Russo Spena). The authors would also like to thank Giovanni Marchiandi (Politecnico di Torino), which supported the cross tension tests.

Author Contributions: Pasquale Russo Spena, Manuela De Maddis and Franco Lombardi conceived and designed the experiments; Pasquale Russo Spena and Manuela De Maddis performed the welding tests; Pasquale Russo Spena carried out the metallographic analysis, shear tension tests and microhardness measurements; Manuela De Maddis and Franco Lombardi performed the cross tension tests; the authors with the contribution of Gianluca D'Antonio analyzed the data of welding monitoring; and Pasquale Russo Spena wrote the paper with the support of Manuela De Maddis and Franco Lombardi.

References

1. De Moor, E.; Speer, J.G.; Matlock, D.K.; Kwak, J.H.; Lee, S.B. Effect of carbon and manganese on the quenching and partitioning response of CMnSi steels. *ISIJ Int.* **2011**, *51*, 137–144. [CrossRef]
2. Sun, J.; Yu, H. Microstructure development and mechanical properties of quenching and partitioning (Q&P) steel and an incorporation of hot-dipping galvanization during Q&P process. *Mater. Sci. Eng. A* **2013**, *586*, 100–107. [CrossRef]
3. Rossini, M.; Russo Spena, P.; Cortese, L.; Matteis, P.; Firrao, D. Investigation on dissimilar laser welding of advanced high strength steel sheets for the automotive industry. *Mater. Sci. Eng. A* **2015**, *628*, 288–296. [CrossRef]
4. Casalino, G.; Campanelli, S.L.; Ludovico, A.D. Laser-arc hybrid welding of wrought to selective laser molten stainless steel. *Int. J. Adv. Manuf. Tech.* **2013**, *68*, 209–216. [CrossRef]
5. Peterson, W.; Borchelt, J. Maximizing cross tension impact properties of spot welds in 1.5 mm low carbon, dual-phase, and martensitic steels. In *SAE Technical Paper 2000-01-2680*; SAE International: Warrendale, PA, USA, 2000.
6. Donders, S.; Brughmans, M.; Hermans, L.; Tzannetakis, N.O. The effect of spot weld failure on dynamic vehicle performance. *Sound Vibrat.* **2005**, *39*, 16–25.

7.   Wang, B.; Duan, Q.Q.; Yao, G.; Pang, J.C.; Li, X.W.; Wang, L.; Zhan, Z.F. Investigation on fatigue fracture behaviors of spot welded Q&P980 steel. *Int. J. Fract.* **2014**, *66*, 20–28. [CrossRef]
8.   Russo Spena, P.; Maddis, M.D.; Lombardi, F.; Rossini, M. Dissimilar resistance spot welding of Q&P and TWIP steel sheets. *Mater. Manuf. Process.* **2016**, *31*, 291–299. [CrossRef]
9.   Russo Spena, P.; Cortese, L.; Maddis, M.D.; Lombardi, F. Effects of process parameters on spot welding of trip and quenching and partitioning steels. *Steel Res. Int.* **2016**, in press. [CrossRef]
10.   Zhang, H. Expulsion and its influence on weld quality. *Weld. J.* **1999**, *78*, 373-s–380-s.
11.   Zhang, H.; Senkara, J. *Resistance Welding: Fundamentals and Applications*, 2nd ed.; CRC Press: London, UK, 2011.
12.   *AWS D8.9M, Test Methods for Evaluating the Resistance Spot Welding Behavior of Automotive Sheet Steel Materials*; American Welding Society (AWS): Miami, FL, USA, 2012.
13.   International Organization for Standardization. *Resistance Welding-Weldability-Part 2: Alternative Procedures for the Assessment of Sheet Steels for Spot Welding*; ISO Copyright Office: Geneva, Switzerland, 2004.
14.   CWT Specification, Power Electronic Measurement Ltd. Available online: www.pemuk.com (accessed on 7 November 2016).
15.   *AWS D8.1M, Specification for Automotive Weld Quality—Resistance Spot Welding of Steel*; American Welding Society (AWS): Miami, FL, USA, 2013.
16.   Dickinson, D.W.; Franklin, J.E.; Stanya, A. Characterization of spot welding behavior by dynamic electrical parameter monitoring. *Weld. J. Weld. Res. Suppl.* **1980**, *59*, 170–176.
17.   Garza, F.; Das, M. On real time monitoring and control of resistance spot welds using dynamic resistance signatures. In Proceedings of the 44th IEEE 2001 Midwest Symposium on Circuits and Systems MWSCAS, Dayton, OH, USA, 14–17 August 2001.
18.   Pouranvari, M.; Asgari, H.T.; Mosavizadch, S.M.; Marashi, P.H.; Goodarzi, M. Effect of weld nugget size on overload failure mode of resistance spot welds. *Sci. Technol. Weld. Join.* **2007**, *12*, 217–225. [CrossRef]
19.   Chao, Y.J. Failure modes of spot welds: Interfacial versus pullout. *Sci. Technol. Weld. Join.* **2003**, *8*, 133–137. [CrossRef]
20.   Biro, E.; Cretteur, L.; Dupuy, T. Higher than expected strengths from dissimilar Configuration advanced high strength steel spot welds. In Proceedings of the Sheet Metal Welding Conference XV, Livonia, MI, USA, 2–5 October 2012.
21.   Huin, T.; Dancette, S.; Fabrègue, D.; Dupuy, T. Investigation of the failure of advanced high strength steels heterogeneous spot welds. *Metals* **2016**, *6*, 111. [CrossRef]

Article

The Interfacial Microstructure and Mechanical Properties of Diffusion-Bonded Joints of 316L Stainless Steel and the 4J29 Kovar Alloy Using Nickel as an Interlayer

Tingfeng Song [1], Xiaosong Jiang [1,*], Zhenyi Shao [1,2], Defeng Mo [3], Degui Zhu [1] and Minhao Zhu [1]

[1] School of Materials Science and Engineering, Southwest Jiaotong University, Chengdu 610031, Sichuan, China; tfsong@yeah.net (T.S.); zysao_10227@163.com (Z.S.); dgzhu@home.swjtu.edu.cn (D.Z.); zhuminhao@home.swjtu.edu.cn (M.Z.)
[2] Department of Material Engineering, Chengdu Technological University, Chengdu 611730, Sichuan, China
[3] Key Laboratory of infrared imaging materials and detectors, Shanghai Institute of Technical Physics, Chinese Academy of Sciences, Shanghai 200083, China; modefeng@163.com
* Correspondence: xsjiang@home.swjtu.edu.cn; Tel.: +86-28-8760-0779

Academic Editor: Giuseppe Casalino
Received: 28 August 2016; Accepted: 19 October 2016; Published: 3 November 2016

Abstract: 316L stainless steel (Fe–18Cr–11Ni) and a Kovar (Fe–29Ni–17Co or 4J29) alloy were diffusion-bonded via vacuum hot-pressing in a temperature range of 850–950 °C with an interval of 50 °C for 120 min and at 900 °C for 180 and 240 min, under a pressure of 34.66 MPa. Interfacial microstructures of diffusion-bonded joints were characterized by optical microscopy (OM), scanning electron microscopy (SEM), X-ray diffraction (XRD), and energy dispersive spectroscopy (EDS). The inter-diffusion of the elements across the diffusion interface was revealed via electron probe microanalysis (EPMA). The mechanical properties of the joints were investigated via micro Vickers hardness and tensile strength. The results show that an Ni interlayer can serve as an effective diffusion barrier for the bonding of 316L stainless steel and the 4J29 Kovar alloy. The composition of the joints was 316L/Ni s.s (Fe–Cr–Ni)/remnant Ni/Ni s.s (Fe–Co–Ni)/4J29. The highest tensile strength of 504.91 MPa with an elongation of 38.75% was obtained at 900 °C for 240 min. After the width of nickel solid solution (Fe–Co–Ni) sufficiently increased, failure located at the 4J29 side and the fracture surface indicated a ductile nature.

Keywords: 316 stainless steel; 4J29 Kovar alloy; nickel; diffusion bonding

1. Introduction

316L stainless steel (Fe–18Cr–11Ni) is widely used for its low corrosion rate, which is attributed to its inner chromium oxide region and outer mixed iron–nickel oxide region [1,2]. Kovar (Fe–29Ni–17Co) alloys possess the advantages of low-temperature constant expansion, and thermal expansion properties and a good thermal matching performance similar to Si, Ge, and glass, thus obtaining wide application in the electronics industry [3–5]. Joining stainless steel and Kovar alloys has been widely done in aerospace, nuclear, and electronic industries for technical and economic reasons [6–8]. However, due to the differences in thermo-mechanical and metallurgical properties, many obstacles to achieving good dissimilar joints have been confronted, such as precipitates, intermetallics, and distortions of the weld interface, which is detrimental to joint properties [1,6–9].

Baghjari et al. [6] investigated the laser welding of AISI 420 stainless steel to a Kovar alloy, and the result showed that sulfur and phosphor were segregated at the austenitic boundary, which led to cracks formed in the weld and chromium carbide precipitation formed in the ferrite grain boundary. Mai et al. [10] reported on the welding of a Kovar alloy and steel via laser welding and confirmed that elliptically shaped pores formed during the welding process, and the number and size of these pores could be controlled by the welding speed. Inconel 718 and 316 stainless steel were joined via electron beam melting additive manufacturing technology, and the results showed that precipitates of niobium carbide, and the Laves phase formed in the fusion zone, generally led to solidification cracking [1]. Nekouie et al. [11] reported that the microstructure of dissimilar joints could be controlled by adjusting the specific point energy and beam offset when low carbon steel and austenitic stainless steel were welded via laser welding. Wu et al. [12] conducted laser welding between ferritic stainless steel and carbon steel. Lath martensite, upper bainite and widmanstatten structure ferrite were formed in the weld bead. A tensile strength similar to a base metal was achieved, and the fracture position located at the carbon steel base metal were far from the weld bead. Verena et al. [13] investigated the microstructure and the properties of dissimilar joints between high temperature steels PM91 and PM2000 welded via electron beam. Precipitation of an Al phase and the formation of a soft dendritic microstructure in the fusion were attributed to a worse performance. In order to further obtain a high strength and high stability dissimilar joints, a filler material that usually possessed good plasticity or an intermediary coefficient of thermal expansion between parent materials was adopted [14–16]. In this context, nickel was considered a suitable intermediate material because the thermal expansion coefficient of nickel was situated between 316 stainless steel and the 4J29 Kovar alloy, and other suitable properties, such as density, melting point, and crystal type, were similar to them. Wang et al. [17] employed a nickel–phosphorus alloy as an interlayer and achieved a good connection between the Kovar alloy and low-carbon low-alloy steel by using parallel seam welding. Sathiskumar et al. [18] studied a dissimilar joint between pulse-plated nickel and Inconel 718 manufactured via electron beam welding. The grain structure, sizes, orientation, and grain boundary characteristic distribution of the weld seam were investigated to reveal the influence of hydrogen induced cold cracking. Precipitated secondary carbides were observed on the intergranular and intragranular γ matrix. Madhusudhan et al. [19] applied nickel as an interlayer to connect maraging steel and low alloy steel via friction welding and revealed that nickel could be an effective barrier for the diffusion of elements such as carbon and manganese, and ductile fracture, higher tensile strength, and higher elongation was obtained. Compared with fusion welding, diffusion welding is a near net shape forming process and more suitable for joining dissimilar materials. It is a kind of welding method where the contact surface achieves porosity closure via creep and diffusion under high temperature and certain pressure [20–22].

In this work, efforts were made to accomplish the solid state diffusion bonding of 316 stainless steel and the 4J29 Kovar alloy using pure nickel as an interlayer. The effect of a nickel interlayer at different temperatures and bonding times was investigated. The research focused on the evolution of the interface microstructure, element diffusion, and the bond strength of joints.

2. Materials and Methods

2.1. Materials and Processing Parameters

The chemical compositions and physical properties of 316 stainless steel and the 4J29 Kovar alloy are presented in Tables 1 and 2, respectively. Cylindrical specimens with a 30 mm diameter and a 25 mm length were machined from the base metals. A pure nickel (99.9 wt. %) interlayer with a thickness of 70 µm was used as an intermediate material. Prior to bonding, the grinding surfaces were cleaned in an ultrasonic bath containing acetone, cleaned by ethyl alcohol, and dried in air. The experiments were conducted in a temperature range of 850–950 °C with an interval of 50 °C, a bonding time of 120 min, and at 900 °C with a bonding time of 120 min, 180 min, and 240 min in $(1–6) \times 10^{-1}$ Pa vacuum. 2.5 T (34.66 MPa) uniaxial pressure was applied via hot pressing. During processing, heating was started at a constant rate of 10 °C/min; after the joining operation, the furnace was cooled at a rate of 5 °C/min to 600 °C and then naturally cooled to room temperature.

Table 1. Chemical composition of base metals (wt. %).

Mn	Si	C	Fe	Co	Ni	Cr	S	P	Mo	Alloy
2.00	1.00	0.03	Bal.	-	11.05	18.17	0.03	0.04	2.00	316L
0.4	0.2	0.02	Bal.	17.17	28.67	-	0.02	0.02	-	4J29

Table 2. Physical properties of base metals (wt. %).

Alloy	Density (g/cm³)	Melting Point (°C)	Expansion Coefficient (10^{-6} K^{-1})	Ultimate Tensile Strength (MPa)
316L	8.9	1375	16	573
4J29	8.1	1460	4.7	580

2.2. Interface Microstructure Characterization

The bonded joints were cut longitudinally, grounded, polished, and etched with a separate reagent for metallographic observations. Stainless steel sides were etched by aqua regia. Kovar alloy sides were etched by a mixture of 2.5 g of $CuCl_2$, 50 mL of ethyl alcohol, and 50 mL of HCl. The change in microstructure owing to diffusion was revealed via light microscopy (AxioCam MRc 5, ZEISS, Wetzlar, Germany). Line scanning of the interface was carried out with a scanning electron microscope (SEM, JEOL Ltd., Tokyo, Japan JSM-7001F at 15 kV) equipped with an X-ray energy-dispersive spectrometer (EDS, JEOL Ltd., Tokyo, Japan) and an electron probe microanalyzer (EPMA, JEOL Ltd., Tokyo, Japan JXA-8530F field-emission hyperprobe at 15 kV) for elemental mapping using a wavelength-dispersive spectrometer (WDS, JEOL Ltd., Tokyo, Japan) to evaluate the diffusion of elements at the interface.

2.3. Evaluation of Mechanical Properties

Tensile properties of the transition joints were evaluated using a microcomputer-controlled electronic universal testing machine (WDW-3100, Rui Machinery, Changchong, China) at a loading rate of 0.5 mm/min at room temperature. The hardness measurements were carried out using a micro-Vickers hardness tester (HXD-100TM/LCD, TaiMing, Shanghai, China). The test load was 200 gf, and the dwell time was 15 s.

3. Result and Discussion

3.1. Optical Microstructure

Figure 1 shows an optical micrograph of the diffusion-bonded joints between the 316 stainless steel and the 4J29 Kovar alloy with Ni as an interlayer for 120 min at different temperatures. Sound diffusion interfaces were achieved, and interface lines were clearly visible. The microstructure of 316L stainless steel is equiaxed grain, and the grain size remains unchanged. However, the grain size of the 4J29 side with equiaxed austenitic grains and some twins gradually increased. Via the quantitative metallography method, the average grain size of the Kovar alloy bonded at 850 °C, 900 °C, and 950 °C was about 24 μm, 37 μm, and 69 μm, respectively. It has been illustrated that the thickness of the diffusion layer between the 316L stainless steel and the Ni interlayer (Ni–316L) was apparently extended, with a temperature varying from 850 to 950 °C. According to Figure 2, it was about 4–8 μm at 850 °C, 6–9 μm at 900 °C, and 8–14 μm at 950 °C, but that of the 4J29 Kovar alloy and the Ni interlayer side (Ni–4J29) cannot be seen obviously, as shown in Figure 3. The influence of the diffusion time on the microstructure of the joints is shown in Figure 2; as bonding time increases, the width of the diffusion layer at the joint interface of 316L stainless steel and the Ni interlayer simultaneously extends. From Figures 1 and 2, it can be seen that the diffusion layer was not uniformly and smoothly distributed, which is attributed to some untight contact between the surface of the base metals, and a zigzagged interface was formed.

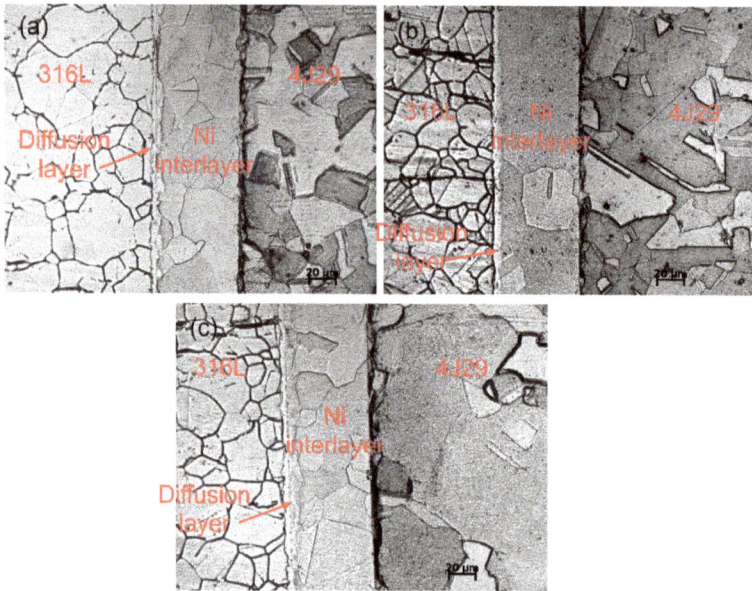

Figure 1. Optical micrograph of diffusion-bonded joints for 120 min at (**a**) 850 °C, (**b**) 900 °C, and (**c**) 950 °C.

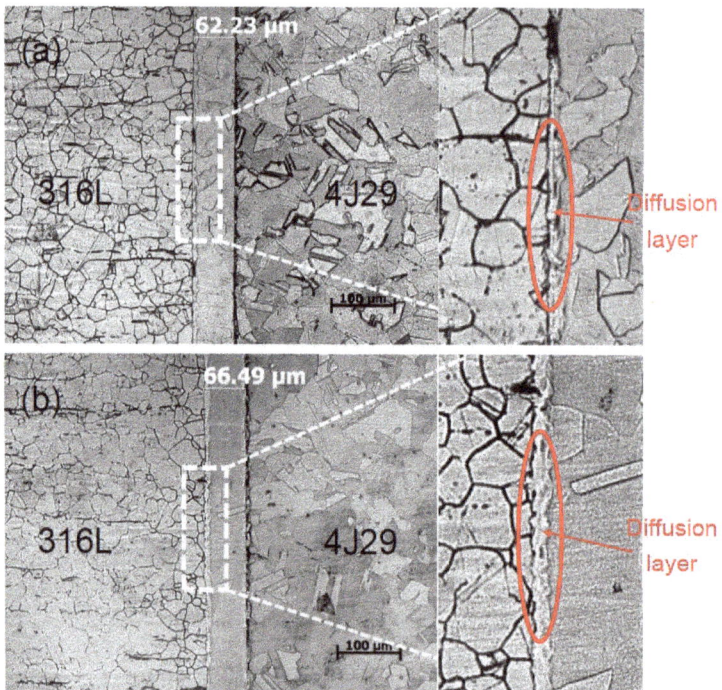

Figure 2. Optical micrograph of diffusion-bonded joints at 900 °C for (**a**) 180 min and (**b**) 240 min.

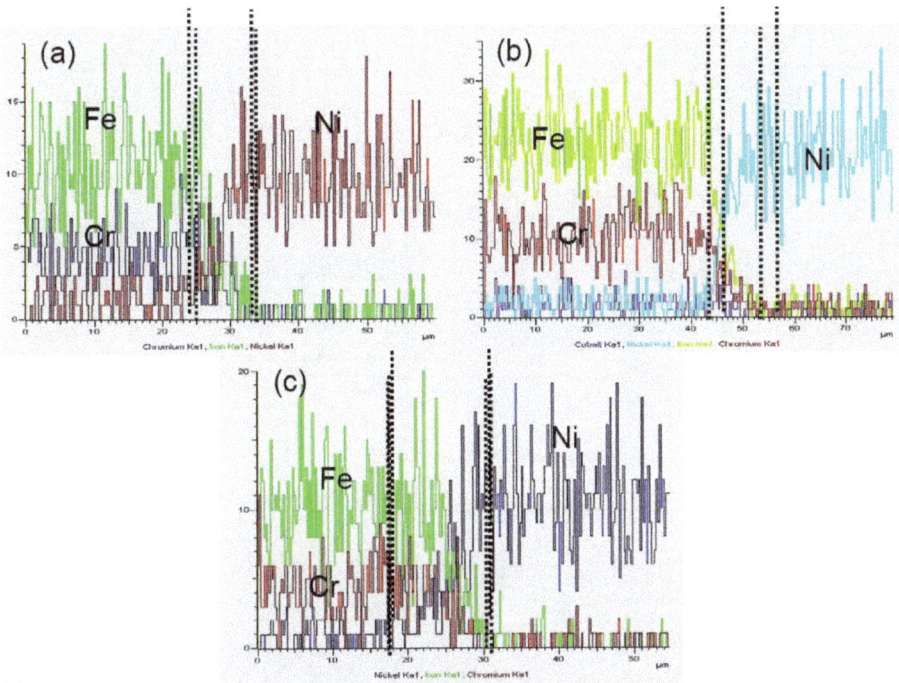

Figure 3. Concentration profiles of elements at the interface of Ni interlayer and the 316L side bonded at (**a**) 850 °C, (**b**) 900 °C, and (**c**) 950 °C for 120 min.

3.2. Element Distribution

The concentration profiles of the elements, namely, Fe, Cr, and Ni, at the interface of the Ni interlayer and the 316L side bonded at different temperatures were measured with EDS line scans, as shown in Figure 3a–c. It can be clearly seen that, as bonding temperature varied from 850 to 950 °C, the width of the diffusion area had a tendency to increase, as shown between the dotted lines. Moreover, the element concentration gradually decreased, and no apparent element aggregation was found at the interface of the Ni–316L side, which means that the intermetallic compounds did not form in the interface. The diffusion bonding of joints mainly depends on the formation of the nickel solid solution of Fe–Cr–Ni, and this is confirmed by the XRD measurements.

Figure 4a–c presents the concentration profiles of the elements at the interface of the Ni interlayer and the 4J29 side bonded at 850 °C, 900 °C, and 950 °C for 120 min. The concentration of Fe, Co, and Ni at the interface markedly declined. By comparison, the width of the diffusion layers did not evidently change as temperature increased, in accordance with metallurgical research. The different widths of the diffusion layers between the Ni–316L side and the Ni–4J29 side is partly attributed to the different grain sizes of the parent materials. The fine grains of the 316L stainless steel side possessed more grain boundary than the coarse grains of the 4J29 Kovar alloy side, which benefits the diffusion of atoms [23]. Thus, wider diffusion layers were obtained at the Ni–316L side.

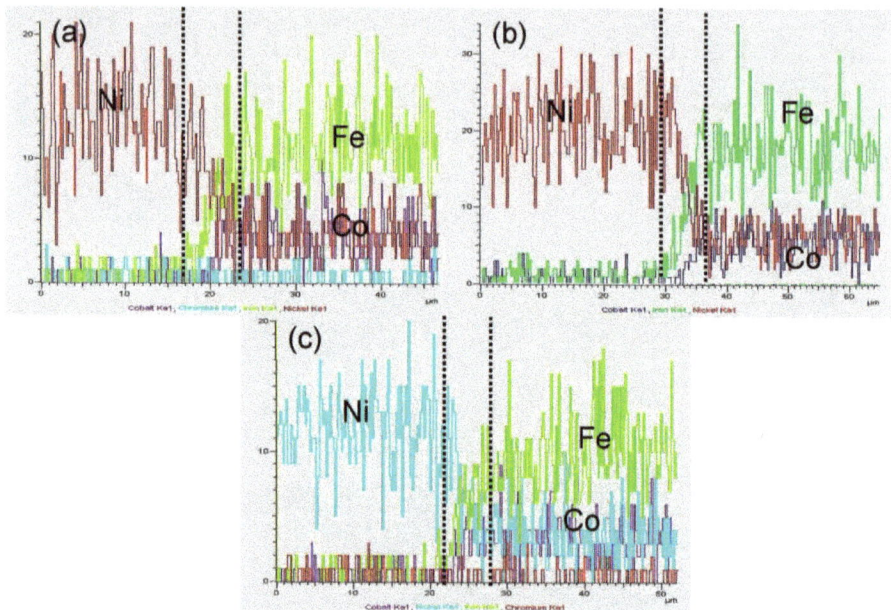

Figure 4. Concentration profiles of elements at the interface of the Ni interlayer and the 4J29 side bonded at (**a**) 850 °C, (**b**) 900 °C, and (**c**) 950 °C for 120 min.

Figure 5 shows the EPMA analysis of the 316L stainless steel and 4J29 joint interface bonded at 950 °C, which displays the diffusion of elements at the interface via elemental maps. Different areas in Figure 5a were detected. It can be seen in Figure 5b that the Ni of the interlayer diffused into both 4J29 and 316L sides. However, the extent of Ni into the 316L side was greater than that into the 4J29 side, which was due to a greater difference in the concentration of Ni between the Ni interlayer and the 316L stainless steel. The Co element of the 4J29 did not present evident diffusion into the Ni interlayer, as shown in Figure 5c, which may be attributed to a higher activation energy for diffusion. The diffusion extent of Fe at the 4J29 and 316L side into the Ni interlayer is revealed in Figure 5d. By comparison, Fe of the 316L side diffused further into the Ni interlayer than that of the 4J29 side due to a higher concentration. Moreover, Cr diffused into the Ni interlayer is further than that of Co, and both Fe and Cr promoted a wider diffusion layer of the 316L side than the 4J29 side. Figure 5e–f shows the concentration of Cr and C in the same position, which shows the phase of chromium carbide at the 316L side, as shown in [6]. According to the element composition of the Ni interlayer and the 4J29 side, the microstructure of the interface was a nickel solid solution of Fe–Co–Ni.

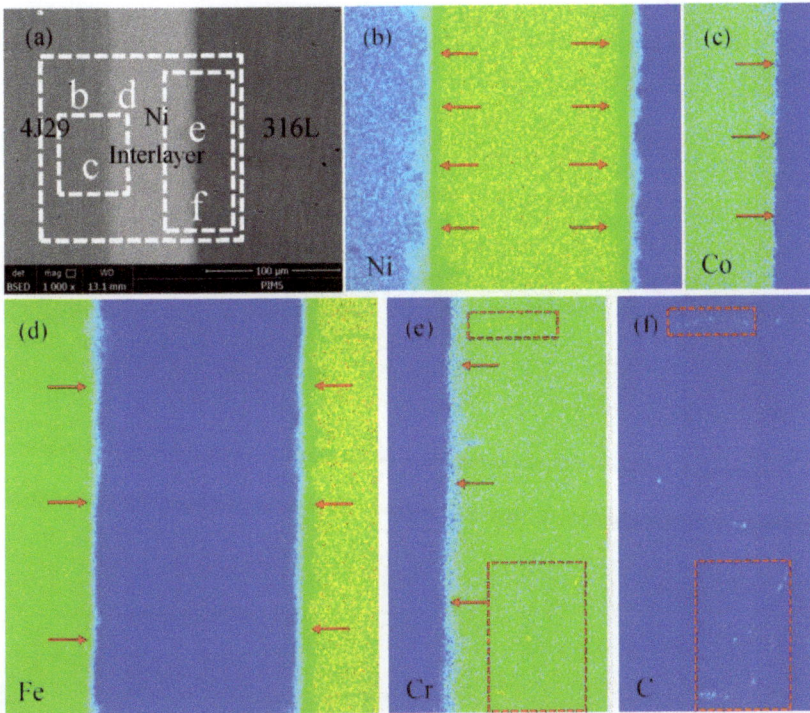

Figure 5. Concentration profiles from electron probe microanalysis (EPMA) analysis of (a) back-scattered electron (BSE) image and elements (b) Ni, (c) Co, (d) Fe, (e) Cr, and (f) C.

3.3. Diffusion Coefficient

According to Fick's second law, in infinite length diffusion, the relationship between elemental concentration and distance from the interface can be written as follows:

$$\frac{\partial C}{\partial t} = D\frac{\partial^2 C}{\partial x^2} \tag{1}$$

When the boundary conditions are conformed, the solution of Equation (1) can be written as follows:

$$C(x,t) = \frac{C_1 + C_2}{2} + \frac{C_1 - C_2}{2}\mathrm{erf}\left(\frac{x}{2\sqrt{Dt}}\right) \tag{2}$$

where C_1 and C_2 are initial concentrations of the element in both materials, x is the distance from the interface, t is diffusion time, and D is diffusion coefficient [24]. Based on the element concentration of a certain distance from the interface, the diffusion coefficient can be calculated by Equation (2).

Table 3 exhibits the diffusion coefficients of Ni on both sides of Ni–316L and Ni–4J29 at temperatures varying from 850 °C to 950 °C. The diffusion coefficients of Ni into both sides gradually increased as temperature increased, which conforms to the regulation of Arrhenius Equation (3), but the increase of the 316L side was more obvious. The calculated diffusion coefficients of Ni into the 316L and 4J29 sides were smaller than the reported literature value, 1.15×10^{-15} m^2/s at 850 °C into pure iron [23]. The lower values in the current work are attributed to a higher concentration of Ni in the 316L stainless steel and the 4J29 Kovar alloy compared with that of pure iron. In addition, the presence of greater amounts of alloying elements such as Cr, Si, Mo and Mn in 316L stainless steel and Co, Si, and Mn in the 4J29 Kovar alloy may cause a low diffusion of nickel due to the change in crystal structure. This result also confirms that the concentration gradient oriented is more obvious than the chemical activation on the diffusion of Ni across the interface. The Arrhenius equation can be written as Equation (4):

$$D = D_0 \exp\left(-\frac{Q}{RT}\right) \tag{3}$$

$$\ln D = \ln D_0 - \frac{Q}{RT} \tag{4}$$

where D_0 is the diffusion constant (m^2/s), Q is the activation energy for diffusion (kJ/mol), R is the real gas constant (8.314 J/mol), and T is the bonding temperature (K). Based on Equation (4), the activation energy for the diffusion of Ni can be calculated for the guidance of the parameter controlling the extent of diffusion. Curves of the diffusion coefficient vs. the temperature are exhibited in Figure 6. According to the slope of the plot in Figure 6a,b, the activation energy for the diffusion of Ni into 316L and 4J29 is 173.68 kJ/mol and 133.27 kJ/mol, respectively.

Table 3. Diffusion coefficients of Ni into the 316L and 4J29 sides.

Sample (°C, min)	Diffusion of Ni into 316L ($\times 10^{-16}$ m^2/s)	Diffusion of Ni into 4J29 ($\times 10^{-16}$ m^2/s)
850, 120	2.03	2.37
900, 120	6.62	5.60
950, 120	9.23	7.58

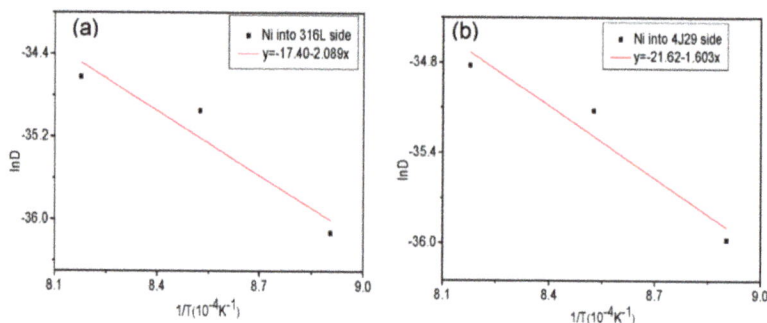

Figure 6. Curves of diffusion coefficient vs. temperature: (**a**) Ni into 316L side; (**b**) Ni into 4J29 side.

3.4. XRD Analysis

X-ray diffraction patterns of the parent materials and the diffusion layer on both sides of the bonded joints at 950 °C are shown in Figure 7. The microstructure of 4J29 consists of the austenite phase (γ-Fe) and the ferrite phase (α-Fe). It was found that, compared with the parent material of 4J29, more γ-Fe was formed owing to the existence of more γ-stabilizing Ni in the Ni–4J29 diffusion layer. Moreover, a solid solution of nickel (Ni s.s) was observed on the 4J29 and Ni interlayer sides, which was due to the high concentration of Ni and the sufficient solid solubility of Ni, Fe, and Co. On the other hand, a Ni s.s was also formed on the 316L and Ni interlayer sides, which promoted a good bond between the 316L stainless steel and the Ni interlayer. Compared with other literature reports [1,6,12], there has been no formation of the detrimental phase and segregation of sulfur and phosphor to result in hot cracks at interfaces. It is concluded here that a Ni interlayer can serve as an effective diffusion barrier for the bonding of 316L and 4J29. The composition of the joints was 316L/Ni s.s (Fe–Cr–Ni)/remnant Ni/Ni s.s (Fe–Co–Ni)/4J29.

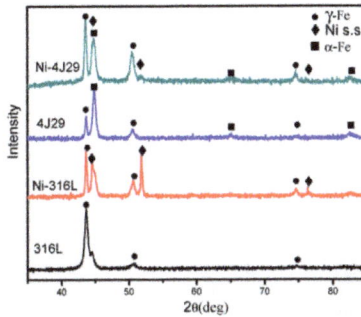

Figure 7. X-ray diffraction patterns of the parent materials and the diffusion layer on both sides of the bonded joints at 950 °C.

3.5. Hardness

Figure 8 shows the variation in micro-hardness of the bonded joint at different parameters. The hardness of the interface distinctly decreased on both sides of the 316L stainless steel and the 4J29 Kovar alloy, and the lowest hardness appeared at the Ni interlayer because of its excellent plasticity, which is a benefit for releasing the stress derived from the significant difference of expansion coefficients between parent metals. By comparison, the hardness distribution and the tendency of the bonded interface did not apparently change with parameters. It was observed that the hardness of the 316L side was a small reduction, which may be due to the aging and tempering treatment of the austenitic stainless steel at 900 °C for 240 min.

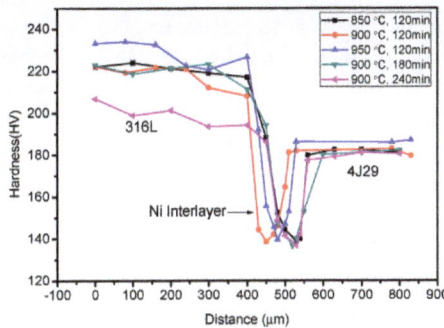

Figure 8. Micro-hardness of 316L stainless steel and the 4J29 Kovar alloy, diffusion-bonded with nickel as an interlayer at different parameters.

3.6. Tensile Strength and Fracture Analysis

Variation in the mechanical properties of the bonded joints processed at different parameters is shown in Table 4 and Figure 9. At 850 °C and 900 °C for 120 min, some joints were fractured at the interface on account of the insufficient creep deformation to form a compact diffusion layer. When the joints were bonded for 120 min, tensile strength gradually increased as temperature increased. At the same time, the stability of the joints significantly improved, as shown in Figure 9a, which was due to a more sufficient diffusion and bonding between the parent metals and the Ni interlayer. When the joints were processed at 900 °C, the mechanical properties rarely changed after heat preservation beyond 180 min, which means that the width of the diffusion layer was enough to form an excellent joint. Moreover, the highest tensile strength of 504.91 MPa with an elongation of 38.75% was obtained at 900 °C for 240 min.

Table 4. Mechanical properties of 316L stainless steel and the 4J29 Kovar alloy with Ni as an interlayer bonded at different conditions.

Sample (°C, min)	Ultimate Tensile Strength (MPa)	Elongation (%)	Failure Location
850, 120	215.86	6	Interface
900, 120	451.99	25.55	Interface/4J29 side
950, 120	490.62	31.25	4J29 side
900, 180	501.84	38.75	4J29 side
900, 240	504.91	38.75	4J29 side

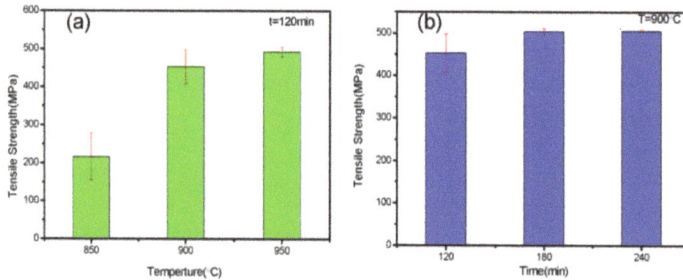

Figure 9. Mechanical properties of diffusion-bonded joints processed at different temperatures (**a**) and times (**b**).

Fracture morphologies on the 4J29 Kovar alloy side of the diffusion-bonded joints at different temperatures are shown in Figure 10. At a lower bonding temperature of 850 °C, a large amount of the terrace area was presented on the fracture surface, as shown in Figure 10a. According to EDS analysis, these areas were mainly not well bonded to the Kovar alloy, which demonstrates insufficient bonding between the Ni interlayer and the 4J29 Kovar alloy. Apart from the terrace, some dimples were formed at the bonded area, which was a nickel solid solution with a composition of 10.78 wt. % Fe, 16.17 wt. % Co, and 73.05 wt. % Ni. Figure 10b shows that, when temperature increased to 900 °C, the area of the terrace sharply decreased, and dimples increased, which significantly promoted the increase in tensile strength. Figure 10c shows a typically ductile fracture of the joint failed at the 4J29 side, which means an excellent bond was achieved at the interface. On applying axial force on the samples, the deformation of the remnant Ni interlayer was restricted by a mechanical constraint formed by the relatively non-deforming 316L and 4J29, which caused a state of triaxial stresses that decreased the effective stress on the remnant Ni interlayer [16]. This resulted in the joint failure at the 4J29 side.

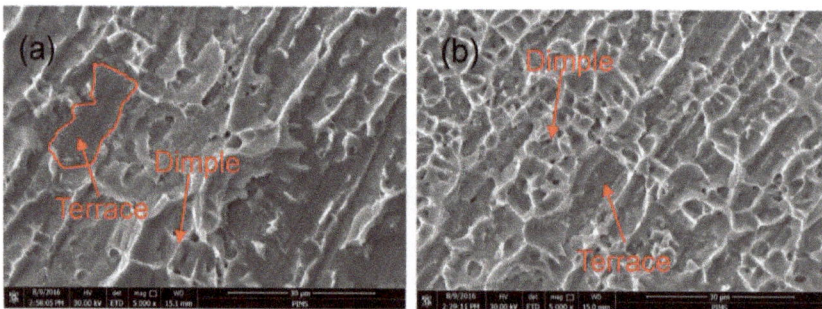

Figure 10. *Cont.*

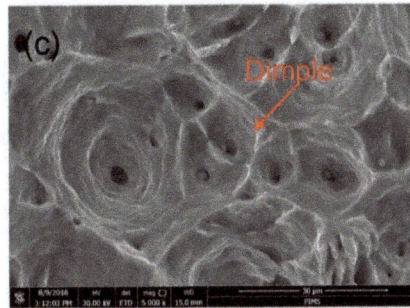

Figure 10. Fracture morphology analysis of diffusion-bonded joints at (**a**) 850 °C, (**b**) 900 °C, and (**c**) 950 °C for 120 min.

4. Conclusions

The diffusion bonding of 316L stainless steel and 4J29 Kovar alloys with nickel as an interlayer was carried out in vacuum in a temperature range of 850–950 °C for 120 min and at 900 °C with a bonding time of 120 min, 180 min, and 240 min under 2.5 T (34.66 MPa) uniaxial pressure, which was applied along the longitudinal direction of the specimen.

1.   The Ni interlayer can serve as an effective diffusion barrier for the bonding of stainless steel (316L) and the Kovar alloy (4J29). The composition of the joints was 316L/Ni s.s (Fe–Cr–Ni)/remnant Ni/Ni s.s (Fe–Co–Ni)/4J29.
2.   Growth of the diffusion layer was determined with the diffusion coefficient and activation energy, and the activation energy for the diffusion of Ni into 316L and 4J29 is 173.68 kJ/mol and 133.27 kJ/mol, respectively.
3.   At lower bonding temperatures and times, fracture takes place at the interface of the Ni–4J29 side due to insufficient bonding. After the width of the nickel solid solution (Fe–Co–Ni) increased, failure located at the 4J29 side and the fracture surface indicated a ductile nature. The highest tensile strength of 504.91 MPa with an elongation of 38.75% was obtained at 900 °C for 240 min.

Acknowledgments: This work was supported by National Natural Science Foundation of China (No. 51201143), Fundamental Research Funds for the Central Universities (No. 2682015CX001), the China Postdoctoral Science Foundation (No. 2015M570794), the Key Laboratory of Infrared Imaging Materials and Detectors, Shanghai Institute of Technical Physics, Chinese Academy of Sciences (No. IIMDKFJJ-14-04), the Sichuan Science and Technology Support Program (No. 2016FZ0079), and R & D Projects Funding from the Research Council of Norway (No. 263875/H30).

Author Contributions: Tingfeng Song implemented and conducted the diffusion bonding experiments and characterized the interfaces. Xiaosong Jiang and Zhenyi Shao analyzed and discussed the results. All authors participated in the design of the experiments and cooperated in the writing of this paper.

Conflicts of Interest: The authors declare no conflict of interest.

References

1.   Alejandro, H.; Jorge, M.; Ashley, R.; Pedro, F.; Peter, H.; Lawrence, E.M.; Ryan, B.W. Joining of Inconel 718 and 316 Stainless Steel using electron beam melting additive manufacturing technology. *Mater. Des.* **2016**, *94*, 17–27.
2.   Casalino, G.; Campanelli, S.L.; Ludovico, A.D. Laser-arc hybrid welding of wrought to selective laser molten stainless steel. *Int. J. Adv. Manuf. Technol.* **2013**, *68*, 209–216. [CrossRef]
3.   Chen, Y.C.; Tseng, K.H.; Wang, H.C. Small-scale projection lap-joint welding of Kovar alloy and SPCC steel. *J. Chin. Inst. Eng.* **2012**, *35*, 211–218. [CrossRef]
4.   Zhu, W.W.; Chen, J.C.; Jiang, C.H. Effects of Ti thickness on microstructure and mechanical properties of alumina-Kovar joints brazed with Ag-Pd/Ti filler. *Ceram. Int.* **2014**, *40*, 5699–5705. [CrossRef]

5. Wei, J.H.; Deng, B.H.; Gao, X.Q. Interface structure characterization of Fe36Ni alloy with ultrasonic soldering. *J. Alloy. Compd.* **2013**, *576*, 386–392. [CrossRef]
6. Baghjari, S.H.; Akbari, M.S. Experimental investigation on dissimilar pulsed Nd:YAG laser welding of AISI 420 stainless steel to Kovar alloy. *Mater. Des.* **2014**, *57*, 128–134. [CrossRef]
7. Akbari, M.S.; Sufizadeh, A.R. Metallurgical investigations of pulsed Nd:YAG laser welding of AISI 321 and AISI 630 stainless steels. *Mater. Des.* **2009**, *30*, 3150–3157. [CrossRef]
8. Rossini, M.; Spena, P.R.; Cortese, L.; Matteis, P.; Firrao, D. Investigation on dissimilar laser welding of advanced high strength steel sheets for the automotive industry. *Mater. Sci. Eng. A* **2015**, *628*, 288–296. [CrossRef]
9. Casalino, G.; Mortello, M.; Peyre, P. Yb–YAG laser offset welding of AA5754 and T40 butt joint. *J. Mater. Process. Technol.* **2015**, *223*, 139–149. [CrossRef]
10. Mai, T.A.; Spowage, A.C. Characterization of dissimilar joints in laser welding of steel-Kovar, copper-steel and copper-aluminum. *Mater. Sci. Eng. A* **2004**, *374*, 224–233. [CrossRef]
11. Nekouie, E.M.; Coupland, J.; Marimuthu, S. Microstructure and mechanical properties of a laser welded low carbon-stainless steel joint. *J. Mater. Process. Technol.* **2014**, *214*, 2941–2948.
12. Wu, W.Y.; Hu, S.S.; Shen, J.Q. Microstructure, mechanical properties and corrosion behavior of laser welded dissimilar joints between ferritic stainless steel and carbon steel. *Mater. Des.* **2015**, *65*, 855–861. [CrossRef]
13. Verena, W.; Bernhard, D.; Silvia, H.; Michael, R. Investigations of dissimilar welds of the high temperature steels P91 and PM2000. *Fus. Eng. Des.* **2013**, *88*, 2539–2542.
14. Wang, T.; Zhang, B.G.; Chen, G.Q.; Feng, J.C.; Tang, Q. Electron beam welding of Ti-15-3 titanium alloy to 304 stainless steel with copper interlayer sheet. *Trans. Nonferr. Met. Soc. China* **2010**, *20*, 1829–1834. [CrossRef]
15. Yuan, X.J.; Tang, K.L.; Deng, Y.Q.; Luo, J.; Sheng, G.M. Impulse pressuring diffusion bonding of a copper alloy to a stainless steel with/without a pure nickel interlayer. *Mater. Des.* **2013**, *52*, 359–366. [CrossRef]
16. Deng, Y.Q.; Sheng, G.M.; Xu, C. Evaluation of the microstructure and mechanical properties of diffusion bonded joints of titanium to stainless steel with a pure silver interlayer. *Mater. Des.* **2013**, *46*, 84–87. [CrossRef]
17. Wang, J.D.; He, X.Q.; Li, X.P.; En, Y.F. Hermetic Packaging of Kovar Alloy and Low-carbon Steel Structure in Hybrid Integrated Circuit (HIC) System Using Parallel Seam Welding Process. In Proceedings of the 15th International Conference on Electronic Packaging Technology, Chengdu, China, 12–15 August 2014.
18. Sathiskumar, J.; Torsten, S.; Davies, H.M.; Eggert, D.R.; Brown, S.G.R. Localized microstructural characterization of a dissimilar metal electron beam weld joint from an aerospace component. *Mater. Des.* **2016**, *90*, 101–114.
19. Madhusudhan, R.G.; Venkata, R.P. Role of nickel as an interlayer in dissimilar metal friction welding of maraging steel to low alloy steel. *J. Mater. Process. Technol.* **2012**, *212*, 66–77. [CrossRef]
20. Sam, S.; Kundu, S.; Chatterjee, S. Diffusion bonding of titanium alloy to micro-duplex stainless steel using a nickel alloy interlayer: Interface microstructure and strength properties. *Mater. Des.* **2012**, *40*, 237–244. [CrossRef]
21. Kundu, S.; Sam, S.; Mishra, B.; Chatterjee, S. Diffusion bonding of microduplex stainless steel and Ti alloy with and without interlayer: Interface microstructure and strength properties. *Metall. Mater. Trans. A* **2014**, *45*, 371–383. [CrossRef]
22. Kundu, S.; Sam, S.; Chatterjee, S. Interface microstructure and strength properties of Ti-6Al-4V and microduplex stainless steel diffusion bonded joints. *Mater. Des.* **2011**, *32*, 2997–3003. [CrossRef]
23. Sun, J.C. Investigation of Surface Nanocrystallinzation and Alloying of Commercial Iron and Diffusion Behavior. Ph.D. Thesis, Chongqing University, Chongqing, China, March 2012.
24. Vigraman, T.; Ravindran, D.; Narayanasamy, R. Effect of phase transformation and intermetallic compounds on the microstructure and tensile strength properties of diffusion-bonded joints between Ti-6Al-4V and AISI304L. *Mater. Des.* **2012**, *36*, 714–727. [CrossRef]

![metals logo] *metals* | MDPI

Article

Microstructure and Mechanical Properties of Friction Stir Welded Dissimilar Titanium Alloys: TIMET-54M and ATI-425

Kapil Gangwar [1], M. Ramulu [1,*], Andrew Cantrell [1] and Daniel G. Sanders [2]

[1] Department of Mechanical Engineering, University of Washington, Seattle, WA 98195, USA; kapildg@uw.edu (K.G.); werdna73@mac.com (A.C.)
[2] The Boeing Company, Seattle, WA 98124-2207, USA; daniel.g.sanders@boeing.com
* Correspondence: ramulum@uw.edu; Tel.: +1-206-543-5349

Academic Editor: Giuseppe Casalino
Received: 14 September 2016; Accepted: 17 October 2016; Published: 24 October 2016

Abstract: Weight reduction in automobiles and in aerospace industries can profoundly register for the behemoth change in the consumption of the fossil fuels and, in turn, CO_2 emission. With a promising hope in hindsight for weight reduction, we have successfully produced butt joints of friction stir welded (FSWed) dissimilar, and rather novice, α-β titanium alloys—ATI-425 and TIMET-54M. The study presented in this article encompasses the microstructural and mechanical properties of the joints for two cases, (1) ATI-425 on the advancing side; and (2) TIMET-54M on the advancing side. The evolution of microstructure and concomitant mechanical properties are characterized by optical microscopy, microhardness, and tensile properties. A detailed description of the microstructural evolution and its correlation with the mechanical properties have been presented in this study. Our investigations suggest that mixing patterns are dependent on the location (advancing, or retreating) of the alloying sheet. However, the microstructure in the weld nugget (WN) is quite similar (grain boundary α, and basket weave morphology consisting of α + β lamellae) in both cases with traces of untransformed β. The thermo-mechanically affected zone (TMAZ) on the either side of the weld is primarily affected by the microstructure of the base material (BM). A noticeable increase in the hardness values in the WN is accompanied by significant deflection on the advancing and retreating sides. The tensile properties extracted from the global stress strain curves are comparable with minimal difference for both cases. In both cases, the fracture occurred on the retreating side of the weld.

Keywords: dissimilar α-β titanium alloys; FSW; microstructure; mechanical properties

1. Introduction

By tradition and technique, fusion welding is well suited for the joining of similar and dissimilar alloys. The widespread span of this method encircles combinations of metallic alloys ranging from lower melting temperatures (of aluminum) to high meting temperatures (steel and titanium). However, the formation of intermetallics in the weld pool, and an inevitable porosity as the molten metal solidifies, are the potential threats for the failure of the metallic structure. In industries, such as aerospace and automotive, where even a slightest leeway of error is blasphemous, discovery of an alternate joining method with a distinct focus on weight reduction is essential to not only strengthen the joint but also to recover from the additional weight that filler materials add during fusion welding. Progressing through intense research, and variations in its kind as it emerged over the decades since 1991 [1], friction stir welding (FSW), as shown in the Figure 1, has clinched its spot in the territory of joining methods around almost all the metallic alloys, of magnesium, aluminum, and even steel and titanium.

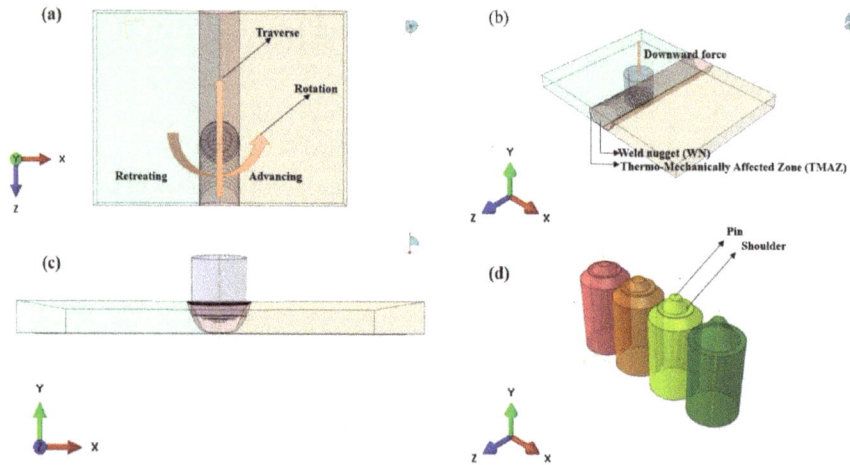

Figure 1. FSW and tool schematics. (**a**) Top; (**b**) side; (**c**) front; (**d**) tool designs. Images not to scale.

The means of FSW are a rotating and traversing tool, with a specified geometry and element composition, and the frictional heat that is being generated between the tool and the worksheets. As the tool comes in contact with the worksheets, the frictional heat aids in the temperature rise, the material underneath softens, and whirs fervently. The rotation of the tool helps in mixing of the plasticized material and, as a result of the traverse speed, the softened material gets deposited from the retreating (RET) side to the advancing (ADV) side, in the wake of the tool. Figure 1 shows the schematic of the process and the emergence of the different zones. While the presence of the thermo-mechanically affected zone (TMAZ) or heat affected zone (HAZ) remains a questionable pursuit in titanium alloys; magnesium, and aluminum alloys lead the path with three different welding zones, as are shown the Figure 1c.

Aerospace industries, whether involved in commercial aircraft or supersonic ones nearing Mach 2, are determined to reduce weight in their current and upcoming fleet of airplanes. On one hand metal matrix composites provides solution for larger structure, such as the fuselage, the present and future of the jet engines solely depends on high-temperature titanium alloys [2–4]. Although the ninth most abundant element, the efficient use of titanium is often endangered by the difficulty in machining, constantly pondered under the shadow of the buy-to-fly ratio. Titanium sheets, typically available in limited sizes, need to be joined, and subsequently formed, in order to produce a larger structure. Once the defect-free joint is achieved, the path to glorification of the final product is hurdled by its superplastic forming ability, the anisotropic nature of the weld, and the deformation mechanism (either grain boundary sliding (GBS) or phase boundary sliding (PBS)) [5].

The potential of FSW has been realized for aluminum, magnesium, and copper since its inception in 1991. With precocious tool design, and well-sought welding processing parameters, even steel and titanium alloys have savored the fruition of this novice joining process. Commercially pure (CP)-titanium, and alloys of titanium, Ti_6Al_4V, Ti-5111, and β-21S have been studied extensively for the microstructural and mechanical properties of the weld [6–14]. Fonda et al. have studied the effect of deformation corresponding with the traverse speed in near-α titanium alloy, Ti-5111. The presence of hcp $P(1)$ and bcc $D(1)$ texture is dependent on the welding speed [6]. Reynolds et al. have studied the formation of torsion texture during welding in β-21S [9].

An encyclopedia of FSW α + β titanium alloy, Ti$_6$Al$_4$V, is well versed in terms of the microstructural properties [12,13], mechanical properties, texture formation [10,14], and even superplastic forming. In a study of identification of the welding parameters, Edwards et al. have predicted the optimum welding parameters along with referencing several other parameters adopted by the researchers corresponding to different thicknesses of Ti$_6$Al$_4$V sheets [15]. The similar study [15] by Edwards et al. and by Rai et al. [16] have summarized the nature and the properties of the tools used in welding of titanium alloys. Ti$_6$Al$_4$V, commonly known as the workhorse of the industry, presents a variety of challenges for the research and development of new near-α, α + β, or β alloys that could potentially match, or surpass, the properties of Ti$_6$Al$_4$V. Material flow in and around the WN for FSW of Ti$_6$Al$_4$V has been studied by Edwards et al. [17] by computerized tomography using a tracer technique. The microstructure and texture development in FSW of investment cast Ti$_6$Al$_4$V has been studied by Pilchak et al. [18]. A series of different microstructure in and around the WN has been observed for the temperature profiles in WN (however, not measured accurately) fluctuate significantly. The fatigue and crack propagation rates in the FSWed Ti$_6$Al$_4$V have been studied for different thicknesses [19–23]. Not only has it been found that the weld has an excellent fatigue crack propagation rate, but welded joints in various shapes (L, and T shape) are far more superior when stress (S) - number of cycles (N) to failure or S-N curves are examined [20]. Although an example of the fusion welding technique, electron beam welding (EBW) processed joints for 24 mm thick Ti$_6$Al$_4$V sheets were comparable with FSWed sheets of same material, and of the same thickness [22]. The microstructure and hardness values of the FSWed titanium sheets of Ti$_6$Al$_4$V with varying rotation or traverse speed have also been studied extensively, demonstrating higher values of hardness in the WN. However, as the rotation speed increases, a drop in the hardness values has been recorded due to coarsening of the prior β grains, in addition to the increase in the width of α lamella (Widmanstätten structure) [11–13].

Steps have been taken in order to FS weld dissimilar alloys. Titanium and steel [24–27], aluminum and magnesium [28,29], and titanium and aluminum, [30–33] have been successfully welded with appropriate welding parameters for their surface and subsurface properties analysis. However, the FSW of dissimilar titanium alloys remain untouched, with the exception of a few studies of the laser welding of Ti$_6$Al$_4$V and Ti$_6$Al$_6$V$_2$Sn by Hsieh et al. [34]. Although, Jata et al. have successfully joined 2 mm thick sheets of Ti$_{17}$, and Ti$_6$Al$_4$V by FSW and presented the microstructural and mechanical properties [35], the detailed analysis of the material flow, study of migration of elements, effect of interchanging of alloying sheets, and corresponding mechanical properties have not yet been investigated so far. This article presents the detailed description of the FSW of dissimilar, rather novice α + β titanium alloys—ATI-425 and TIMET-54M—for their mechanical and microstructural properties and their correlation with the evolving microstructure with the material flow. The evolving microstructure and the material flow observed and analyzed in this study show dependence on the initial microstructure and the β transus temperature (for ATI-425, 1780 °F ± 25 °F (971 °C ± 14 °C) [36] and for TIMET-54M, 1720 °F–1770 °F (938 °C–966 °C) [37]).

2. Materials and Methods

The test plates of α + β titanium alloys used in this study were fabricated from standard, 4 mm thick, ATI-425 and TIMET-54M sheets. The chemical composition of the alloying sheets is given in Table 1.

Table 1. Mass fraction (wt. %) composition of the alloys.

Alloys	wt. %	Al	Mo	V	Fe	O	C	N	H	Ti
ATI-425	Min	3.5		2	1.2	0.2				Bal.
	Max	4.5		3	1.8	0.3	0.08	0.03	0.015	
TIMET-54M	Min	4.5	0.4	3	0.2				0.015	Bal.
	Max	5.5	1	5	0.8	0.2	0.1			

2.1. Processing

Prior to welding, the longer edges that would be the abutting faces of the square groove butt joint were machined and chemically etched to achieve adequate fit up and to remove any contaminates (e.g., oil, grease, etc.) in the resulting weld. All welds were made at 300 rpm and 75–100 mm/min. These parameters were selected based on a previous study focused on the identification of optimal process parameters for FSW in various thicknesses of Ti_6Al_4V [15]. The FSW welding tool used was a tungsten lanthanum (W-La) alloy with a small shoulder and a large tapered pin, more details of which can be found in [38]. All welds were made on a W-La backing anvil. Thermal management via water-cooling of the pin tool and anvil was also utilized. Welding parameters are shown in the Table 2.

Table 2. Friction stir welding Process parameters adopted in this study.

Parameters	Values
Spindle Speed	300 rpm
Tool Traverse Speed	~75–100 mm/min
Forging Load	~15.6 kN
Tool Plunging Depth	~1.6 mm
Tool Tilt	3° from direction of traverse
Tool Material	W-La
Tool Pin Length	~1.4 mm
Tool Pin Diameter	~8.6 mm
Tool Shoulder Diameter	~15.9 mm

Welds produced in this study were a square groove butt joint configuration as shown in Figure 1c. Two sheets of ATI-425 and TIMET-54M of 4 mm thickness, 600 mm length, and 100 mm width were butt joined by FSW. The welding direction and the pre-milled rolling direction of the base materials are parallel to each other. Sections for metallurgical analysis and tensile coupons were extracted from the welded sheets by water-jet machining. Micrographs in this study are from the transverse cross-section of the weld, Figure 2. The tensile coupons were perpendicular to the welding direction with a gauge length encompassing the base materials on both sides of the WN.

Figure 2. Typical schematic diagram appearance of the sectioned friction stir welded specimen.

2.2. Mettalurgical Characterization

For metallurgical characterization, the FSWed sheets were water-jetted perpendicular to the welding direction. The cut section was then mounted and polished, first on silicon carbide (SiC) paper (grit size from 240–1200), and secondly on 6 μm, 3 μm, 1 μm, and finally 0.3 μm alumina-oxide polishing wheels. To reveal the optical microstructure the polished specimens were etched using using Kroll's reagent (distilled water, 92 mL; HNO_3, 6 mL; HF, 2 mL). The macrographs with were captured with a stereomicroscope (Nikon SMZ1000, Nikon Corporation, Tokyo, Japan) and analyzed using NIS element software, to observe the quality of the weld produced with the adopted set of welding parameters.

2.3. Microhardness

The specimens were re-polished after microstructural analysis for microhardness evaluation. Each microhardness (as per ASTM E384-06) profile consisted of a grid pattern covering the entire weld with indents spaced at 254 μm in each row and column (matrix of 12 rows and 105 columns for ATI-425 on the advancing side, and of 11 rows and 90 columns for TIMET-54M on the advancing side); please refer to Figure 3. All of the microhardness indentations were conducted on a LECO AMH43 Automatic

Hardness Testing System (LECO Corporation, St. Joseph, MI, USA) using a Vickers indenter with a 500-g load applied for 13 s.

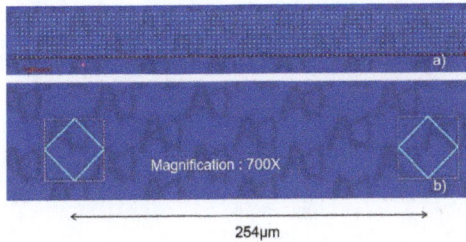

Figure 3. (a) Indention pattern on the transverse cross-section of the weld; and (b) distance between two indent at 700× magnification.

2.4. Tensile Specimens

Four tensile coupons were extracted from the as-welded condition. The dimensions of the specimens were in accordance with ASTM E8 (Figure 4). For all specimens, the welds were oriented transversely to the loading direction and the WN was centered in the specimen gage length. All tests were conducted under displacement control at a constant cross-head extension speed of 1.27 mm/min. Every sample was tested to failure. The load data was monitored by a load cell built into the load frame. The axial strain data was monitored by a clip-on extensometer (Model number: 2650-562, Instron Industrial Products, Grove City, PA, USA), which was removed at the onset of yielding to prevent damage at failure and the test was allowed to run to failure with elongations in the plastic portion of the stress strain curve measured by the cross-head of the load frame. In each test, the yield and ultimate strengths were recorded in addition to the elongation at failure and failure location.

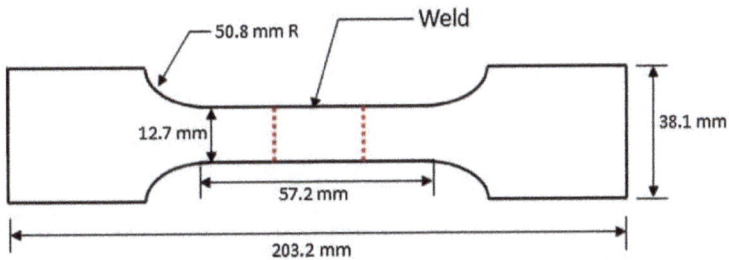

Figure 4. Schematic diagram of tensile specimens.

3. Results

3.1. Metalography

Figure 5 shows the typical appearance of the FSWed butt weld. The macrographs of dissimilar FS welds of ATI-425 and TIMET-54M are shown in the Figure 5. It can be observed that that there are no inclusions or voids present in the WN. Both welds were produced with identical process parameters. For the case of P4, when TIMET-54M is on the advancing side, a prominent mixing pattern has been observed in the WN in comparison to P1, where ATI-425 is on the advancing side.

Figure 5. Typical macrographs of butt welds, (P1) ATI-425 on the advancing side, and (P4) TIMET-54M on the advancing side. RET: Retreating; and ADV: Advancing side.

The microstructures were taken at the locations as prescribed in Figure 6. Position a, and g correspond to the base material. Locations b, and f correspond with the TMAZ/HAZ on the retreating and the advancing side, respectively. Location g corresponds with the center of the WN. Locations c, and e are in the vicinity of the boundary inside the WN on the retreating and the advancing side, respectively. Subscripts 1 and 4 in the microstructures corresponds with the specimen P1 and P4, respectively.

Figure 6. Prescribed locations for the microstructure taken for comparison. a, and g represent base material; b, and f represent TMAZ on RET, and ADV side; c, and e represent locations inside the weld nugget, however, closer to TMAZ on RET, and ADV side; d, represents locations in WN.

Please refer to Figure 7 for micrographs. The microstructure of the base material of TIMET-54M (a_1 and g_4) is a bi-modal microstructure, containing both equiaxed α grains and an acicular, or plate-like, α phase. The microstructure of the base material ATI-425, (a_4 and g_1), consists primarily of an α phase with infused intergranular β. The boundaries on the advancing sides, f_1 and f_4, consist of refined equiaxed α grains in a matrix of β; and prior β grains with grain boundary α, respectively. The boundaries on the retreating sides, b_1 and b_4, comprises of, b_1: equiaxed α (bottom left), untransformed α (streak in the middle), and β grains surrounded by grain boundary α (top right); and b_4: refined equiaxed α, and needle-like α surrounded by grain boundary α. The observed microstructure in the vicinity of the boundary inside the WN on the advancing side (e_1 and e_4) comprises of a basket-weave morphology with plate-like α surrounded by grain boundary α. No visible differences in the grain size has been observed, however, e_1 appears to be richer in α phase. Furthermore, in the case of e_4, some traces of untransformed β have been observed; see the bottom left of e_4. For c_1 and c_4, in the vicinity of the boundary inside the WN on the retreating side, the microstructure is comprised of the following: c_1: prior β grains delineated with grain boundary α (bottom left), and acicular plate-like α surrounded by grain boundary α; c_4: β grains surrounded by grain boundary α (bottom left), untransformed α, and acicular plate-like α surrounded by grain boundary α with an uneven distribution of the β phase. In the center of the weld, d_1, the microstructure consists of a basket-weave morphology with flakes of untransformed β. For the case of d_4, the microstructure is quite similar to d_1; nonetheless, a rather high amount of untransformed β with grain boundary α has been observed.

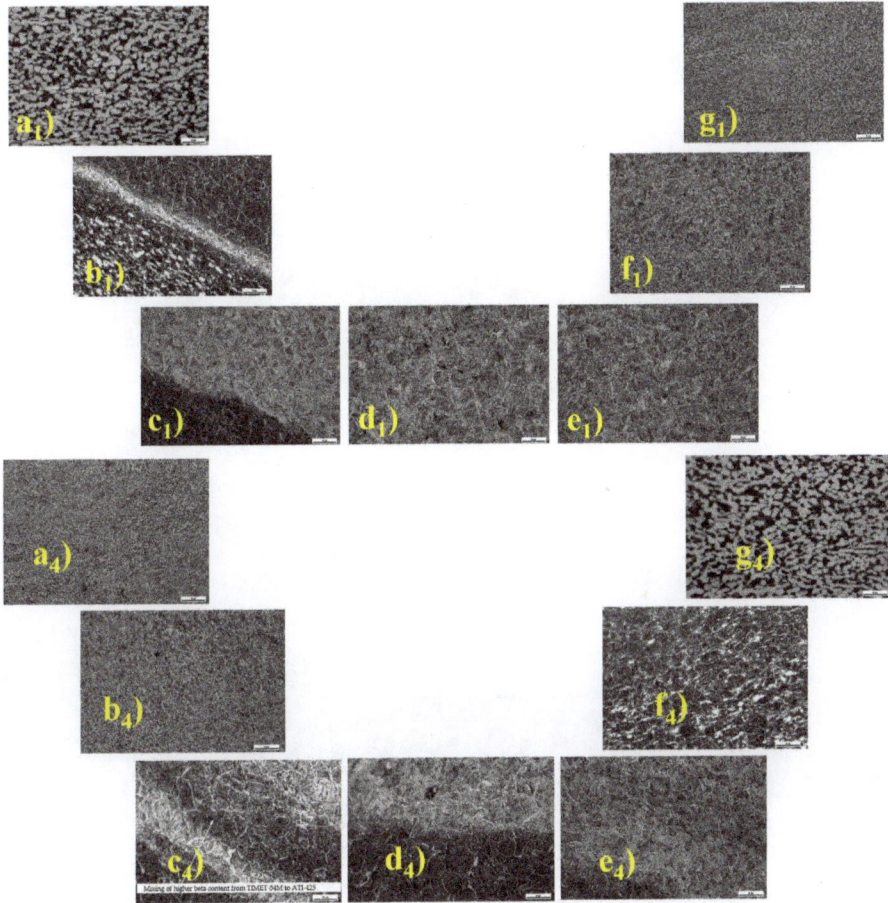

Figure 7. Optical microstructure of the specimens, P1 (marked a_1 to g_1), and P4 (marked a_4 to g_4). With $a_{1,4}$ being on the retreating side; and $g_{1,4}$ being on the advancing side. The reader is recommended to the web version for enhanced clarity of the text. Scale: 20 µm for all the micrographs except c_4. For c_4, the scale is 10 µm.

3.2. Microhardness

The microhardness contours on the traverse cross section of the weld are shown in the Figure 8. The approximate location of the weld boundaries are identified, however, the width of the TMAZ cannot be determined from the hardness maps. The hardness profiles in the center of each specimen are also plotted in Figure 9. The approximate center line, marked by the dotted line, is the reference line for hardness variations (Figure 9) in the center of the transverse cross-section of the weld.

(a)

(b)

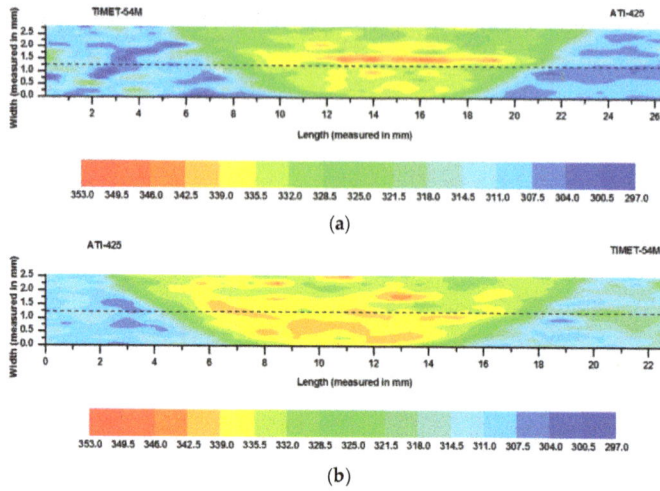

Figure 8. Color-coded microhardness profiles on the transverse cross-sections of the weld. (a) P1; and (b) P4.

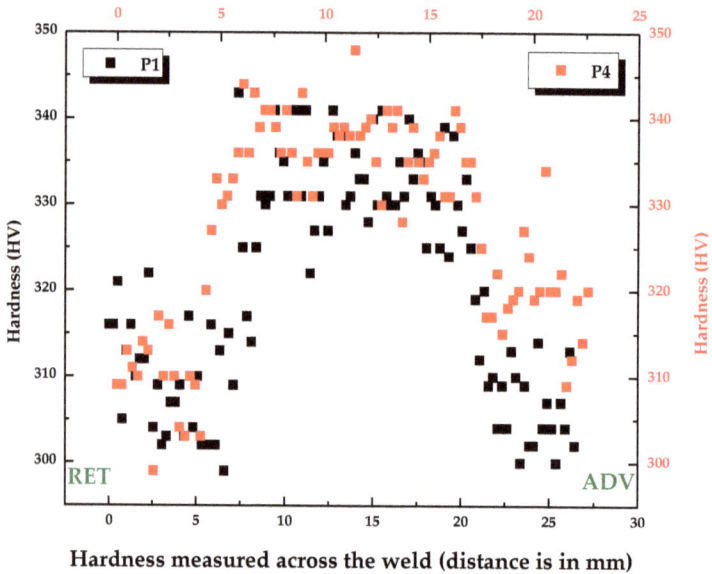

Hardness measured across the weld (distance is in mm)

Figure 9. Hardness profile in the center of the weld.

From the hardness profiles and the maps it can be observed that the hardness values are higher in the WN. A discernible boundary between WN and base material can be observed on either side of the WN. Furthermore, the hardness values in the center of the weld are higher for P4, in comparison with P1. The hardness for TIMET-54M is slightly lower than ATI-425. When plotted on the center of the weld (as shown in the Figure 9), a significant amount of deflection is observed in the hardness values on either side of the WN. A rather uniform hardness profile is observed for P4 in the center of the WN. In both cases the hardness appears to be slightly higher on the retreating side.

3.3. Tensile Specimens

The average yield strength (YS), ultimate tensile strength (UTS), and percent elongation (% Elong.) to failure for four tensile coupons were tested and are shown in Figure 10. From the typical global stress strain data collected for dissimilar FS welds, it can be observed that the values of YS, UTS, and % Elong. for P4 are slightly higher in comparison to P1. The average values of the YS, UTS, and % Elong. are presented in Table 3.

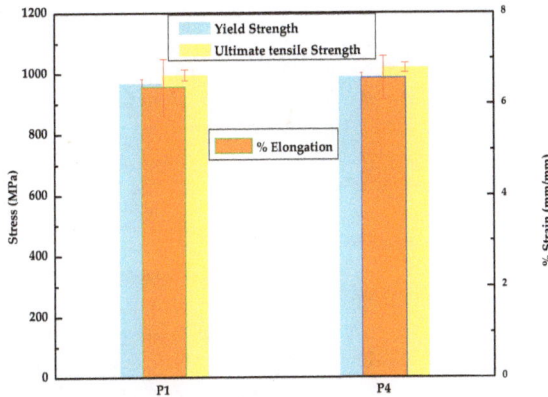

Figure 10. Tensile test results.

Table 3. Average tensile test values. Yield strength (YS); Ultimate tensile strength (UTS); percent elongation (% Elong.) and standard deviation (St. Dev.).

Specimen	YS (in MPa)	St. Dev. in YS (MPa)	UTS (MPa)	St. Dev. in UTS (MPa)	% Elong.	St. Dev. in % Elong.
P1	966.48	15.87	995.64	18.29	6.37	0.62
P4	989	12.33	1019.07	16.1	6.57	0.48

4. Discussion

4.1. Morphology of the Welds

From the macrographs it can be observed that the defect-free welds were made under the adopted processing conditions. A clear distinction of BM, TMAZ (not easily discernible at this level of resolution), and WN can be seen in all of the welds. Based on the macrographs it can be observed that mixing of the material is quite different in the bottom of the weld when TIMET-54M is kept on the advancing side (P4). For the case when ATI-425 is kept on the advancing side (P1), a rather more uniform pattern can been observed. Based on the macrographs it can be deduced that the material flow is primarily dependent on the location of the material; either on the advancing side or on the retreating side.

4.2. Microstructural Evolution

The microstructures evolved in the TMAZ and in the WN are addressed in terms of the grain morphology, phase fractions, and distribution and presence of adiabatic shear bands (ASB). The microstructure in the TMAZ and in the WN are dependent on the total strain, strain rate, stacking fault energy of α (hcp), and β (bcc) phases, and temperature evolving during FSW. During severe plastic deformation imposed by the tapered and tilted tool, a large fraction of the plastic work is converted into heat, and due to low thermal conductivity of the titanium, the heating rate at higher strain rate could be dominating over the heat loss by conduction, resulting in a localized zone of higher temperature, hence, the formation of adiabatic shear bands [39]. The formation of these adiabatic shear bands running across the weld boundary (Figure 7, P1-b_1 and P4-c_4) is quite distinct in both cases (P1 and P4).

4.2.1. P1: TIMET-54M on the Retreating Side

ImageJ analysis, shown in Figure 11, of the BM of TIMET-54M and ATI-425, has shown that TIMET-54M has slightly higher β content (43%–45%) and, hence, a lower α content, in comparison with ATI-425 (38%–39%). For the case of P1 (see Figure 7a_1–g_1) when TIMET-54M is present on the retreating side, a severe plastic deformation, along with slightly lower temperature in comparison with the advancing side is experienced. Due to this, a significant amount of α (without being transformed into β) is sheared and migrating from the retreating side in the form of untransformed band of α along the WN, Figure 7b_1. As we approach closer to the weld center, c_1, a significant amount of the grain boundary α with a rather coarser acicular α has been observed adjacent to the prior β grain boundaries containing a fine acicular α. At the center of the weld d_1, and at e_1 where temperatures are rather uniform, the cooling rate also appears to be quite adequate in the proximity of the center resulting in a basket-weave morphology, with a grain boundary α with coarser acicular α in the prior β grains. The region which is closer to the advancing side, e_1, is significantly deformed, however, unlike the retreating side, no crescent-like structure has been formed on the advancing side. The microstructure at e_1 is quite similar to that observed at d_1, due to nearly similar temperature profiles. The microstructure at f_1, which is closer to ATI-425, and on the advancing side, results in refined α grains in a matrix of β, suggesting that, even on the advancing side, the temperature are not par β transus of ATI-425 (β transus temperature of ATI-425 is 1780 °F).

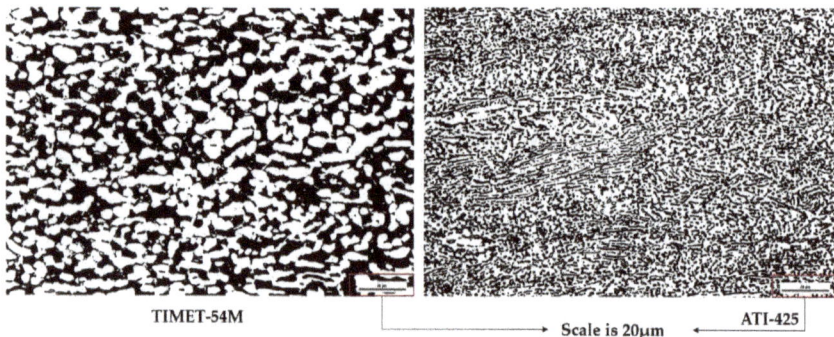

TIMET-54M   Scale is 20μm   ATI-425

Figure 11. ImageJ analysis of TIMET-54M and ATI-425, for P1. TIMET-54M and ATI-425 have β percentages of 45.3099% and 41.5035%, respectively.

Briefly, for this case it can be summarized that the region which appears darker in the macrographs in the WN contains prior β grains and fine acicular α. The lighter region, on the other hand, consists of grain boundary α with coarser acicular α in β. Based on the observation it can be concluded that heating rates at high strain rate and cooling rate (by conduction, and by constant argon gas flow) are sufficient enough to avoid formation of any localized heatsink in the WN, hence, resulting in a rather uniform microstructure (grain boundary α with acicular α in β). However, on the retreating side, some regions of α remain untransformed suggesting a sub-β transus temperature profile in that region.

4.2.2. P4: ATI-425 on the Retreating Side

For this case, Figure 7a_4–g_4, as can be observed that due to the higher β transus temperature of the ATI-425 (1780 °F ± 25 °F (971 °C ± 14 °C)) in comparison with TIMET-54M (1720 °F–1770 °F

(938 °C–966 °C)), the crescent-like structure is not formed in the proximity of the weld boundary on the retreating side. The microstructural observation suggests that only refined α grains in β matrix have been observed, at b_4. As we move closer to WN inside the weld, at c_4, most of the α that is being migrated from the retreating side (ATI-425), due to the higher β transus, is untransformed and appears as a wider streak. In the center of the weld, d_4, a chaotic mixing pattern is observed, due to which the top section of weld microstructure (d_4) results in the formation of coarser acicular α in β grains that are delineated with grain boundary α. The bottom section of the of the WN microstructure results in the formation of prior β grains with fine acicular α. The microstructure inside the WN close to the weld boundary on the advancing side, at e_4, results in coarser acicular α in β. This can be attributed to the fact that although the temperature profiles are nearly similar, due to differences in the β transus temperature and the cooling rates (faster cooling rates at the bottom) the bottom of the weld has finer acicular α and, at e_4, rather coarser acicular α has been observed. It can be said that d_4, and e_4 are mainly made of TIMET-54M. The boundary on the advancing side is severely deformed and due to temperatures reaching on par with the β transus, refined and sheared equiaxed α grains have been observed along with β grains delineated with grain boundary α.

As a general observation, by looking at the microstructures of both cases, it can be said that temperatures on the retreating side are slightly lower due to the tangential vector of the tool rotation and the axial vector of the traverse direction being in the opposite directions. As a result of rather slightly lower temperatures on the retreating side, the material with lower β transus temperature (TIMET-54M) has resulted in the evolution of a Widmanstätten microstructure with fine acicular α in the transformed β grains at the weld boundary. Although, material with higher β transus temperature (ATI-425) resulted in the deformation of α grain at the boundary, yet, as we approach closer to the WN, a similar microstructure (untransformed α, and grain boundary α with acicular α) has been observed. For the combination of alloys considered in this study it is safe to assume that although temperatures on the retreating side are lower than the advancing side, they are still sufficient enough for reaching above the β transus for TIMET-54M and not for ATI-425.

4.3. Microhardness

From Figure 8, it can be seen that TIMET-54M has shown relatively higher hardness in comparison with ATI-425 for BM; the reason being, the base material microstructure of the TIMET-54M contains equiaxed α grain presenting a higher flow stress in comparison with the ATI-425, where α is infused with primary β. Hardness in the WN increases for both cases. The hardness values are significantly fluctuating inside, and around, the WN. In the case of P1, when TIMET-54M, with a lower β transus temperature, was kept on the retreating side, the increase in the hardness on the boundary of the retreating side is a result of finer acicular α in refined prior β grains in comparison with refined α grains on the advancing side of P1. The higher values of hardness on the retreating side for P1 are in correspondence with the microstructure observed (Figure $7b_1$). Upon close inspection one can observe a very distinctive fine streak in the center of the weld in the macrograph of P1 (Figure 5). The higher values of hardness in that region are also observed in Figure 8a.

For the case of P4, when ATI-425 was kept on the retreating side, the increase in the hardness at the boundary on the retreating side is marked by the grain refinement indicating that temperatures produced on the retreating side have not crossed the β transus of ATI-425; rather, the microstructure is severely deformed into refined equiaxed α grains. As we move closer to the weld center, the temperatures increase, and an increase in the hardness values is registered by the evolution of the finer acicular α in the β grains. At the bottom of the specimen there appears a significant variation in the hardness values owing to the prominent mixing of the two alloys and the formation of refined β grains with finer acicular α in it (Figure $7d_4$). Based on the hardness pattern the following can be concluded:

(1) If a material with lower β transus is kept on the retreating side, this results in the formation of a Widmanstätten microstructure with coarser α lamellae for most of the WN and slightly lower hardness values.

(2) If a material with higher β transus is kept on the retreating side, an increase in the hardness values on the retreating side is marked by the deformation of the α phase. In the weld center the microstructure mostly consists of a basket-weave morphology with finer α lamellae. As we approach the boundary on the advancing side, the hardness increase is a result of both grain refinement and of acicular α in prior β grains.

(3) As such, no uniformity can be conformed based on the evolution of the microstructure but, for the case of P4, higher values of hardness have been observed on the retreating side suggesting that grain refinement is more dominant in comparison with the finer acicular α formation.

4.4. Mechanical and Microstructural Relationship

From tensile data, as plotted in Figure 10, extracted from global stress strain curves for four specimens of each case, it can be observed that P4, when ATI-425 is on the retreating side, showed higher values of YS and UTS in comparison to P1, when TIMET-54M is on the retreating side. In all of the cases, out of four specimens tested for the analysis, the fracture always occurred on the retreating side. For the case of P4, where the hardness in the WN is slightly higher in comparison with that of P1, it can be said that residual stresses are slightly more compressive [40] giving rise to higher values of YS and UTS. The residual stresses are not measured, however, due to migration of the elements, Al, V, and Fe, the titanium atoms in the bcc (β) and hcp (α) phases are replaced by these elements causing the shrinkage or expansion of the lattice spacing. Further research with transmission electron microscopy (TEM) needs to be done in order to confirm the results as presented during our findings in order to verify the detailed phase transformation and the true nature of deformation. As it appears from the microstructure, it can also be said that if a titanium alloy of lower β transus is kept on the advancing side (P4, in our case, with TIMET-54M on the advancing side) it results in the formation of a basket-weave morphology (refined β grains with finer acicular α) inside the WN that, as a part of entire gage length (if measured in a transverse tensile test for global stress strain curve), assists in achieving superior values of mechanical properties.

5. Conclusions

In the present study, we have successfully joined two dissimilar titanium alloys, ATI-425 and TIMET-54M, with FSW at 300 rpm and 75–100 mm/min traverse speed by using W-La tool with a specified geometry. Weld joints with no visible defects (as shown in Figure 5) were obtained by exchanging the locations (advancing or retreating) of the sheets. The following conclusions were drawn:

(1) The typical microstructure as it evolved inside, and around, the WN, is dependent on the initial BM microstructure. If a material with a lower β transus temperature (TIMET-54M) is kept on the advancing side(P4), the majority of the microstructure inside the WN is characterized by the refined prior β grains and finer acicular α. On the other hand, if same material with a lower transus temperature (TIMET-54M) is kept on the retreating side (P1), the microstructure in the WN is characterized by coarser α lamellae with grain boundary α.

(2) Higher values of hardness have been observed for the case when TIMET-54M was kept on the advancing side (P4).

(3) The global stress strain curve showed an increase in the mechanical properties when TIMET-54M was kept on the advancing side.

A trivial conclusion that can be drawn from this study is that when a titanium alloy with a lower β transus temperature is kept on the advancing side it results in better mechanical properties. However, further study needs to be conducted in order to confirm the concluding remarks.

Acknowledgments: We sincerely thank The Boeing Company for financial support of Titanium Component Manufacturing Research Project at University of Washington. The Boeing Company for support & encouragement.

Metals **2016**, *6*, 252

Author Contributions: Daniel G. Sanders's lab produced the welded samples, Kapil Gangwar and Andrew Cantrell conducted the experiments and microstructural evaluation. M. Ramulu, Kapil Gangwar analyzed the data in consultation with Daniel G. Sanders and contributed to writing and editing the manuscript.

Conflicts of Interest: The authors declare no conflict of interest.

References

1. Thomas, W.M.; Nicholas, E.D.; NeedHam, J.C.; Murch, M.G.; Templesmith, P.; Dawes, C.J. *Friction Stir Welding*; The Welding Institure: Cambridge, UK, 1991.
2. Boyer, R.R. An overview on the use of titanium in the aerospace industry. *Mater. Sci. Eng. Struct. Mater. Prop. Microstruct. Process.* **1996**, *213*, 103–114. [CrossRef]
3. Peters, J.O.; Lutjering, G. Comparison of the fatigue and fracture of alpha plus beta and beta titanium alloys. *Metall. Mater. Trans. Phys. Metall. Mater. Sci.* **2001**, *32*, 2805–2818. [CrossRef]
4. Lutjering, G. Influence of processing on microstructure and mechanical properties of (alpha + beta) titanium alloys. *Mater. Sci. Eng. Struct. Mater. Prop. Microstruct. Process.* **1998**, *243*, 32–45. [CrossRef]
5. Kim, J.S.; Kim, J.H.; Lee, Y.T.; Park, C.G.; Lee, C.S. Microstructural analysis on boundary sliding and its accommodation mode during superplastic deformation of Ti-6Al-4V alloy. *Mater. Sci. Eng. Struct. Mater. Prop. Microstruct. Process.* **1999**, *263*, 272–280. [CrossRef]
6. Fonda, R.W.; Knipling, K.E. Texture development in near-alpha Ti friction stir welds. *Acta Mater.* **2010**, *58*, 6452–6463. [CrossRef]
7. Fonda, R.W.; Knipling, K.E. Texture development in friction stir welds. *Sci. Technol. Weld. Join.* **2011**, *16*, 288–294. [CrossRef]
8. Pao, P.S.; Fonda, R.W.; Jones, H.N.; Feng, C.R.; Moon, D.W. Fatigue crack growth in friction stir welded Ti-5111. In *Friction Stir Welding and Processing V*; Rajiv, M.W.M., Mishra, S., Lienert, T.J., Eds.; TMS (The Minerals, Metals & Materials Society): Warrendale, PA, USA, 2009.
9. Reynolds, A.P.; Hood, E.; Tang, W. Texture in friction stir welds of Timetal 21S. *Scr. Mater.* **2005**, *52*, 491–494. [CrossRef]
10. Yoon, S.; Ueji, R.; Fujii, H. Microstructure and texture distribution of Ti-6Al-4V alloy joints friction stir welded below beta-transus temperature. *J. Mater. Process. Technol.* **2016**, *229*, 390–397. [CrossRef]
11. Zhang, Y.; Sato, Y.S.; Kokawa, H.; Park, S.H.C.; Hirano, S. Stir zone microstructure of commercial purity titanium friction stir welded using pcBN tool. *Mater. Sci. Eng. Struct. Mater. Prop. Microstruct. Process.* **2008**, *488*, 25–30. [CrossRef]
12. Zhang, Y.; Sato, Y.S.; Kokawa, H.; Park, S.H.C.; Hirano, S. Microstructural characteristics and mechanical properties of Ti-6Al-4V friction stir welds. *Mater. Sci. Eng. Struct. Mater. Prop. Microstruct. Process.* **2008**, *485*, 448–455. [CrossRef]
13. Zhou, L.; Liu, H.J.; Liu, Q.W. Effect of process parameters on stir zone microstructure in Ti-6Al-4V friction stir welds. *J. Mater. Sci.* **2010**, *45*, 39–45. [CrossRef]
14. Zhou, L.; Liu, H.J.; Wu, L.Z. Texture of friction stir welded Ti-6Al-4V alloy. *Trans. Nonferr. Met. Soc. China* **2014**, *24*, 368–372. [CrossRef]
15. Edwards, P.; Ramulu, M. Identification of Process Parameters for Friction Stir Welding Ti-6Al-4V. *J. Eng. Mater. Technol. Trans. Asme* **2010**, *132*. [CrossRef]
16. Rai, R.; De, A.; Bhadeshia, H.K.D.H.; DebRoy, T. Review: Friction stir welding tools. *Sci. Technol. Weld. Join.* **2011**, *16*, 325–342. [CrossRef]
17. Edwards, P.D.; Ramulu, M. Material flow during friction stir welding of Ti-6Al-4V. *J. Mater. Process. Technol.* **2015**, *218*, 107–115. [CrossRef]
18. Pilchak, A.L.; Williams, J.C. Microstructure and Texture Evolution during Friction Stir Processing of Fully Lamellar Ti-6Al-4V. *Metall. Mater. Trans. Phys. Metall. Mater. Sci.* **2011**, *42*, 773–794. [CrossRef]
19. Edwards, P.; Ramulu, M. Fracture toughness and fatigue crack growth in Ti-6Al-4V friction stir welds. *Fatigue Fract. Eng. Mater. Struct.* **2015**, *38*, 970–982. [CrossRef]
20. Edwards, P.; Ramulu, M. Fatigue performance of Friction Stir Welded titanium structural joints. *Int. J. Fatigue* **2015**, *70*, 171–177. [CrossRef]
21. Edwards, P.; Ramulu, M. Fatigue performance of Friction Stir Welded Ti-6Al-4V subjected to various post weld heat treatment temperatures. *Int. J. Fatigue* **2015**, *75*, 19–27. [CrossRef]

22. Edwards, P.D.; Ramulu, M. Comparative study of fatigue and fracture in friction stir and electron beam welds of 24mm thick titanium alloy Ti-6Al-4V. *Fatigue Fract. Eng. Mater. Struct.* **2016**, *39*, 1226–1240. [CrossRef]
23. Sanders, D.G.; Edwards, P.; Cantrell, A.M.; Gangwar, K.; Ramulu, M. Friction Stir-Welded Titanium Alloy Ti-6Al-4V: Microstructure, Mechanical and Fracture Properties. *JOM* **2015**, *67*, 1054–1063. [CrossRef]
24. Fazel-Najafabadi, M.; Kashani-Bozorg, S.F.; Zarei-Hanzaki, A. Dissimilar lap joining of 304 stainless steel to CP-Ti employing friction stir welding. *Mater. Des.* **2011**, *32*, 1824–1832. [CrossRef]
25. Liao, J.S.; Yamamoto, N.; Liu, H.; Nakata, K. Microstructure at friction stir lap joint interface of pure titanium and steel. *Mater. Lett.* **2010**, *64*, 2317–2320. [CrossRef]
26. Ishida, K.; Gao, Y.; Nagatsuka, K.; Takahashi, M.; Nakata, K. Microstructures and mechanical properties of friction stir welded lap joints of commercially pure titanium and 304 stainless steel. *J. Alloy. Compd.* **2015**, *630*, 172–177. [CrossRef]
27. Fazel-Najafabadi, M.; Kashani-Bozorg, S.F.; Zarei-Hanzaki, A. Joining of CP-Ti to 304 stainless steel using friction stir welding technique. *Mater. Des.* **2010**, *31*, 4800–4807. [CrossRef]
28. Buffa, G.; Baffari, D.; di Caro, A.; Fratini, L. Friction stir welding of dissimilar aluminium-magnesium joints: Sheet mutual position effects. *Sci. Technol. Weld. Join.* **2015**, *20*, 271–279. [CrossRef]
29. Yuan, W.; Mishra, R.S.; Carlson, B.; Verma, R.; Mishra, R.K. Material flow and microstructural evolution during friction stir spot welding of AZ31 magnesium alloy. *Mater. Sci. Eng. Struct. Mater. Prop. Microstruct. Process.* **2012**, *543*, 200–209. [CrossRef]
30. Chen, Y.H.; Ni, Q.; Ke, L.M. Interface characteristic of friction stir welding lap joints of Ti/Al dissimilar alloys. *Trans. Nonferr. Met. Soc. China* **2012**, *22*, 299–304. [CrossRef]
31. Hong, J.K.; Lee, C.H.; Kim, J.H.; Yeom, J.T.; Lee, C.G. Friction Stir Welding of Dissimilar Al 5052 to Ti-6Al-4V Alloy with WC-Co Tool. *Steel Res. Int.* **2010**, *81*, 1092–1095.
32. Zhang, Z.H.; Li, B.; Feng, X.M.; Shen, Y.F.; Hu, W.Y. Friction-stir welding of titanium/aluminum dissimilar alloys: Joint configuration design, as-welded interface characteristics and tensile properties. *Proc. Inst. Mech. Eng. Part B J. Eng. Manuf.* **2014**, *228*, 1469–1480. [CrossRef]
33. Vaidya, W.V.; Horstmann, M.; Ventzke, V.; Petrovski, B.; Kocak, M.; Kocik, R.; Tempus, G. Structure-property investigations on a laser beam welded dissimilar joint of aluminium AA6056 and titanium Ti6Al4V for aeronautical applications Part I: Local gradients in microstructure, hardness and strength. *Mater. Werkst.* **2009**, *40*, 623–633. [CrossRef]
34. Hsieh, C.T.; Chu, C.Y.; Shiue, R.K.; Tsay, L.W. The effect of post-weld heat treatment on the notched tensile fracture of Ti-6Al-4V to Ti-6Al-6V-2Sn dissimilar laser welds. *Mater. Des.* **2014**, *59*, 227–232. [CrossRef]
35. Jata, K.V.; Subramanian, P.R.; Reynolds, A.P.; Trapp, T.; Helder, E. Friction stir welding of titanium alloys for aerospace application: Microstructure and mechanical behavior. In Proceedings of the Fourteenth (2004) International Offshore and Polar Engineering Conference, Toulon, France, 23–28 May 2004.
36. ATI Technical Data Sheet. Available online: https://www.atimetals.com/Documents/ati_425_alloy_tds_en_v5.pdf (accessed on 21 October 2016).
37. Kosaka, Y.; Gudipati, P. Supreplastic Forming Prperties of TIMETAL 54M. Available online: http://c.ymcdn.com/sites/www.titanium.org/resource/resmgr/2010_2014_papers/KosakaYogi_2010_AerospaceMat.pdf (accessed on 21 October 2016).
38. Edwards, P.; Ramulu, M. Surface Residual Stresses in Ti-6Al-4V Friction Stir Welds: Pre- and Post-Thermal Stress Relief. *J. Mater. Eng. Perform.* **2015**, *24*, 3263–3270. [CrossRef]
39. Lienert, T.J. Microstructure and mechanical properties of friction stir welded titanium alloys. In *Friction Stir Welding and Processing*; Rajiv, M.W.M., Mishra, S., Eds.; ASM International: Geauga County, OH, USA, 2007; p. 128.
40. Frankel, J.; Abbate, A.; Scholz, W. The Effect of Residual-Stresses on Hardness Measurements. *Exp. Mech.* **1993**, *33*, 164–168. [CrossRef]

![metals logo] *metals*

MDPI

Article

Microstructure and Mechanical Properties of Dissimilar Friction Stir Welding between Ultrafine Grained 1050 and 6061-T6 Aluminum Alloys

Yufeng Sun [1,2,*], Nobuhiro Tsuji [2] and Hidetoshi Fujii [1]

[1]   Joining and Welding Research Institute, Osaka University, Ibaraki 5670047, Japan; fujii@jwri.osaka-u.ac.jp
[2]   Department of Materials Science and Engineering, Kyoto University, Kyoto 6068501, Japan; nobuhiro-tsuji@mtl.kyoto-u.ac.jp
*   Correspondence: yfsun@jwri.osaka-u.ac.jp; Tel.: +81-6-68798663; Fax: +81-6-687-986-53

Academic Editor: Giuseppe Casalino
Received: 25 August 2016; Accepted: 15 October 2016; Published: 21 October 2016

Abstract: The ultrafine grained (UFGed) 1050 Al plates with a thickness of 2 mm, which were produced by the accumulative roll bonding technique after five cycles, were friction stir butt welded to 2 mm thick 6061-T6 Al alloy plates at a different revolutionary pitch that varied from 0.5 to 1.25 mm/rev. In the stir zone, the initial nano-sized lamellar structure of the UFGed 1050 Al alloy plate transformed into an equiaxial grain structure with a larger average grain size due to the dynamic recrystallization and subsequent grain growth. However, an equiaxial grain structure with a much smaller grain size was simultaneously formed in the 6061 Al alloy plates, together with coarsening of the precipitates. Tensile tests of the welds obtained at different welding speeds revealed that two kinds of fracture modes occurred for the specimens depending on their revolutionary pitches. The maximum tensile strength was about 110 MPa and the fractures were all located in the stir zone close to the 1050 Al side.

Keywords: ultrafine grained structure; dissimilar friction stir welding; dynamic recrystallization; aluminum alloy

---

1. Introduction

As a solid state joining process, friction stir welding (FSW) was invented by the welding insitute TWI in the UK with the original purpose to weld light metals such as aluminum and magnesium alloys, which has been demonstrated to be very difficult by traditional fusion welding methods [1,2]. Since FSW is performed at a temperature lower than the melting point of the materials to be welded, FSW can produce joints with fewer defects or porosity, low residual stresses, etc., compared with other fusion welding methods [3,4]. The FSW technique has developed very rapidly since its emergence and it has now been expanded to many high melting point metallic materials, including Cu, Ti, Fe, stainless steels, and even high carbon steels, which were considered to be unweldable materials by fusion welding methods because of the formation of the brittle martensite phase [5–8]. Therefore, with the increasing effort to improve fuel efficiency in industry, the use of the FSW technique must be in strong demand in the near future for the welding of light materials, especially aluminum and magnesium alloys.

Recently, the FSW of certain types of ultrafine grained (UFGed) materials has been investigated. UFGed materials are those that generally have an average grain size of less than 1 μm. UFGed materials have an increased length of grain boundaries acting as obstacles for moving dislocation and therefore exhibit superior properties at ambient temperature to their coarse grained counterparts, which also makes them attractive candidates for a range of potential applications in the automotive, aerospace and biomedical industries. There are many severe plastic deformation (SPD) techniques like accumulative roll bonding (ARB), equal channel angular pressing (ECAP), and high pressure torsion (HPT) to produce UFGed metallic materials, among which the ARB technique can be used to produce UFGed materials in bulk plate form [9]. The welding or joining of UFGed materials is therefore becoming more and more important. Obviously, the UFGed materials cannot be welded by the general fusion welding methods, since the molten pool generated during the welding process will inevitably destroy the UFGed structure and result in a much coarser grain size in the solidified butt. The UFGed alloys like the 1050, 1100, 6016 Al alloys and IF steel prepared by the ARB technique have also been successfully subjected to the FSW process [10–15]. It was revealed that a hardness reduction did take place in the FSW processed materials with the UFGed structure due to the coarsening of the grain size. Although the initial UFGed structure is still very difficult to be retained in the stir zone, the degradation of the mechanical properties was significantly reduced.

However, the dissimilar FSW involved with UFGed materials has never been reported according to the best of our knowledge. The FSW of dissimilar alloys has been significantly studied including dissimilar aluminum alloys, Al/Mg alloys, Al/Steel pairs, etc. The main salient feature of FSW of dissimilar metals and alloys is thought to be the variation in asymmetry or the degree of symmetry with reference to the weld centerline [16]. For example, Lee et al. evaluated the joint microstructure of the dissimilar welds between cast A356 and wrought 6051 aluminum alloys produced at various welding speeds [17]. Palanivel et al. [18] studied the effect of the tool rotational speed and pin profile on the microstructure and tensile strength of dissimilar FSW between the AA5083-H111 and AA6351-T6 aluminum alloys. They found that joint strength was affected due to the variations in the materials behavior. It is evident that an important aspect in the FSW of dissimilar materials is the selection of the appropriate alloys for the advancing and the retreating sides to obtain the optimum mixing and weld properties due to the asymmetric material flow in the joints. It was found that the maximum tensile strength was achieved for the dissimilar FSW AA2024/AA7075 aluminum alloy joints only when the 2024 Al alloy was located on the advancing side [19]. Kwon et al. successfully obtained Al/Mg dissimilar FSW joints when the AZ31 alloy and Al alloy were located on the RS and AS, respectively. However, the reason of the work-piece configuration was not explained in detail [20]. According to the investigation of dissimilar FSW between Al and Cu alloys, the suitable configuration and even the amount of offset of the tool from the joint centerline were considered to play an important role in obtaining high joints properties [21–23]. More recently, Sun et al. conducted the dissimilar spot FSW between the UFGed 1050Al and 6061-T6 aluminum alloys [24]. However, the UFGed materials have not been reported to be dissimilar FSW processed with other materials.

In this study, the dissimilar friction stir butt welding was carried out between the UFGed 1050 Al and 6061-T6 aluminum alloys in order to expand the application of UFGed materials. After welding, the microstructure and mechanical properties of the joints were characterized and discussed.

2. Materials and Methods

In this study, the dissimilar FSW was performed between 2 mm thick UFGed 1050 Al and commercial 6061-T6 Al alloy plates. The UFGed 1050 Al alloy plates were fabricated by the ARB process after 5 cycles. The 6061-T6 Al alloy has a yielding strength of about 266 MPa and the UFGed 1050 Al has a yielding strength of about 200 MPa, much higher than the 75 MPa of coarse grained 1050 Al alloys. Prior to the dissimilar FSW, the two kinds of Al plates were cleaned with acetone to remove any impurities on the surface such as dirt and oil. To optimize the welding conditions, the UFGed Al plates were separately placed on the advancing side (AS) and retreating side (RS). The rotating tools were made of tool steel, which had a concave-shaped shoulder geometry with a diameter of 12 mm and a threaded probe with a diameter of 4 mm and a length of 1.8 mm. The tool axis was tilted by 3° with respect to the normal direction of the sample surface. During welding, the rotating tool exactly penetrated into the butt interface between the two dissimilar materials at a speed of 0.5 mm/s. After the shoulder of the tool touched the plate surface, the tool started to travel along the butt interface and left a weld seam behind. The welding was carried out at a constant load of 8000 kN and a constant rotation speed of 800 rpm, while the welding speed was varied at 400, 600, 800 and 1000 mm/min, corresponding to the revolutionary pitch (welding speed/rotation speed) of 0.5, 0.75, 1 and 1.25 mm/rev, respectively.

After welding, optical microscopy (OM, Olympus Microscopy, Tokyo, Japan, BX51M) and scanning electronic microscopy (SEM, JEOL Microscopy, Osaka, Japan, JEOL-7001FA) with an electron backscattered diffraction (EBSD, TSL Solutions, Tokyo, Japan) system were used to characterize the microstructure of the joints. For the OM observations, the specimens were first mechanically polished and then chemically etched with Keller's reagent. The specimens for the EBSD measurement were electro-polished using a solution of $HNO_3:CH_4O = 3:7$ at 15 V and $-30\,^\circ C$. The Vickers micro-hardness tests on the cross-sectional plane of the joints were carried out at an interval of 0.5 mm using a testing machine (AAV500, Akashi, Tokyo, Japan). The tensile tests of the joints were carried out using a testing machine (5500, Instron, Norwood, MA, USA) with a cross-head speed of 1 mm/min.

3. Results and Discussion

Figure 1 shows the microstructure of the two types of base metals. The UFGed 1050 aluminum alloys have a lamellar structure with an average lamellar width of about 300 nm as shown in Figure 1a, which is the typical microstructure of the ARB processed materials. These lamellar shaped grains normally have a boundary spacing or grain width much smaller than 1 μm. In addition, the elongated grains in the ARB processed materials are generally surrounded by high angle boundaries, a grain misorientation angle larger than 15°, and high fraction of high angle boundaries of more than 70%. The 6061-T6 aluminum alloys have a coarse equiaxial grain structure with an average grain size of about 18 μm as shown in Figure 1b. As a typical precipitate hardened aluminum alloy, the mechanical properties of the 6061 Al alloy slightly depend on the grain size of the alloy, but are strongly dominated by the volume fraction, size and distribution of the strengthening precipitates. The 6061 alloy under T6 heat treatment consists of a high density of needle-shape precipitates and a low density of β-Mg_2Si precipitates [25].

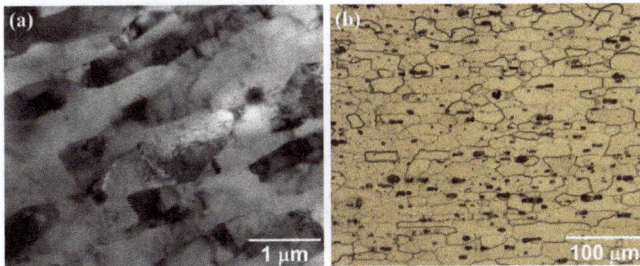

Figure 1. The microstructure of the base metal of (**a**) ultrafine grained (UFGed) 1050 aluminum and (**b**) 6061-T6 aluminum alloys.

Figure 2 is the photos of the samples produced by FSW at different revolutionary pitches, in which the UFGed 1050 and 6061-T6 aluminum plates were placed on the AS and RS side, respectively. It was found that the dissimilar FSW between these two materials could never been successful. During every welding process, a large defect was formed in the UFGed 1050 aluminum plates and the defect became larger when decreasing the welding speed, as shown in the figures. It was noted that the dissimilar FSW between two aluminum alloys might be difficult since they have quite different deformation characteristics at high temperature. Because the rotating tool on the AS shows the same direction of self-rotation and traveling along the butt interface, the material in the AS generally experiences a higher strain rate and higher temperature than that in the RS [26,27]. Therefore, the materials with a low melting point are usually placed on the RS during the dissimilar welding process between two materials with quite different melting points, for example, dissimilar FSW between aluminum and steel alloys [28,29]. Sometimes the rotating tool also needs to be offset from the butt interface toward the lower melting temperature materials in order to prevent tool wear or overheating of the lower melting materials. In this study, although the 1050 and 6061 aluminum alloys had very similar melting points, the UFGed 1050 aluminum had a very unstable microstructure, which easily resulted in the dynamic recrystallization and subsequent grain growth at elevated temperature during the welding process. When the UFGed 1050 Al plate was placed on the AS, the materials became very soft and would flow into the RS with the rotation of the tool. However, the 6061 Al on the RS was still hard to be plasticized due to the relatively lower temperature and strain rate. The softened 1050 Al could not be stirred into the 6061 Al on the RS and finally was extruded outside stir zone. As a result, a defect was formed on the AS due to the insufficient amount of the materials. Therefore, placing the UFGed materials on the AS was not suitable for the dissimilar FSW in this study.

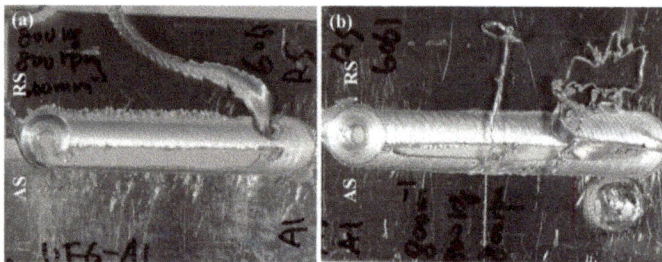

Figure 2. Photo showing the appearance of the dissimilar friction stir welding (FSW) joints with UFGed 1050Al on the advancing side (AS) and 6061-T6 on the retreating side (RS) produced at a revolutionary pitch of (**a**) 0.75 mm/rev; and (**b**) 1 mm/rev.

Figure 3 shows photos of the samples produced at different welding speeds, in which the UFGed 1050 Al and 6061-T6 Al alloys were placed on the RS and AS, respectively. The welding was successfully done at the wide revolutionary pitches from 0.5 to 1.25 mm/rev and no obvious defects could be observed based on the appearance of the joints as shown in Figure 3a,b. However, it was found that at a larger revolutionary pitch of speed of 1 or 1.25 mm/rev, the width of the welding seam became smaller and smaller with the increasing welding distance. It was proposed that the heat input was not sufficient at the higher welding speed during welding process, and the rotating tool could not penetrate deep enough to make the tool shoulder touch the sample surface completely. As a result, a lack of contact between the tool shoulder and the surface of the work-piece was formed, as shown by the arrow in Figure 3d. Due to the very different material flow of the two materials, the interface could still be observed on the surface of the joint centerline. At the lower revolutionary pitch of less than 0.75 mm/rev, the welding process became very stable and a homogeneous welding seam was obtained, except that a slightly more flash formed on the retreating side of the joint welded at a revolutionary pitch of 0.5 mm/rev. It was revealed that the FSW of dissimilar aluminum alloys should be conducted with suitable locations and in this study the UFGed aluminum plates should be put on the RS.

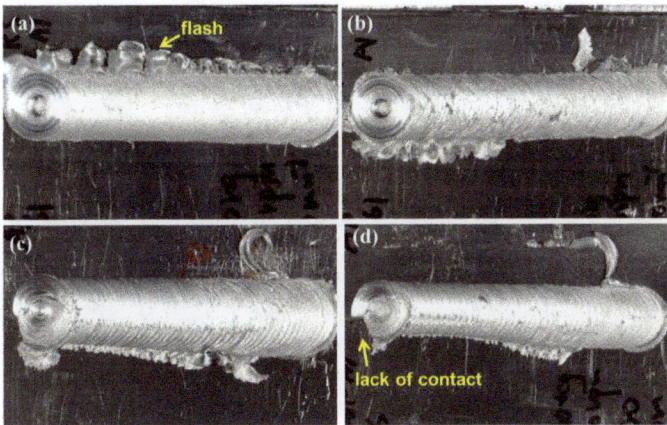

Figure 3. Photo showing the appearance of the dissimilar FSW joints with UFGed 1050Al on the RS side and 6061-T6 on the AS produced at a revolutionary pitch of (**a**) 0.5; (**b**) 0.75; (**c**) 1; and (**d**) 1.25 mm/rev.

Figure 4 shows the cross-sectional macrostructure of the dissimilar FSW joints between the UFGed 1050 Al and 6061-T6 Al alloy plates, which were obtained at different revolutionary pitches ranging from 0.5 to 1.25 mm/rev. Since the 6061-T6 Al alloy was etched black using Keller's reagent, the interface and mixing status of the two materials could be readily discerned in the joints. The entire stir zone of the FSW joint consists of the shoulder-affected zone and probe-affected zone. The shoulder affected zone becomes larger at higher rotation speed or lower welding speed, while the probe affected zone shows less sensitive to the welding condition. This has been recently confirmed by the investigation with an adjustable rotating tool [30]. As a result, a crown-like stir zone is generally formed along the transverse direction, especially for thin plates like in this study. It can be found that the area of the stir zone became larger with the decreasing welding speed for all the samples. At higher revolutionary pitch of 1 and 1.25 mm/rev, the bending of the interface between the alternative layers of the aluminum sheets was observed in the UFGed 1050 Al side near the stir zone as shown by the arrow in Figure 4c. Generally, a well chemically etched cross-section of the friction stir weld reveals an onion-ring structure in the stir zone with a round flow pattern formed by bright and dark lamellae in the dissimilar aluminum joints [31,32]. However, in this study the round flow pattern containing bright and dark lamellae was not observed in the stir zone, probably due to the relatively smaller thickness of the plates. In contrast, roughly two kinds of mixing types could be classified. One was the mixing of dissimilar materials caused by the lower heat input at the high revolutionary pitches of 1 and 1.25 mm/rev. In this case, the size of the stir zone was small and only several prismatic blocks of the 6061 alloy were found in the stir zone. Most of the stir zone was composed of the 1050 Al alloys probably due to its lower viscosity than 6061 alloy at high temperature. In addition, the TMAZ/SZ boundary close to the 6061 Al side was very sharp and nearly vertical, which indicated that the probe-affected zone was very limited due to the fast traveling speed. The other kind of mixing was caused by the higher heat input at the lower revolutionary pitches of 0.75 and 0.5 mm/rev. The shoulder affected zone became much larger and expand downward to the probe affect zone. The material flow of the two materials in the stir zone was stronger and the boundary between the TMAZ/SZ near the 6061-T6 Al side declined more to the base metal. As a result, more material of the 6061-T6 Al alloy was plastically deformed and pushed into the stir zone. Especially, at the low revolutionary pitch of 0.5 m/rev, the two materials could not be easily discerned any more in the stir zone due to the severe mechanical mixing by the rotating tool.

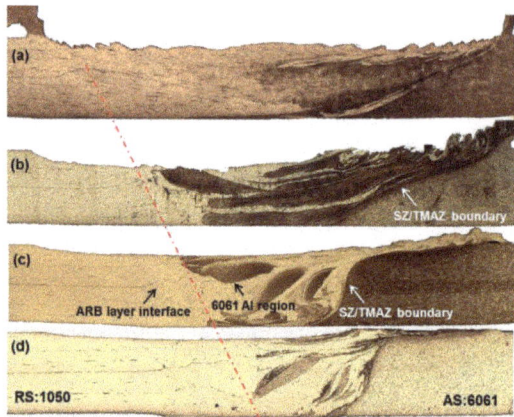

Figure 4. Optical microscopy (OM) images showing the cross-sectional macrostructure of the UFGed 1050/6061-T6 Al alloys dissimilar FSW joints produced at the different revolutionary pitches of (**a**) 0.5; (**b**) 0.75; (**c**) 1; and (**d**) 1.25 mm/rev.

Figure 5 shows the enlarged OM images of the stir zone of the dissimilar FSW joints produced at the different welding speeds. In all the figures except for Figure 5d, the white and black areas indicated the 1050 Al and the 6061 Al alloys, respectively. As shown in Figure 4, only several large blocks of the 6061 Al alloys were mixed in the stir zone at the high welding speed. Figure 5a,b show the boundary between the 6061 Al blocks and the surrounding 1050 Al produced at the revolutionary pitches of 1.25 and 1 mm/rev, respectively. A couple of small welding defects could be distinguished at the corner of the 6061-T6 region. The formation of the defects might be caused by the insufficient plastic deformation at the lower heat input. In addition, it is interesting to note that deformed UFGed 1050 Al layers were observed to be mixed in the region of the 6061-T6 aluminum alloy. In most of the large 6061-T6 blocks, a large number of small segments of white color, which corresponded to the 1050 Al alloy were observed. However, the 6061-T6 aluminum materials were never found to be distributed inside the 1050 Al area. When the revolutionary pitch decreased to 0.75 mm/rev, the mixing of the two materials became much stronger. However, large stripe-like areas of the two different materials were still alternatively distributed together as shown in Figure 5c. When the revolutionary pitch further decreased to 0.5 mm/rev, the mixing became quite complete and the two different materials could not be easily discerned under by OM observation, as shown in Figure 5d.

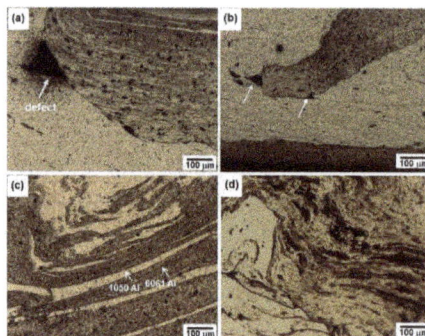

Figure 5. OM images showing the microstructure of the stir zone of the dissimilar FSW joints obtained at the different revolutionary pitches of (**a**) 1.25; (**b**) 1; (**c**) 0.75; and (**d**) 0.5 mm/rev.

To reveal how the two materials were mixed in the stir zone, SEM observations of the cross-sectional microstructure of the joints were carried out. As a typical example, Figure 6 shows the microstructure of the joints produced at a revolutionary pitch of 0.75 mm/rev. In most parts of the area it showed a very high density of white particles, while in the other parts particles were hardly observed. From the element mapping by Energy Dispersive X-ray Spectroscopy EDS measurement as shown in Figure 7, the particles were confirmed to be the Mg_2Si phase. The coarsening of the Mg_2Si particles in the 6061 Al alloy area due to the quite high heat input generated during the FSW process, was also observed in the FSW of similar 6xxx aluminum alloys [33]. The distribution of particles in the 6061-T6 Al alloy made it easy to be distinguished from the area of the 1050 Al, which contained no particles. The 1050 Al part was like a long stripe with a width of about 10 μm in the mixture of the two materials, which was very likely caused by the flow stress during welding. Figure 6b shows the boundary between the 6061 and 1050 Al alloys. The Mg_2Si particles were found distributed on the boundary, which played a role in pinning of the migration of the grain boundary during the welding process. In addition, sub-grains with the average size of several nanometers were formed inside the 6061 Al grains, which were also found in the dissimilar friction stir spot welding between UFGed 1050 and 6061-T6 Al alloys [24].

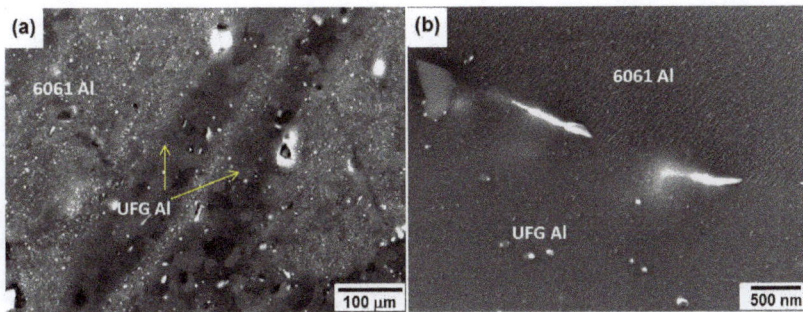

Figure 6. SEM (scanning electronic microscopy) images showing the mixing of the two kinds of aluminum alloys in the stir zone. (**a**) Alternative layered structure of the two materials; (**b**) interfacial microstructure between the UFGed 1050/6061-T6 aluminum alloys.

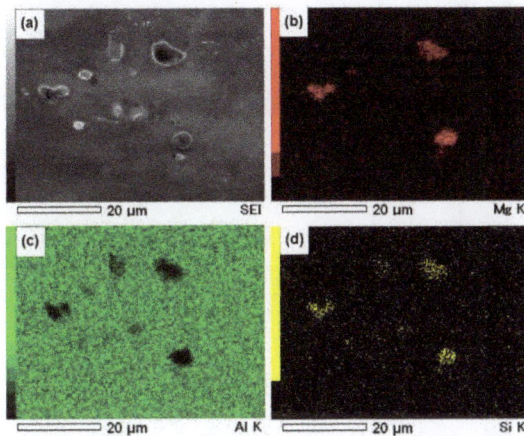

Figure 7. The EDS mapping showing the Mg_2Si particles distributed in the area of the 6061 Al alloy. (**a**) SEM image and corresponding distribution of (**b**) Mg; (**c**) Al; and (**d**) Si elements.

Figure 8 shows the EBSD maps measured in the mixed area of the stir zone, in which only the high angle boundary larger than 15° was plotted. The black area in the maps indicated the distribution of the coarsened Mg_2Si precipitates due to the generated Kikuchi pattern with a low confidence index (CI) value. Figure 8a shows the EBSD-IPF map of the mixed area in the stir zone of the joint produced at a revolutionary pitch of 0.75 mm/rev. An obvious bimodal structure containing two types of grains with quite different grain size could be observed. One type of grain structure had a larger average grain size of about 2.6 µm, while the other type of grain structure had a smaller average grain size of about 1.1 µm. Figure 8b shows the corresponding image quality (IQ) map, in which the matrix of the 6061 Al alloy contains additional elements like Mg and Cu and therefore its crystalline lattice deviated from that of the pure Al. Another reason for the low IQ value of the 6061 Al region might be caused by the formation of a high density of subgrain, in which the density of dislocation might be higher. Therefore, the area with the larger grain size corresponded to the 1050 Al alloys, while the area with the smaller grain size corresponded to the 6061 Al alloys. Figure 8c,d shows the IPF and IQ map of the joint produced at a revolutionary pitch of 0.5 mm/rev, when a higher heat was generated during welding. The mixing of the two materials became stronger; however, it still can be distinguished due to their quite different grain size and IQ value. In this case, the average grain size of the 1050 Al and 6061 Al were 4.5 and 1.5 µm, respectively. With the increase in the heat input during welding, the grain growth of the 1050 Al increases faster than that of the 6061 Al alloys.

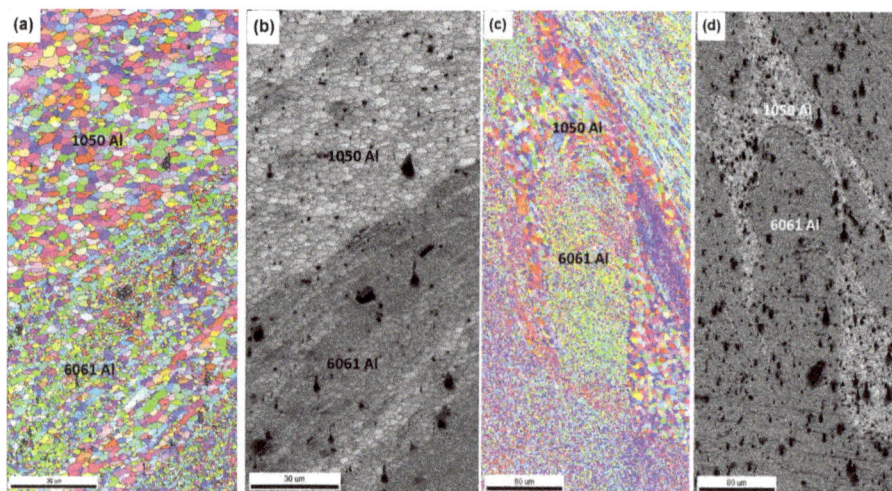

Figure 8. Electron backscattered diffraction (EBSD) map showing the microstructure of the stir zone of the dissimilar FSW joint produced at the revolutionary pitches of (**a**) and (**b**) 0.75 mm/rev; (**c**) and (**d**) 0.5 mm/rev.

It is interesting to note that in the stir zone the 1050 Al showed an average grain size quite larger than 300 nm of the BM, while the 6061 Al showed a fairly refined microstructure compared with that 18 µm of the BM. In addition, both 1050 Al and 6061Al had equiaxial grain structure and large fraction of high angle grain boundary, indicating the occurrence of dynamic recrystallization. Because FSW is a kind of high strain rate plastic deformation at high temperature, the evolution of the grain structure in the stir zone during the FSW strongly depends on the initial microstructure such as the grain boundary structure, grain size, dislocation density, etc., of the base metal [34]. Usually, when the FSW process is applied to the conventional metals or alloys, a very refined microstructure will be formed in the stir zone of the joints. The grain refinement process is generally believed to be driven by the grain subdivision or the continuous dynamic recrystallization, termed geometric dynamic recrystallization which was first recognized by Humphreys and McQeen [35–37]. The continuous dynamic recrystallization is characterized by a strain-induced progressive rotation of the subgrains with little boundary migration during the FSW process and is prone to occur at grain boundary in aluminum alloys with a high level of solute like Mg and Zn, by progressive lattice rotation.

The pure Al alloy has a high stacking fault energy and the UFGed 1050 Al produced by the SPD process generally showed a fairly unstable microstructure upon heat treatment. That is because unlike conventional metals or alloys, the as-ARB processed materials contain a large quantity of vacancies and dislocations generated by the severe plastic deformation. Therefore, the UFGed alloys are very likely to be structurally recovered at a relatively low temperature to decrease the defect density within the grains. At the same time, continuous grain growth took place and the mean grain size increased. Similarly, during the dissimilar FSW process in this study, the area of the 1050 Al showed a fairly coarse grain structure, which resulted in an obvious hardness reduction in the stir zone.

Figure 9 shows the microhardness profile for the dissimilar FSW joints produced at different welding speeds. The variation in the hardness value corresponded to the typical microstructural zones. For all the specimens, the base metal of both the UFGed 1050 and 6061-T6 Al alloys showed the highest value. From the BM to HAZ, a gradual decrease in microhardness was observed for both materials. For the 6061-T6 alloy, the decrease was due to the accelerated solid solution of precipitates and the simultaneous occurrence of coarsening of particles caused by the weld thermal cycles. However, the stir zone also showed some regions with a high hardness value similar to that of the base metal, which was caused by the significantly refined grain size, while in the UFG 1050 Al side, the decrease in the hardness was mainly caused by the grain growth and the dislocation density.

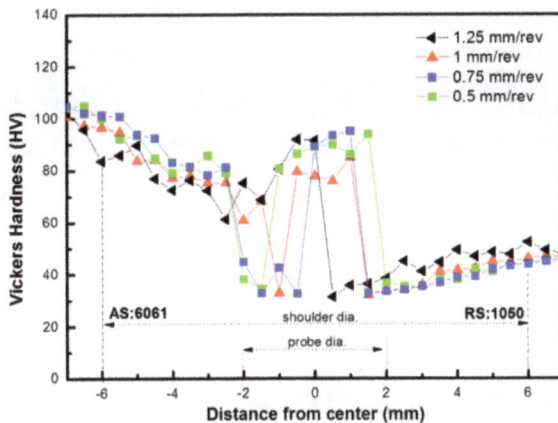

Figure 9. Hardness profile along center line of the cross-sectional plane of the dissimilar FSW joints produced at various welding speeds.

Figure 10 shows the tensile strain-stress curves of the specimens produced at different welding speed. For the specimens produced at the revolutionary pitches of 1.25 and 1 mm/rev, both the tensile strength and elongation were lower than the other specimens produced at the smaller revolutionary pitches. It is proposed that the mixing of the two materials was not sufficient and several large blocks of different materials were formed in the stir zone. In addition, a couple of defects with several micrometers in size were found at the corner of the 6061-T6 aluminum block that may more or less decrease the strength of the joints. When the revolutionary pitch decreased to 0.75 or 0.5 mm/rev, the high heat input enhanced the plastic deformation and therefore led to further mixing of the two materials in the stir zone. The microstructure of the stir zone also became more homogeneous. As a result, the tensile strength and elongation increased to about 110 MPa and 13% for the sample welded at 0.75 mm/rev and 110 MPa and 22.5% for the sample welded at 0.5 mm/rev. The largest joint efficient was therefore about 55% with respect to the UFG 1050 Al base metal. However, the joints strength was still much larger than the ultimate strength of the commercial coarse grain structured 1050 Al alloy. However, too much heat input in the joints produced at 0.5 mm/rev resulted in serious grain growth of both materials, especially the 1050 Al alloy parts. The tensile strength slightly decreased again, however, with compensation of a much increased elongation.

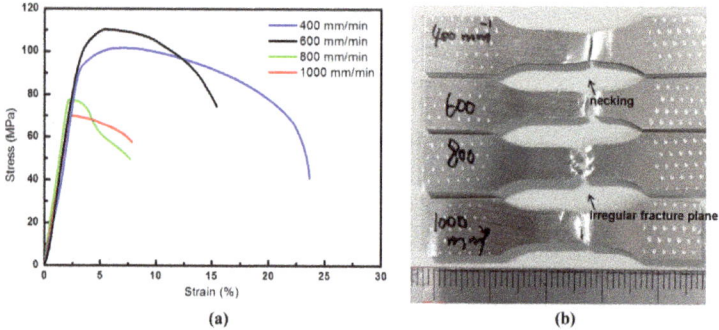

Figure 10. (**a**) Tensile strain-stress curves of the dissimilar FSW joints produced at different welding speeds; and (**b**) photo of the fractured specimens after the tensile tests.

From the photos showing the appearance of the fractured tensile specimen, the joints produced at high welding speeds showed a brittle fracture, and even some irregular zigzag edges near the fracture plane could be found. In contrast, the joints produced at a revolutionary pitch of 0.5 mm/rev showed obvious necking near the fractured plane, indicative of ductile failure of the specimen.

Figure 11 shows the morphologies of the fracture plane after tensile testing of the dissimilar FSW joints produced at 1 and 0.5 mm/rev, corresponding to the brittle and ductile kinds of fracture mode. For the specimen produced at 1 mm/rev, several irregular fractured planes were observed as shown in Figure 11a. Some dimple patterns could be found in some of the fracture planes, while some other planes showed tearing of the specimens. For the specimen produced at a welding speed of 0.5 mm/rev, the fracture plane had typical dimple patterns as shown in Figure 11b, indicating ductile failure of the specimen.

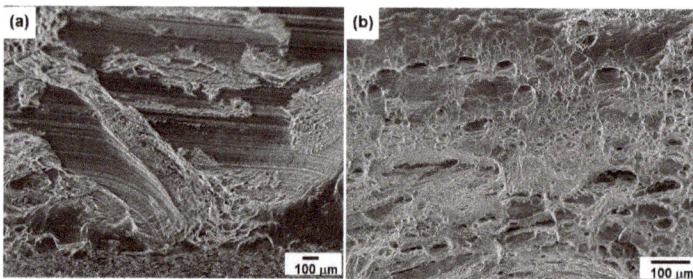

Figure 11. SEM images showing the morphologies of the fractured surfaces of the joints produced at a revolutionary pitch of (**a**) 1; and (**b**) 0.5 mm/rev.

4. Conclusions

The dissimilar FSW the UFGed 1050 Al to 6061-T6 alloy was successfully performed at the wide revolutionary pitches from 0.5 to 1.25 mm/min; however, this took place only when the 6061-T6 Al alloy was put on the AS. Otherwise, sound welds could not be obtained due to the large defects formed in the softened 1050 Al side.

The size of the stir zone became larger with the decrease in the revolutionary pitch and the mixing between the two dissimilar Al alloys became much more homogeneous. However, in the entire stir zone of the dissimilar FSW, it was found that the 1050 Al was mixed into the 6061 side. However, no 6061 Al was mixed into the 1050 Al area.

In the stir zone, the two dissimilar Al alloys could still be distinguished. The nano-sized lamellar structure of the UFGed 1050 Al alloy could not be distinguished any more. Finally, the 1050 Al showed dynamic recrystallization and had an average grain size larger than that of the 6061 Al alloy.

All the dissimilar joints fractured in the stir zone during the tensile tests. For the joints produced at 1.25 and 1 mm/rev, the fracture strength was low and showed the brittle fracture mode. In contrast, the joints produced at 0.75 and 0.5 mm/rev showed a higher tensile strength and large plastic elongation.

Acknowledgments: The authors wish to acknowledge the financial support of the Collaborative Research Based on Industrial Demand "Heterogeneous Structure Control: Towards Innovative Development of Metallic Structural Materials" by Japan Science and Technology Agency (JST), the Global COE Programs from the Ministry of Education, Sports, Culture, Science, and a Grant-in-Aid for Science Research from the Japan Society for Promotion of Science and Technology of Japan, ISIJ Research Promotion Grant.

Author Contributions: Yufeng Sun designed the experiments, performed the experiments, analyzed the data and wrote the paper. Nobuhiro Tsuji and Hidetoshi Fujii provided the alloy sheets, directed the research and contributed to the discussion and interpretation of the results.

Conflicts of Interest: The authors declare no conflict of interest.

References

1. Thomas, W.M.; Nicholas, E.D.; Needhman, J.C.; Church, M.G.; Templesmith, P.; Dawes, C.J. Friction Stir Butt Welding. International Patent Application No. PCT/GB92/02203, 1991.
2. Mishra, R.S.; Ma, Z.Y. Friction stir welding and processing. *Mater. Sci. Eng. R* **2005**, *50*, 1–78. [CrossRef]
3. Luijendijk, L. Welding of dissimilar aluminum alloys. *J. Mater. Process. Technol.* **2003**, *103*, 29–35. [CrossRef]
4. Nandan, R.; Debroy, T.; Bhadeshia, H.K.K.H. Recent advances in friction-stir welding-Process, weldment structure and properties. *Prog. Mater. Sci.* **2008**, *53*, 980–1023. [CrossRef]
5. Sun, Y.F.; Fujii, H. Investigation of the welding parameter dependent microstructure and mechanical properties of friction stir welded pure copper. *Mater. Sci. Eng. A* **2010**, *527*, 6879–6886. [CrossRef]
6. Fujii, H.; Sun, Y.F.; Kato, H. Investigation of welding parameter dependent microstructure and mechanical properties in friction stir welded pure Ti joints. *Mater. Sci. Eng. A* **2010**, *527*, 3386–3391. [CrossRef]
7. Sun, Y.F.; Fujii, H.; Imai, H.; Kondoh, K. Suppression of hydrogen-induced damage in friction stir welded low carbon steel joints. *Corros. Sci.* **2015**, *94*, 88–98. [CrossRef]
8. Fujii, H.; Cui, L.; Maeda, M.; Nogi, K. Effect of tool shape on mechanical properties and microstructure of friction stir welded aluminum alloys. *Mater. Sci. Eng. A* **2006**, *419*, 25–31. [CrossRef]
9. Saito, Y.; Utsunomiya, H.; Tsuji, N.; Sakai, T. Novel ultra-high straining process for bulk materials-development of the accumulative roll-bonding (ARB) process. *Acta Mater.* **1999**, *47*, 579–583. [CrossRef]
10. Lipinska, M.; Olejnik, L.; Pietras, A.; Rosochowski, A.; Bazarnik, P.; Goliński, J.; Brynk, T.; Lewandowska, M. Microstructure and mechanical properties of friction stir welded joints made from ultrafine grained aluminum 1050. *Mater. Des.* **2015**, *88*, 22–31.
11. Topic, I.; Hoppel, H.W.; Goken, M. Friction stir welding of accumulative roll-bonded commercial-purity aluminum AA1050 and aluminum alloy AA6016. *Mater. Sci. Eng. A* **2009**, *503*, 163–166. [CrossRef]
12. Sabooni, S.; Karimzadeh, F.; Enayati, M.H.; Ngan, A.H.W. Recrystallization mechanism during friction stir welding of ultrafine-and coarse-grained AISI 304L stainless steel. *Sci. Tech. Weld. Join.* **2016**, *21*, 287–294. [CrossRef]
13. Sato, Y.S.; Urata, M.; Kokawa, H.; Ikeda, K. Hall-Petch relationship in friction stir welds of equal channel angular-pressed aluminium alloys. *Mater. Sci. Eng. A* **2003**, *354*, 298–305. [CrossRef]
14. Fujii, H.; Ueji, R.; Takada, Y.; Kitahara, H.; Tsuji, N.; Nakata, K.; Nogi, K. Friction stir welding of ultrafine grained interstitial free steel. *Mater. Trans.* **2006**, *47*, 239–242. [CrossRef]
15. Sun, Y.F.; Fujii, H.; Takada, Y.; Tsuji, N.; Nakata, K.; Nogi, K. Effect of initial grain size on the joint properties of friction stir welded aluminum. *Mater. Sci. Eng. A* **2009**, *527*, 317–321. [CrossRef]

16. Murr, L.E. A review of FSW research on dissimilar metal and alloy systems. *J. Mater. Eng. Perform.* **2010**, *19*, 1071–1089. [CrossRef]
17. Lee, W.B.; Yeon, Y.M.; Jung, S.B. The joint properties of dissimilar formed Al alloys by friction stir welding according to the fixed location of materials. *Scr. Mater.* **2003**, *49*, 423–428. [CrossRef]
18. Palanivel, R.; Mathews, P.K.; Murugan, N.; Dinaharan, I. Effect of tool rotational speed and pin profile on microstructure and tensile strength of dissimilar friction stir welded AA5083-H111 and AA6351-T6 aluminum alloys. *Mater. Des.* **2012**, *40*, 7–16. [CrossRef]
19. Khodir, S.A.; Shibayanagi, T. Friction stir welding of dissimilar AA2024 and AA7075 aluminum alloys. *Mater. Sci. Eng. B* **2008**, *148*, 82–87. [CrossRef]
20. Kwon, Y.J.; Shigematsu, I.; Saito, N. Dissimilar friction stir welding between magnesium and aluminum alloy. *Mater. Lett.* **2008**, *62*, 3827–3829. [CrossRef]
21. Tolephih, M.H.; Mahmood, H.M.; Hashem, A.H.; Abdullah, E.T. Effect of tool offset and tilt angle on weld strength of butt joint friction stir welded specimens of AA2024 aluminum alloy welded to commercial pure copper. *Chem. Mater. Res.* **2013**, *3*, 49–58.
22. Carlone, P.; Astarita, A.; Palazzo, G.S.; Paradiso, V.; Squillace, A. Microstructural aspects in Al-Cu dissimilar joining by FSW. *Int. J. Adv. Manu. Technol.* **2015**, *79*, 1109–1116. [CrossRef]
23. Sahu, P.K.; Pal, S.; Pal, S.K.; Jain, R. Influence of plate position, tool offset and tool rotational speed on mechanical properties and microstructures of dissimilar Al/Cu friction stir welding joints. *J. Mater. Proc. Technol.* **2016**, *235*, 55–67. [CrossRef]
24. Sun, Y.F.; Fujii, H.; Tsuji, N. Microstructure and mechanical properties of spot friction stir welded ultrafine grained 1050 Al and conventional grained 6061-T6 Al alloys. *Mater. Sci. Eng. A* **2013**, *585*, 17–24. [CrossRef]
25. Sato, Y.S.; Urata, M.; Kokawa, H. Parameters controlling microstructure and hardness during friction stir welding of precipitation-Hardenable Aluminum alloy 6063. *Metall. Mater. Trans. A* **2002**, *33*, 625–635. [CrossRef]
26. Maeda, M.; Liu, H.J.; Fujii, H.; Shibayanagi, T. Temperature Field in the vicinity of FSW-Tool during friction stir welding of aluminum alloys. *Weld. World* **2005**, *49*, 69–75. [CrossRef]
27. Liu, H.; Fujii, H.; Maeda, M.; Nogi, K. Heterogeneity of mechanical properties of friction stir welded joints of 1050-H24 aluminum alloy. *J. Mater. Sci. Lett.* **2003**, *22*, 441–444. [CrossRef]
28. Yasui, T.; Ishii, T.; Shimoda, Y.; Tsubaki, M.; Fukumoto, M.; Shinoda, T. Friction stir welding between aluminum and steel with high welding speed. In Proceedings of the 5th International Symposium on Friction Stir Welding, Metz, France, 14–16 September 2004.
29. Tanaka, T.; Morishige, T.; Hirata, T. Comprehensive analysis of joint strength for dissimilar friction stir welds of mild steel to aluminum alloys. *Scr. Mater.* **2009**, *61*, 756–759. [CrossRef]
30. Takeoka, N.; Fujii, H.; Morisada, Y.; Sun, Y.F. Clarification of formation mechanism of stri zone using adjustable tool. In Proceedings of the 11th International Symposium on Friction Stir Welding, Cambridge, UK, 17–19 May 2016.
31. Murr, L.E. Intercalation vortices and related microstructural features in the friction stir welding of dissimilar metals. *Mater. Res. Innov.* **1998**, *2*, 150–153. [CrossRef]
32. Murr, L.E. A comparative study of friction stir welding of aluminum alloys. *Alum. Trans.* **1999**, *1*, 141–154.
33. Liu, F.C.; Ma, Z.Y. Influence of tool dimension and welding parameters on microstructure and mechanical properties of friction stir welded 6061-T651 aluminum alloy. *Metall. Mater. Trans. A* **2008**, *39*, 2378–2388. [CrossRef]
34. Sato, Y.S.; Kurihara, Y.; Pack, S.H.C.; Kokawa, H.; Tsuji, N. Friction stir welding of ultrafine grained Al alloy 1100 produced by accumulative roll-bonding. *Scr. Mater.* **2004**, *50*, 57–60. [CrossRef]
35. McQueen, H.J.; Knustad, O.; Ryum, N.; Solberg, J.K. Microstructural evolution in Al deformed to strains of 60 at 400 °C. *Scr. Metall.* **1985**, *19*, 73–78. [CrossRef]
36. Humphreys, F.J.; Hatherly, M. *Recrystallization and Related Annealing Phenomenon*; Pergamon Press: Oxford, UK, 2004.
37. McQueen, H.J.; Solberg, J.K.; Ryum, N.; Ness, E. Evolution of flow stress in Al during ultra-high straining at elevated temperature. *Philos. Mag. A* **1989**, *60*, 473–485. [CrossRef]

metals

MDPI

Article

The Optimization of Process Parameters and Microstructural Characterization of Fiber Laser Welded Dissimilar HSLA and MART Steel Joints

Celalettin Yuce *, Mumin Tutar, Fatih Karpat and Nurettin Yavuz

Department of Mechanical Engineering, Uludag University, Bursa 16059, Turkey;
mumintutar@uludag.edu.tr (M.T.); karpat@uludag.edu.tr (F.K.); nyavuz@uludag.edu.tr (N.Y.)
* Correspondence: cyuce@uludag.edu.tr; Tel.: +90-224-294-1919

Academic Editor: Giuseppe Casalino
Received: 22 July 2016; Accepted: 10 October 2016; Published: 18 October 2016

Abstract: Nowadays, environmental impact, safety and fuel efficiency are fundamental issues for the automotive industry. These objectives are met by using a combination of different types of steels in the auto bodies. Therefore, it is important to have an understanding of how dissimilar materials behave when they are welded. This paper presents the process parameters' optimization procedure of fiber laser welded dissimilar high strength low alloy (HSLA) and martensitic steel (MART) steel using a Taguchi approach. The influence of laser power, welding speed and focal position on the mechanical and microstructural properties of the joints was determined. The optimum parameters for the maximum tensile load-minimum heat input were predicted, and the individual significance of parameters on the response was evaluated by ANOVA results. The optimum levels of the process parameters were defined. Furthermore, microstructural examination and microhardness measurements of the selected welds were conducted. The samples of the dissimilar joints showed a remarkable microstructural change from nearly fully martensitic in the weld bead to the unchanged microstructure in the base metals. The heat affected zone (HAZ) region of joints was divided into five subzones. The fusion zone resulted in an important hardness increase, but the formation of a soft zone in the HAZ region.

Keywords: laser welding; dissimilar weld; parameter optimization; microstructural examination

1. Introduction

The automotive sector is focused on developing and manufacturing fuel saving, higher safety vehicles with cost efficient methods. This will be achieved through proper design and using lighter and stronger materials on the auto body parts. Therefore the utilization of advanced high-strength steels (AHSS) is widespread. Due to the higher strength and good formability properties, AHSS can replace conventional thicker materials used in vehicle bodies without comprising crashworthiness. Dual phase (DP), complex phase (CP), martensitic steel (MART) and transformation-induced plasticity (TRIP) steels are the most common types of the AHSS [1]. Among these AHSS types, MART steel is one of the strongest cold-rolled AHSS on the market and has become the preferred material for automotive body applications, such as side impact beams, bumpers and structural components. Although using AHSS steels in the automobile structure is increasing, due to specific mechanical properties, high strength low alloy (HSLA) steel is still mainly used for structural parts, such as cross members, longitudinal beams, chassis components, etc. [2].

Welding is one of the most used and essential joining technique in the fabrication of the auto body and plays a significant role in assessing the final mechanical and metallurgical properties of the joined parts [3]. Due to much superiority over conventional welding methods, such as non-contact and single side access welding, low process cost and suitability of automation, laser welding is becoming an attractive and economically advantageous joining technique in the automotive industry [4]. Joints of dissimilar steel combinations in auto body structures are widely utilized for several applications requiring a special combination of properties besides cost saving and weight reduction. However, due to different metallurgical, thermal and physical properties of the materials, dissimilar material welding is more challenging than similar materials welding. Due to low and concentrated heat input and high speed properties, laser welding has also advantages on joining dissimilar materials over other conventional methods [5]. Thus, reduced distortion and a narrower heat affected zone (HAZ) with limited microstructural changes can be obtained.

There are several studies in the literature concerning the laser welding of similar or dissimilar DP and HSLA steels. Saha et al. [2] examined the mechanical and microstructural properties of laser welded DP980 and HSLA steel sheets. They stated that the tensile strength of the dissimilar welds was lower than DP welds. Xu et al. [6] investigated microstructural and mechanical properties, and Parkes et al. [7] reported the fatigue properties of laser welded DP and HSLA joints with varying weld geometries. Parkes et al. [8] evaluated the tensile properties of laser welded HSLA and DP steels at cryogenic, room and elevated temperatures. They reported that with the temperature increase, the tensile properties were decreased. In addition, several research works investigated laser welding of higher degree DP steels and AHSS. Wang et al. [9,10] investigated the effect of energy input and softening mechanism on the laser butt welded DP1000 steel. They found that the weld bead width and softening zone width become narrowed at lower energy input levels. Additionally, the mechanical properties were increased. The study of Rossini et al. [11], concerned with laser welding of dissimilar AHSS types, has shown that a fully martensitic microstructure was present in the 22MnB5, DP and TRIP steels close to the fusion zone (FZ), while mainly tempered martensite and ferrite zones were close to the base metal.

Although there are many research works about laser welding of DP and HSLA steels, only limited work has been reported on the laser welding of MART steels. Nemecek et al. [12] compared the microstructural and mechanical properties of MART steel joints made by laser and metal active gas (MAG) welding. They stated that the strength of the laser welded joints was higher than arc welding, and the HAZ width and grain coarsening in the HAZ were minimal. Zhao et al. [13] investigated the effect of welding speed on weld bead geometry and the tensile properties of the laser welded MART steel. They observed that, due to the fast cooling rate, the FZ of the joints contained predominantly martensite. Furthermore, the tensile load gradually increased with decreasing welding speed.

Due to welding process parameters directly affecting the quality of the weld joints, it is necessary to work in the suitable range. However, defining the suitable parameters to obtain the required quality welded joints is a time-consuming process. Several optimization methods are utilized in order to solve this problem. The Taguchi method is one of the most common design of experiment (DOE) techniques that allows the analysis of experiments with the minimum number [14,15]. In the literature, several researchers have used DOE methods to optimize quality characteristics in laser welding parameters. Benyounis and Olabi [16] have presented a review of the application of optimization techniques in several welding processes. Anawa and Olabi [17] used the Taguchi method for the purpose of increasing the productivity and decreasing the operation cost of laser welding ferritic-austenitic steel sheets. Another study of the authors [18] analyzed the optimized shape of dissimilar laser welded joints and fusion zone area depending the process parameters. Sathiya et al. [19] carried out the Taguchi method and desirability analysis to relate the parameters to the weld bead dimension and the tensile strength of the joints with various shielding gasses. Fiber laser welding has demonstrated its capability of welding dissimilar steel joint with and without the help of a synergic power source like the arc [20]. Acherjee et al. [21] used Taguchi, response surface methodology (RSM) and desirability function analyses in laser transmission welding, and they investigated the optimal parameter combination for the joint quality.

In addition to these studies, several researchers used other DOE methods to investigate the effect of laser parameters on the mechanical properties and bead geometries of laser welded joints. Benyounis et al. [22] examined the influence of process parameters on the weld bead geometry. They stated that weld bead dimensions were affected by the level of heat input. Ruggiero et al. [23] and Olabi et al. [24] showed the effects of the process parameters on the weld geometry and operating cost for austenitic steel and low carbon steel. The authors developed models and stated that, in terms of weld bead dimensions, the most influential parameter was welding speed. Reisgen et al. [25] optimized the parameters of the laser welded DP and TRIP steels to obtain the highest mechanical strength and minimum operation costs. Zhao et al. [26] investigated the effects of prescribed gap and laser welding parameters on the weld bead profile of galvanized steel sheets in a lap joint format and developed regression models. Benyounis et al. [27] reported the multi-response optimization of laser welded austenitic stainless steel. They developed mathematical models and established relationships between process parameters and responses, such as cost, tensile and impact strength.

As a result of the literature review, laser power, welding speed and focal position were found to be the most important welding parameters for welded joints' quality and mechanical performance. Due to the mechanical properties, especially tensile strength, being dependent on the weld bead geometry, heat input comes to the fore [19]. Although various studies examined the influence of laser parameters on the weld quality of dissimilar HSLA and DP steel joints, the information on fiber laser welding of dissimilar HSLA and MART steel sheets is still not quite clear. Whereas, resolving the issue of reducing vehicle mass while improving crash safety, the use of AHSS and HSLA is increasing. It is essential to investigate the effect of laser welding process parameters on the mechanical performance and quality of these steel types. Therefore, the aim of this work was to evaluate the effects of laser welding parameters of laser power, welding speed and focal position on the response, which was a proportional combination of tensile load (TL) and heat input (HI) using the Taguchi method. In this way, we will be able to find the optimal welding parameters that would maximize TL, while minimizing the HI of the fiber laser welded dissimilar HSLA and MART steel joints. In addition, for the selected samples, the microstructural and microhardness examinations were discussed.

2. Experimental Details

In this study, all experiments were carried out on 1.5 mm-thick cold rolled MART and HSLA steel sheets. The mechanical and chemical properties of the materials are shown in Table 1 [28]. The steel sheets were sheared into 250 mm × 80 mm coupons, which had the sheared edges placed together for running welds in butt joint configuration to make 250 mm × 160 mm, as shown in Figure 1a.

Table 1. Mechanical properties and chemical composition of the steels.

Material	C	Si	Mn	P	S	Al	Nb + Ti	Yield Strength (MPa)	Ultimate Strength (MPa)	Elongation (min %)
Docol 1200M	0.14	0.4	2.0	0.02	0.01	0.015	0.1	950	1200–1400	3
HSLA *	0.1	0.5	1.8	0.025	0.025	0.015	0.15	500	570–710	14

* HSLA: high strength low alloy.

The IPG ytterbium fiber laser attached to a Kuka robotic arm was used for welding experiments. The maximum power of the laser was 3 kW, and the wavelength was 1070 nm. The laser transmitted through the fiber optic cables and then came to a welding head. The fiber laser had a fiber core diameter of 0.2 mm with a laser beam spot diameter of 0.6 mm. The focal length was 300 mm. During the fiber laser welding process, no shielding gas was used.

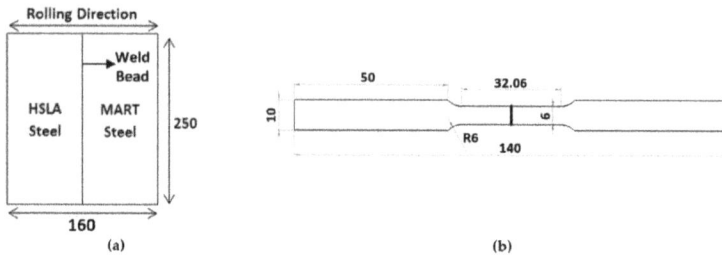

Figure 1. (a) Schematic illustration of the fiber laser welded steel sheets; (b) dimensions of the tensile test specimens. MART, martensitic steel.

In this study, for optimizing the process parameters, the Taguchi method was used. The parameter design is the key step in this method to achieving high quality without increasing cost. Firstly, a suitable orthogonal array should be selected depending on the total degree of freedom (DOF), which can be calculated by summing the individual DOF of each process parameter. The DOF for each parameter is the number of parameter levels minus 1. Then, the experiments were run based on the orthogonal array, analyzing the data and identifying the optimum parameters and, finally, if necessary, conducting confirmation trials with the optimal levels of the parameters. In this study, experiments were designed using an L25 orthogonal array, which means 25 rows and three columns. Five levels were considered for each of the three process parameters, which were laser power, welding speed and focal position. The levels of the parameters were chosen based on previous works in the literature and considering the laser system capabilities. Furthermore, trial experiments were applied to determine the operating range of each process parameter in order to produce an acceptable quality welding. The levels of the process parameters are shown in Table 2. A negative defocus is obtained when the focal point position is below the specimen surface.

Table 2. Laser welding process parameters and levels.

Variables	Unit	Symbol	Level 1	Level 2	Level 3	Level 4	Level 5
Laser Power	W	P	1000	1250	1500	1750	2000
Welding Speed	mm/s	S	5	15	25	35	45
Focal Position	mm	F	0	−0.2	−0.4	−0.6	−0.8

In the data analysis, in order to evaluate the effect of the selected parameters on the response, the signal-to-noise (S/N) ratios are calculated. In addition, S/N ratios are used to reduce the response variability. In this work, the larger-the-better S/N ratio was chosen in order to maximize the responses. The S/N ratio for the larger-the-better for the responses was calculated as follows:

$$S/N = -10\log\left(\frac{1}{n}\sum_{i=1}^{n}\frac{1}{y_i^2}\right) \tag{1}$$

where y_i is the response data from the experiment for the i-th parameter and n is the number of experiments. A higher S/N ratio indicates superior consideration for the optimal parameter combination, since the major signal dominates the noise. Equation (2) is used to calculate the parameter effects:

$$S/N_{i,j} = \frac{1}{n}\sum_{k=1}^{n}S/N_k \tag{2}$$

where $S/N_{i,j}$ is the average S/N value of the j-th level of the i-th parameter and n is the number of the experiment, which includes the j-th level of the i-th parameter. Additionally, the S/N_k is the value of the *k*-th experiment *S/N*. Finally, a statistical analysis of variance (ANOVA) was used to indicate the relative effect of each process parameter on the responses.

At the metallographic examination stage of the study, the samples were cut from the weld cross-section using an electrical discharge cutting machine, then mounted in Bakelite, ground and polished up to 0.25-µm diamond paste. Two different etching procedures were conducted to reveal the grain boundaries and weld zone microstructure. In the first stage of the etching, 3% Nital solution was used. Then, to reveal some microstructures, subsequently, tint etched using 10% $Na_2S_2O_5$ was performed. Then, samples were analyzed for microstructural changes and possible defects using an optic microscope (OM, Nikon DIC, Tokyo, Japan) with the Clemex image analysis system and the scanning electron microscope (SEM, Zeiss EVO 40 XVP, Oberkochen, Germany). Vickers microhardness measurements (DUROLINE-M microhardness tester, Metkon, Turkey) were performed with a 200-g load, and 10-s dwell time. Tensile samples were machined from the perpendicular to the welding direction in accordance with ASTM, E8/E8M (Figure 1b). Tensile tests were performed using a computerized tensile testing machine (UTEST-7014, Ankara, Turkey) using a constant crosshead speed of 5 mm/min.

3. Results and Discussion

3.1. Optimization of the Process Parameters via the Taguchi Method

In this study, a Taguchi orthogonal array, which can handle five levels of the parameters with three columns and 25 rows, was used. The parameter optimization procedure was done in order to get a welded joint that has the maximum TL by minimizing the HI. HI plays a crucial role in the quality of the joint and indirectly the operation cost. The weld joint quality can be defined as weld bead geometry, mechanical properties and distortions [25]. Weld bead geometry, which means the bead width and penetration depth, is an important physical characteristic of a weldment, especially for dissimilar laser welding processes [19]. The appropriate weld bead geometry depends on the HI rate [22]. A shallower and inadequate penetration depth is related to an insufficient HI rate. Thence, the TL of the welded joint will decrease. However, a higher HI gives a slower cooling rate, and so, in the HAZ, large grain sizes can have poor toughness and decrease in TL. Hence, HI and, consequently, weld bead geometry affect the tensile strength of the joints [16,18]. Therefore, in this study, TL and HI were evaluated together as a response variable. Due to the tensile strength being the most important quality indicator of the welded joint, the effect ratio of the TL was determined to be higher, 60%. In determining the effect ratio of the HI, operational cost and weld bead geometry were considered. Namely, this ratio should not be too low because of the insufficient penetration, and also, it should not be too high in terms of cost and decreased strength of the joint. Therefore, it was determined to be 40%. In determining these effect ratios, they have also benefited from operational experience.

The TL of the laser welded joints was experimentally determined using tensile tests. At least three different specimens' tensile test results' average were taken. Additionally, HI was calculated by the laser power divided by the welding speed. Due to the scale of the values of TL and HI being different, a normalization process was applied to these values. Equation (3) was used for the normalization of the TL values.

$$X_n = \frac{X_i}{X_{max}} \tag{3}$$

where X_n is the normalized value, X_i is the value of the relevant row and X_{max} is the maximum value. Since the objective function was a combination of the TL and HI, it is necessary to express it in the same form. Therefore, before applying Equation (3), the reciprocals of the HI values were taken using Equation (4) to convert the values to the larger the better form.

$$X_p = \frac{1}{X_i} \tag{4}$$

where X_p is a pre-normalized value, which was used in Equation (1), and X_i is the HI value of the relevant row.

The experimental layout for the process parameters, average TL, standard deviations (SD), HI values and normalized values are shown in Table 3. The S/N ratios for the response were calculated. The response column represents the sum of 60% normalized TL and 40% normalized HI. The S/N ratios of the process parameters were calculated by using Equation (2), and the effect of each parameter level was determined. As can be seen in Table 4, welding speed was the most important parameter for the response. Laser power and focal position followed this parameter, respectively.

Table 3. Design matrix with experimental results. TL, tensile load; HI, heat input.

Exp. No.	Parameters			Outputs and Calculations						
	Power (W)	Speed (mm/s)	Focal (mm)	TL (kN)	SD	Normalized TL	HI (J/mm)	Normalized HI	Response	S/N Ratio
1	1000	5	0	5.92	0.04	0.995	200.000	0.111	0.642	−3.849
2	1000	15	−0.2	5.49	0.05	0.923	66.667	0.333	0.687	−3.260
3	1000	25	−0.4	4.61	0.22	0.775	40.000	0.556	0.687	−3.260
4	1000	35	−0.6	3.43	0.11	0.578	28.571	0.778	0.658	−3.635
5	1000	45	−0.8	3.18	0.20	0.534	22.222	1.000	0.720	−2.853
6	1250	5	−0.2	5.88	0.02	0.990	250.000	0.089	0.629	−4.026
7	1250	15	−0.4	5.82	0.06	0.978	83.333	0.267	0.694	−3.172
8	1250	25	−0.6	5.32	0.08	0.894	50.000	0.444	0.714	−2.926
9	1250	35	−0.8	4.44	0.08	0.746	35.714	0.622	0.697	−3.135
10	1250	45	0	3.71	0.14	0.625	27.778	0.800	0.695	−3.160
11	1500	5	−0.4	5.73	0.04	0.964	300.000	0.074	0.608	−4.321
12	1500	15	−0.6	5.93	0.05	0.997	100.000	0.222	0.687	−3.260
13	1500	25	−0.8	5.93	0.06	0.997	60.000	0.370	0.746	−2.545
14	1500	35	0	5.82	0.00	0.979	42.857	0.519	0.795	−1.992
15	1500	45	−0.2	4.30	0.10	0.723	33.333	0.667	0.701	−3.085
16	1750	5	−0.6	5.52	0.06	0.929	350.000	0.063	0.583	−4.686
17	1750	15	−0.8	5.90	0.02	0.992	116.667	0.190	0.671	−3.465
18	1750	25	0	5.79	0.10	0.974	70.000	0.317	0.712	−2.950
19	1750	35	−0.2	5.87	0.02	0.987	50.000	0.444	0.770	−2.270
20	1750	45	−0.4	5.95	0.01	1.000	38.889	0.571	0.829	−1.628
21	2000	5	−0.8	5.52	0.07	0.929	400.000	0.056	0.579	−4.746
22	2000	15	0	5.87	0.10	0.987	133.333	0.167	0.659	−3.622
23	2000	25	−0.2	5.58	0.08	0.938	80.000	0.278	0.674	−3.426
24	2000	35	−0.4	5.62	0.04	0.946	57.143	0.389	0.723	−2.817
25	2000	45	−0.6	5.67	0.05	0.953	44.444	0.500	0.772	−2.247

Table 4. Response table for the S/N ratios for the objective.

Level	Laser Power	Welding Speed	Focal Position
1	−3.372	−4.326	−3.349
2	−3.284	−3.356	−3.351
3	−3.041	−3.022	−3.040
4	−3.000	−2.770	−3.214
5	−3.372	−2.595	−3.115
Delta	0.372	1.731	0.311
Rank	2	1	3

The S/N ratios' main effect plot showed how each process parameter affects the response characteristic. The means of the S/N ratios exhibit a good correlation with the main effects of the mean of means (Figure 2). This result indicates that process parameters show higher mean values resulting in higher variability. The response seems to be mainly affected by the process parameters, as shown in Figure 2. It can be seen that the welding speed was the most important process parameter that affected the response. There was a small difference between laser power and focal position; while the focal position plots showed the lowest effect on the response to those parameters.

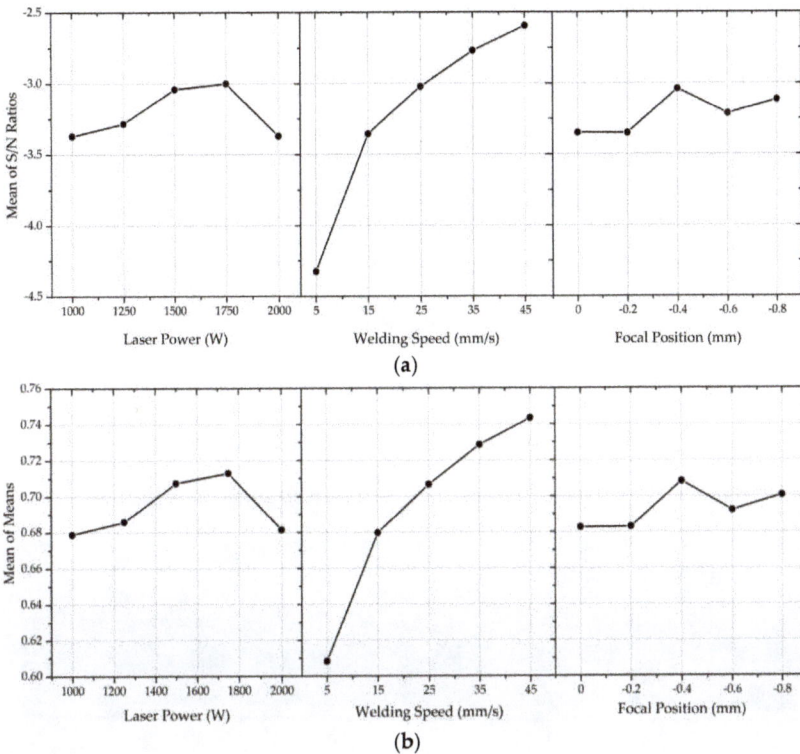

Figure 2. Effects plots of (**a**) S/N ratios; and (**b**) mean of means for the response.

In this study, the optimal parameter combination was found to be 1750 W for laser power, 45 mm/s for welding speed and −0.4 mm for the focal position. This parameter combination was Sample 20 in the orthogonal array in Table 3; thus, no additional confirmation experiments were required.

3.2. Analysis of Variance

The order of importance of the parameters on the response was determined using ANOVA. By comparing the estimation of the experimental errors against the mean square, ANOVA tests the importance of all main factors and their interactions. In this study, The ANOVA shows that for the response (maximum TL and minimum HI), welding speed has the greatest effect with a contribution of 64.01 percent. Laser power and focal position effects were 5.60% and 2.82%, respectively (Table 5). This result is compatible with Table 4, which is the response table for the S/N ratios. Due to the interactions between the processes parameters not being defined, the residual error was large in ANOVA.

Table 5. Analyses of variance table for means.

Source	Degrees of Freedom (DF)	Sum of Squares (Seq SS)	Adjusted Mean Squares (Adj MS)	F	p	Contribution (%)
Laser Power	4	0.004975	0.001244	0.61	0.663	5.60
Welding Speed	4	0.056813	0.014203	6.97	0.004	64.01
Focal Position	4	0.002506	0.000627	0.31	0.867	2.82
Residual Error	12	0.024450	0.002038			27.55
Total	24	0.088745				

3.3. Effects of Process Parameters on the Response

In this study, it was observed that welding speed was a significant parameter that affects the response, which is maximum TL and minimum HI. Although the effect of laser power may seem quite small in ANOVA results, it is an important process parameter due to the associated HI. The increase of laser power causes more heat input. Under the high laser power, if the welding speed were not chosen properly, the weld bead would be broadened and the surface quality of the weld decreased. Therefore, the laser power and welding speed should be considered together to get good weld profiles and TL. When the laser power was kept constant, with increasing welding speed, HI decreased. Due to weld bead geometry related to the HI, weld bead width was increased with increasing HI. In all laser power levels, when the speed was 5 mm/s, the beads were larger due to the excessive heat input (Figure 3a). On the other side, when the speed was 45 mm/s, the beads were found to be narrower (Figure 3b).

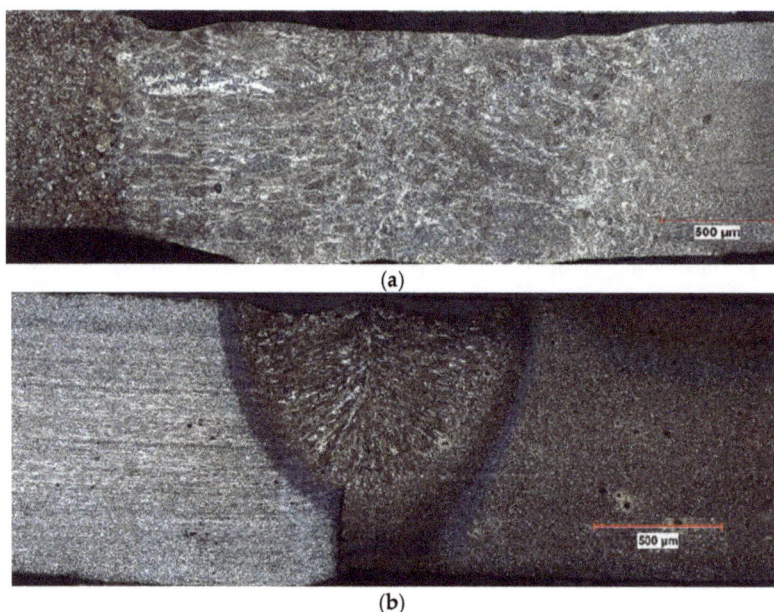

(a)

(b)

Figure 3. Transverse sections of the joints using different heat inputs: (**a**) 300 J/mm, Sample 11; (**b**) 44 J/mm, Sample 25.

As known weld bead dimensions directly affect the TL of the joints [25], at insufficient HI at the low laser power levels or high welding speeds, adequate penetration did not occur, and the TL of the joints was decreased. Besides, at excessive HI levels, the HAZ would be wider, and that causes a decrease in TL. According to the tensile test results, the welding speed in the range between 35 mm/s and 45 mm/s would lead to minimum HI and acceptable TL for the joints. The focal position determines the laser spot size and consequently the power density on the surface, depending on the optical path. In this study, the focal position has the lowest effect on the response. It is believed that the level range of this parameter caused this situation due to the range of the spot diameters being quite small.

3.4. Microstructure and Microhardness Evolution

The microstructural examination and microhardness evolution of the selected welds that have the highest (Sample 20) and lowest (Sample 21) response values were discussed. Three different zones, including FZ, HAZ and base metal (BM), were revealed by examining the selected sample's cross-sections. The BM of the HSLA consisted of a ferrite matrix with carbides dispersed in the grains and at the grain boundaries (Figure 4a). As shown in Figure 4b, MART steels were comprised of martensitic microstructures and a small proportion of ferritic and bainitic grains.

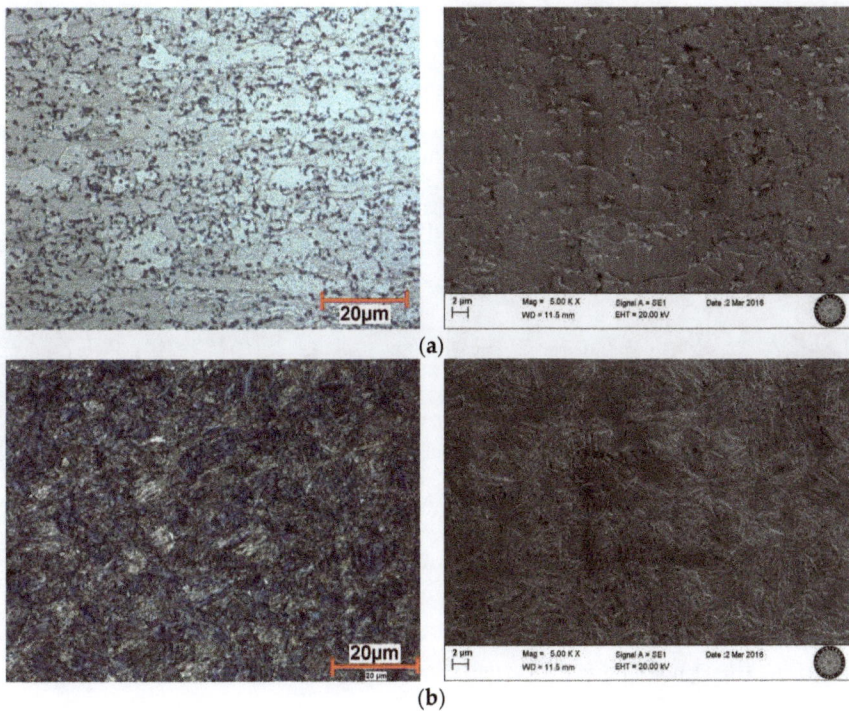

(a)

(b)

Figure 4. Optical micrograph and SEM views of the: (**a**) HSLA base metal (BM); (**b**) MART BM.

In the welding process, final microstructures are affected by peak temperature and the cooling rate of the relevant zones, and carbon equivalent (CE) value resulted from the chemistry of the steels [29–32]. Although there are numerous formulae for calculating CE, Yurioka's formula was used in this study because of its suitability for C-Mn steels [33]. The CE values of steels were calculated using Yurioka's formula given by Equation (5) and shown in Table 6 [34,35]. The Ti element was considered as the Nb element because of their similar effect on the steels' hardenability.

$$CE = C + f(C) \left[\frac{Si}{24} + \frac{Mn}{6} + \frac{Cu}{15} + \frac{Ni}{20} + \frac{(Cr + Mo + Nb + V)}{5} \right] \tag{5}$$

where $f(C)$ is the accommodation factor and is calculated as;

$$f(C) = 0.75 + 0.25\tanh\left[20\left(C - 0.12\right)\right] \tag{6}$$

Table 6. The carbon equivalent (CE) values of the HSLA and MART steels. FZ, fusion zone.

Calculated Zone	HSLA	MART	FZ
CE	0.330	0.453	0.391

The microstructure of the FZ of Sample 20, with a 0.391 CE value (average of MART and HSLA steels), is predominantly martensite with a bainitic structure (Figure 5). With the effect of the heat exchange gradient, in the vicinity of the fusion boundary, grains were elongated towards the weld center. However, in the center of the FZ, equiaxed grains were observed (Figure 5a). Furthermore, due to the lack of shielding gas, as a possible result of the diffusion of some elements, i.e., oxygen and nitrogen from the air, it is thought to be some inclusions in the FZ, which were marked with yellow arrows in Figure 5b.

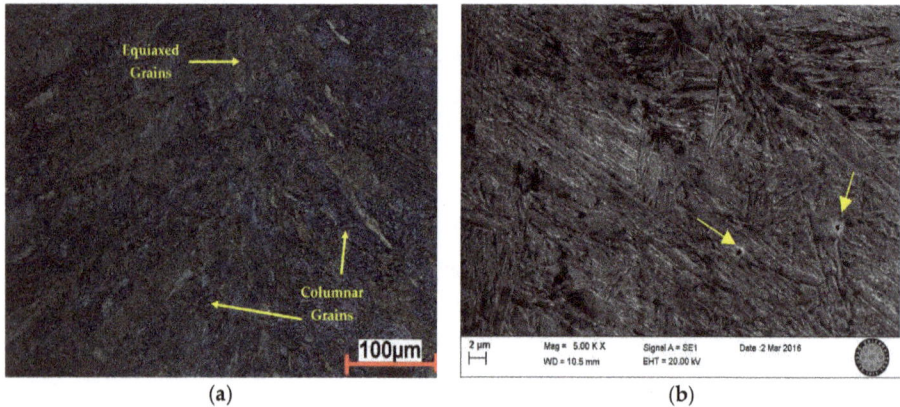

(a)   (b)

Figure 5. (a) Optical micrograph; and (b) SEM micrograph showing the FZ of the Sample 20.

Weld zone microstructures of Sample 21, which have the highest heat input and, of course, slowest cooling rate, are completely different from Sample 20 and not associated with the CE values due to the slow cooling conditions. The FZ of Sample 21 consisted of ferritic microstructures with multiple morphologies, e.g., grain boundary, acicular and Widmanstatten (Figure 6a). Due to the oriented solidification and slow cooling rate, elongated and extremely coarse grains were revealed. In Figure 6b, grain boundaries were dashed with yellow, which contain different ferritic structures. Acicular ferritic microstructures can also be seen in Figure 6. The yellow arrows show the inclusions where acicular ferrites nucleated (Figure 6c).

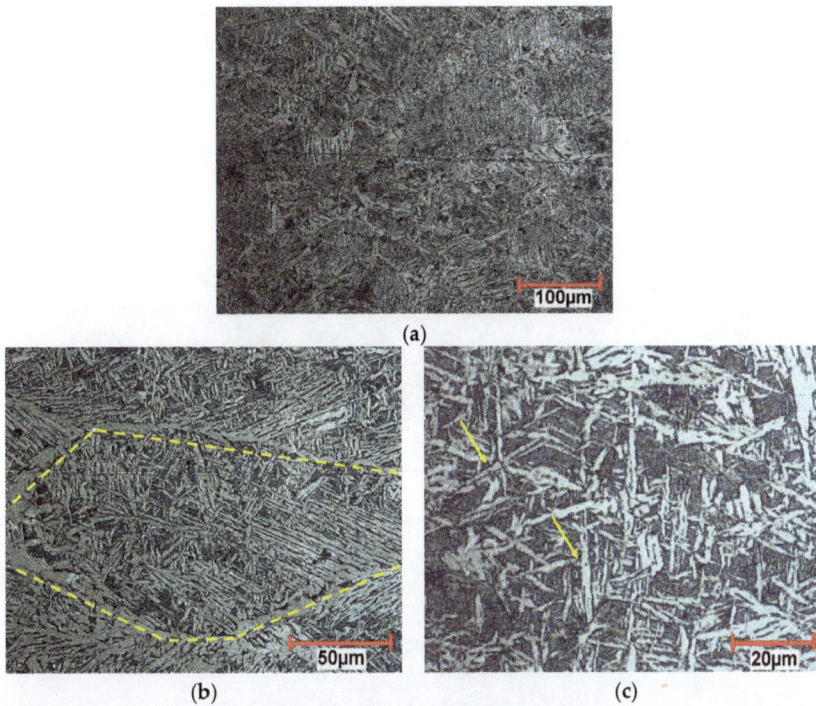

Figure 6. Detailed different magnifications of FZ microstructures of Sample 21: (**a**) FZ at ×100 magnification; (**b**) extremely coarse grains in FZ; and (**c**) inclusions in FZ.

The HAZ of Sample 20 can be divided into five subzones, namely partially molten zone (PMZ), coarse-grained HAZ (CGHAZ), fine-grained (FGHAZ), inter-critical HAZ (ICHAZ) and sub-critical HAZ (SCHAZ). Optical micrographs of these different subzones can be seen in Figures 7 and 8. In the microstructural examinations, PMZ could not be observed. Both MART and HSLA steel, in CGHAZ, consisted of martensitic-bainitic microstructure as a result of the transformation of coarsened austenite grains (Figures 7a and 8a). While the CGHAZ of MART steel shows a higher proportion of martensitic and lower proportion of bainitic microstructures, HSLA steel shows a higher proportion of bainitic and lower proportion of martensitic microstructures. This can be attributed to the CE values of the steels. A higher CE value promoted the formation of martensite, whereas a lower CE value promoted bainitic structures. Although the FGHAZ of MART steel's microstructure is similar to CGHAZ, but consisted of finer grains, this zone could not be observed in HSLA steel. In the ICHAZ, where the peak temperature is between A_3 and A_1, the partial transformation of ferrite to a mixture of ferrite and austenite resulted in martensite islands between the fine-grained ferrite matrix and carbides in HSLA steel (Figure 7b) [2]. Figure 7b shows a transition zone towards SCHAZ.

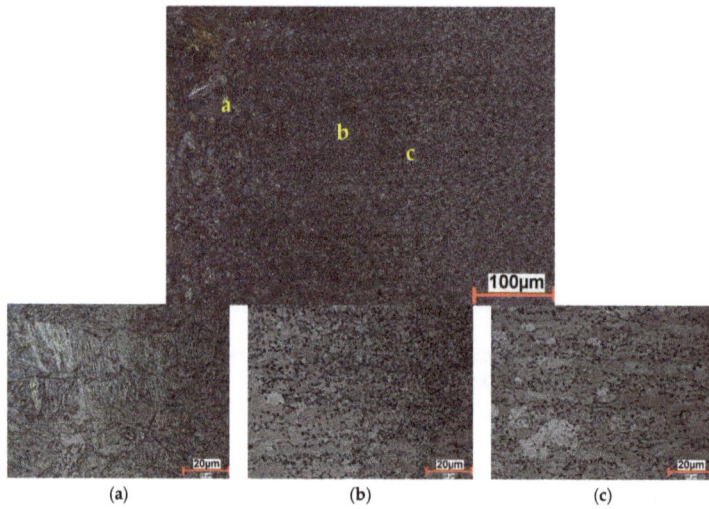

Figure 7. Detailed heat affected zone (HAZ) microstructures and subzones of the HSLA side of Sample 20: (**a**) coarse-grained HAZ (CGHAZ); (**b**) inter-critical HAZ (ICHAZ); and (**c**) sub-critical HAZ (SCHAZ).

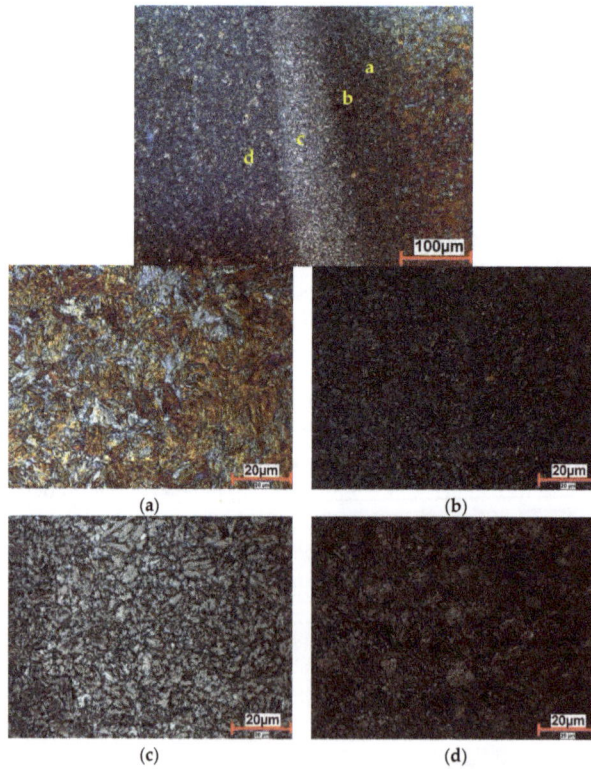

Figure 8. Detailed HAZ microstructures and subzones of the MART side of Sample 20: (**a**) CGHAZ; (**b**) fine-grained (FGHAZ); (**c**) ICHAZ; and (**d**) SCHAZ.

The ICHAZ of MART steel exhibited a dual phase microstructure containing ferrite with fine and well-dispersed martensite. In addition, some portion of the acicular ferritic microstructures can be seen in Figure 8c. Since shielding gas was not used, nitrogen and oxygen absorption could promote titanium base nitrides, carbo-nitrides and oxide inclusions where acicular ferrites can nucleate [2,36–38]. Furthermore, the slow cooling rate of this zone could induce ferritic structures to be formed. Figure 8d shows the SCHAZ of MART steel. In this zone, tempered martensite and bainite formed due to the lower peak temperature than A_1. However, it is expected that the coarsening of the carbides occurs in the HSLA side, and there is no difference identified metallographically. This can be related to the thermal stability of the HSLA, which is greater than MART and, therefore, does not have a microstructure that is distinct from its BM [2].

For Sample 21, the whole weld zone was roughly 11 mm, so only the micrographs of specific zones are presented here. The CGHAZ of HSLA side of Sample 21 consisted of ferritic and bainitic structures and it is shown with dashed lines. The FGHAZ of the HSLA side contains similar, but finer grains with respect to CGHAZ (Figure 9b). Beside the FGHAZ, coarsening of the carbides occurred in the HSLA side.

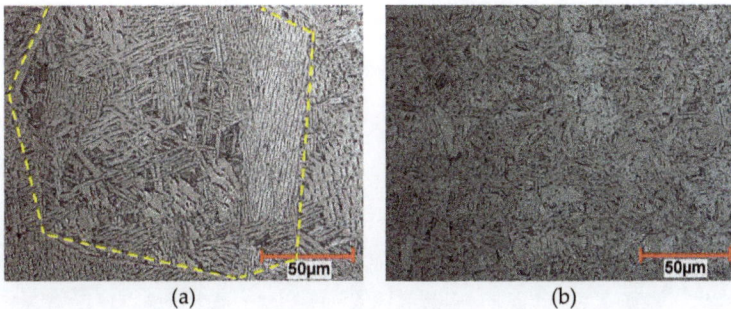

(a)   (b)

Figure 9. Microstructures of the HAZ zone for HSLA side of Sample 21: (a) CGHAZ and (b) FGHAZ.

In the CGHAZ and FGHAZ of the MART side of Sample 21, as a result of the higher CE, coarse baiting, ferritic and martensitic microstructures were identified (Figure 10a,b). The ICHAZ, in accordance with the Fe-Fe$_3$C equilibrium diagram, consisted of fine ferritic structures with small portions of pearlitic structures (Figure 10c). As expected, under the influence of a relatively high temperature, which is in the range of martensite tempering temperatures, tempered martensite formed in SCHAZ of the MART side (Figure 10d).

Microhardness measurements were conducted in the various zones of Samples 20 and 21. The microhardness of the BM of the HSLA and MART steels was measured as 213 and 404 Vickers, respectively. The hardness profile of the welded joint section varies significantly because of the phase transformations during the thermal cycle of the welding process. Figure 11 shows the microhardness map of Sample 20. Figure 11 also presents the microhardness profile across the mid-section of the sample. Due to the rapid cooling of FZ, each material showed an increase in hardness of FZ relative to BM. The average microhardness value in the FZ is 480 Vickers and varies across the section. This fluctuation is attributed to the mixed microstructure of the FZ. Different hardness of the martensitic and bainitic microstructures could cause the fluctuation of the hardness profile. In addition, various morphologies (i.e., columnar and equiaxed) in FZ could be a reason for the various hardness. However, some researchers have focused to determine an empirical formula for FZ hardness using CE values; in the present study, the measured hardness of FZ is higher than the calculated values using the mentioned formulas [31,36]. The calculated hardness values using the formulas given in the literature are 434 HV and 365 HV. In all compared zones, MART steel exhibited higher hardness values due to the higher CE value, which has a significant influence on the hardenability. While the hardness of the

HSLA side exhibits a sharp increase through the HAZ up to the FZ, the MART side shows a softening zone in HAZ. The continuous increase trend in the HSLA side was due to the ferritic microstructure of HSLA steel. The tempering zone and ferritic/martensitic dual phase structures in MART steel caused a decrease in hardness.

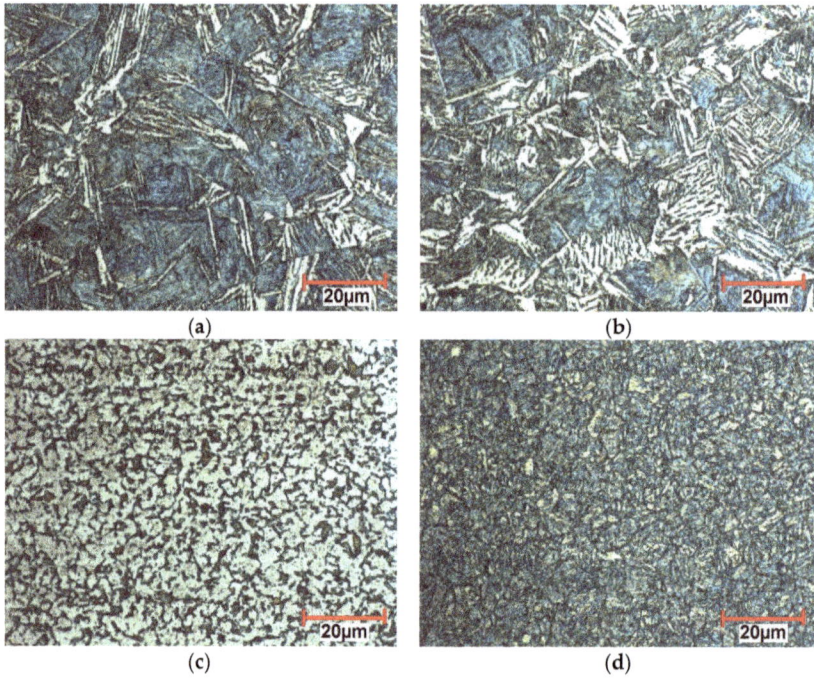

(a)

(b)

(c)

(d)

Figure 10. Detailed HAZ microstructures and subzones of the MART side of Sample 21: (**a**) CGHAZ; (**b**) FGHAZ; (**c**) ICHAZ; and (**d**) SCHAZ.

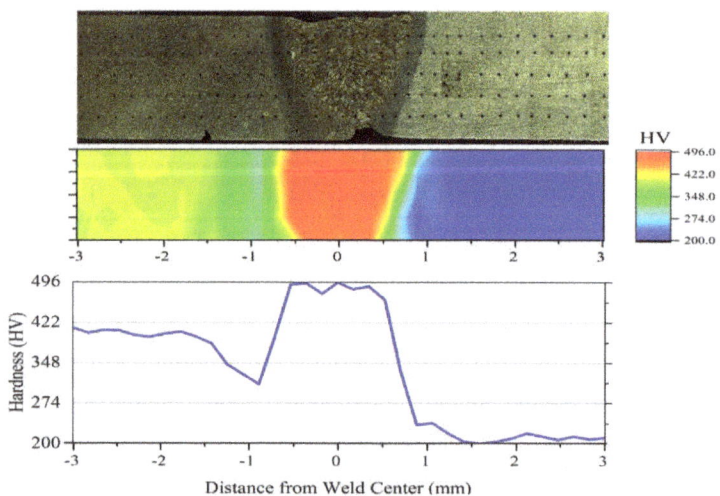

Figure 11. Microhardness map and profile of Sample 20.

The microhardness map and profile of Sample 21 can be seen in Figure 12. The highest and the lowest microhardness values were measured in the BM of MART and HSLA, respectively. The highest value is related to the predominantly martensitic microstructure of the BM of MART steel. Among the weld zone of the MART steel, the ICHAZ showed the lowest microhardness corresponding with the ferritic-pearlitic microstructure. The measured microhardness values through the FZ showed a fluctuation, which can be a result of multiple morphologies of ferritic structures. In the HSLA side, microhardness values showed a decreasing trend up to the BM.

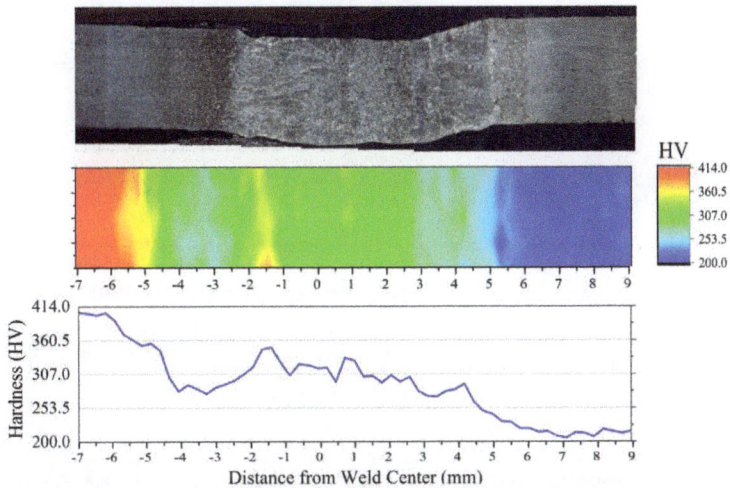

Figure 12. Microhardness map and profile of Sample 21.

4. Conclusions

In this study, fiber laser welded dissimilar MART and HSLA steels have been evaluated with respect to tensile properties, microstructure and hardness profile. As the first step of this study, the process parameters of laser welded dissimilar steel joints have been optimized to maximize the TL and minimize the HI of the welded joints using the Taguchi method. The order of importance of the process parameters on the response was welding speed, laser power and focal position. The welding speed was found to be the most effective process parameter, and its interaction with the laser power should be monitored for the HI and TL of the joints. It was observed that, if the HI was not sufficient due to high speed or low laser power, the weld bead geometry was not formed appropriately. In addition, when applying excessive HI, the HAZ would be wider, and that causes a decrease in TL. The optimum combination of laser welding process parameters was a welding speed of 45 mm/s, a laser power of 1750 W and the focal position of −0.4 mm.

In the second step, the microstructural examination and microhardness evolution of the selected welds that have the highest and lowest response values were discussed. Weld zone microstructures of selected samples were completely distinct due to the different HI and consequently not associated with CE values due to slow cooling rates. The HAZ of the samples was divided into five subzones, namely PMZ, CGHAZ, FGHAZ, ICHAZ and SCHAZ, due to the grain transformations. Due to the phase transformations during the thermal cycle of the process, the hardness profile of the welded sections varies significantly. Due to the rapid cooling of FZ, each sample showed an increase in hardness of FZ relative to BM. While the hardness of the HSLA side exhibits a sharp increase through the HAZ up to the FZ, the MART side shows a softening zone in HAZ.

Acknowledgments: The authors acknowledge the Uludağ University Commission of Scientific Research Projects under Contract No. OUAP (MH)-2016/6 for supporting this research. Additionally, a part of this research was supported by the Coskunoz Holding Research and Development Department at Bursa, Turkey.

Author Contributions: Celalettin Yuce and Fatih Karpat conceived of and designed the experiments. Celalettin Yuce and Mumin Tutar performed the experiments and microstructural examination. Celalettin Yuce, Mumin Tutar and Nurettin Yavuz analyzed the data. All of the authors discussed the results and commented on the manuscript at all stages. All co-authors contributed to the manuscript proofing and submissions.

Conflicts of Interest: The authors declare no conflict of interest.

References

1. Kuziak, R.; Kawalla, R.; Waengler, S. Advanced high strength steels for automotive industry. *Arch. Civ. Mech. Eng.* **2008**, *8*, 103–117. [CrossRef]
2. Saha, D.C.; Westerbaan, D.; Nayak, S.S.; Biro, E.; Gerlich, A.P.; Zhou, Y. Microstructure-properties correlation in fiber laser welding of dual-phase and HSLA steels. *Mater. Sci. Eng. A* **2014**, *607*, 445–453. [CrossRef]
3. Liu, Y.; Dong, D.; Wang, L.; Chu, X.; Wang, P.; Jin, M. Strain rate dependent deformation and failure behavior of laser welded DP780 steel joint under dynamic tensile loading. *Mater. Sci. Eng. A* **2015**, *627*, 296–305. [CrossRef]
4. Benasciutti, D.; Lanzutti, A.; Rupil, G.; Haeberle, E.F. Microstructural and mechanical characterisation of laser-welded lap joints with linear and circular beads in thin low carbon steel sheets. *Mater. Des.* **2014**, *62*, 205–216. [CrossRef]
5. Casalino, G.; Dal Maso, U.; Angelastro, A.; Campanelli, S.L. Hybrid laser welding: A review. *DAAAM Int. Sci. Book* **2010**, *38*, 413–430.
6. Xu, W.; Westerbaan, D.; Nayak, S.S.; Chen, D.L.; Goodwin, F.; Zhou, Y. Tensile and fatigue properties of fiber laser welded high strength low alloy and DP980 dual-phase steel joints. *Mater. Des.* **2013**, *43*, 373–383. [CrossRef]
7. Parkes, D.; Xu, W.; Westerbaan, D.; Nayak, S.S.; Zhou, Y.; Goodwin, F.; Bhole, S.; Chen, D.L. Microstructure and fatigue properties of fiber laser welded dissimilar joints between high strength low alloy and dual-phase steels. *Mater. Des.* **2013**, *51*, 665–675. [CrossRef]
8. Parkes, D.; Westerbaan, D.; Nayak, S.S.; Zhou, Y.; Goodwin, F.; Bhole, S.; Chen, D.L. Tensile properties of fiber laser welded joints of high strength low alloy and dual-phase steels at warm and low temperatures. *Mater. Des.* **2014**, *56*, 193–199. [CrossRef]
9. Wang, J.; Yang, L.; Sun, M.; Liu, T.; Li, H. Effect of energy input on the microstructure and properties of butt joints in DP1000 steel laser welding. *Mater. Des.* **2016**, *90*, 642–649. [CrossRef]
10. Wang, J.; Yang, L.; Sun, M.; Liu, T.; Li, H. A study of the softening mechanisms of laser-welded DP1000 steel butt joints. *Mater. Des.* **2016**, *97*, 118–125. [CrossRef]
11. Rossini, M.; Spena, P.R.; Cortese, L.; Matteis, P.; Firrao, D. Investigation on dissimilar laser welding of advanced high strength steel sheets for the automotive industry. *Mater. Sci. Eng. A* **2015**, *628*, 288–296. [CrossRef]
12. Němeček, S.; Mužík, T.; Míšek, M. Differences between laser and arc welding of HSS steels. *Phys. Proced.* **2012**, *39*, 67–74. [CrossRef]
13. Zhao, Y.Y.; Zhang, Y.S.; Hu, W. Effect of welding speed on microstructure, hardness and tensile properties in laser welding of advanced high strength steel. *Sci. Technol. Weld. Join.* **2013**, *18*, 581–590. [CrossRef]
14. Arslanoglu, N.; Yigit, A. Experimental investigation of radiation effect on human thermal comfort by Taguchi method. *Appl. Therm. Eng.* **2016**, *92*, 18–23. [CrossRef]
15. Tutar, M.; Aydin, H.; Yuce, C.; Yavuz, N.; Bayram, A. The optimisation of process parameters for friction stir spot-welded AA3003-H12 aluminium alloy using a Taguchi orthogonal array. *Mater. Des.* **2014**, *63*, 789–797. [CrossRef]
16. Benyounis, K.Y.; Olabi, A.G. Optimization of different welding processes using statistical and numerical approaches—A reference guide. *Adv. Eng. Softw.* **2008**, *39*, 483–496. [CrossRef]
17. Anawa, E.M.; Olabi, A.G. Optimization of tensile strength of ferritic/austenitic laser-welded components. *Opt. Lasers Eng.* **2008**, *46*, 571–577. [CrossRef]
18. Anawa, E.M.; Olabi, A.G. Using Taguchi method to optimize welding pool of dissimilar laser-welded components. *Opt. Laser Technol.* **2008**, *40*, 379–388. [CrossRef]

19. Sathiya, P.; Jaleel, M.Y.A.; Katherasan, D.; Shanmugarajan, B. Optimization of laser butt welding parameters with multiple performance characteristics. *Opt. Laser Technol.* **2011**, *43*, 660–673. [CrossRef]

20. Casalino, G.; Campanelli, S.L.; Ludovico, A.D. Laser-arc hybrid welding of wrought to selective laser molten stainless steel. *Int. J. Adv. Manuf. Technol.* **2013**, *68*, 209–216. [CrossRef]

21. Acherjee, B.; Kuar, A.S.; Mitra, S.; Misra, D. A sequentially integrated multi-criteria optimization approach applied to laser transmission weld quality enhancement—A case study. *Int. J. Adv. Manuf. Technol.* **2013**, *65*, 641–650. [CrossRef]

22. Benyounis, K.Y.; Olabi, A.G.; Hashmi, M.S.J. Effect of laser welding parameters on the heat input and weld-bead profile. *J. Mater. Process. Technol.* **2005**, *164*, 978–985. [CrossRef]

23. Ruggiero, A.; Tricarico, L.; Olabi, A.G.; Benyounis, K.Y. Weld-bead profile and costs optimisation of the CO_2 dissimilar laser welding process of low carbon steel and austenitic steel AISI316. *Opt. Laser Technol.* **2011**, *43*, 82–90. [CrossRef]

24. Olabi, A.G.; Alsinani, F.O.; Alabdulkarim, A.A.; Ruggiero, A.; Tricarico, L.; Benyounis, K.Y. Optimizing the CO_2 laser welding process for dissimilar materials. *Opt. Lasers Eng.* **2013**, *51*, 832–839. [CrossRef]

25. Reisgen, U.; Schleser, M.; Mokrov, O.; Ahmed, E. Optimization of laser welding of DP/TRIP steel sheets using statistical approach. *Opt. Laser Technol.* **2012**, *44*, 255–262. [CrossRef]

26. Zhao, Y.; Zhang, Y.; Hu, W.; Lai, X. Optimization of laser welding thin-gage galvanized steel via response surface methodology. *Opt. Lasers Eng.* **2012**, *50*, 1267–1273. [CrossRef]

27. Benyounis, K.Y.; Olabi, A.G.; Hashmi, M.S.J. Multi-response optimization of CO_2 laser-welding process of austenitic stainless steel. *Opt. Laser Technol.* **2008**, *40*, 76–87. [CrossRef]

28. SSAB Products. Available online: http://www.ssab.com/products/brands/docol (accessed on 22 August 2016).

29. Guo, W.; Crowther, D.; Francis, J.A.; Thompson, A.; Liu, Z.; Li, L. Microstructure and mechanical properties of laser welded S960 high strength steel. *Mater. Des.* **2015**, *85*, 534–548. [CrossRef]

30. Khan, M.I.; Kuntz, M.L.; Biro, E.; Zhou, Y. Microstructure and mechanical properties of resistance spot welded advanced high strength steels. *Mater. Trans.* **2008**, *49*, 1629–1637. [CrossRef]

31. Oyyaravelu, R.; Kuppan, P.; Arivazhagan, N. Metallurgical and Mechanical properties of Laser welded High Strength Low Alloy Steel. *J. Adv. Res.* **2016**, *7*, 463–472. [CrossRef] [PubMed]

32. Coelho, R.S.; Corpas, M.; Moreto, J.A.; Jahn, A.; Standfuß, J.; Kaysser-Pyzalla, A.; Pinto, H. Induction-assisted laser beam welding of a thermomechanically rolled HSLA S500MC steel: A microstructure and residual stress assessment. *Mater. Sci. Eng. A* **2013**, *578*, 125–133. [CrossRef]

33. Talaş, Ş. The assessment of carbon equivalent formulas in predicting the properties of steel weld metals. *Mater. Des.* **2010**, *31*, 2649–2653. [CrossRef]

34. Yurioka, N. Carbon Equivalents for Hardenability and Cold Cracking Susceptibility of Steels. In Proceedings of the Select Conference on Hardenability of Steels, Derby, UK, 17 May 1990.

35. Santillan Esquivel, A.; Nayak, S.S.; Xia, M.S.; Zhou, Y. Microstructure, hardness and tensile properties of fusion zone in laser welding of advanced high strength steels. *Can. Metall. Q.* **2012**, *51*, 328–335. [CrossRef]

36. Midawi, A.R.H.; Santos, E.B.F.; Huda, N.; Sinha, A.K.; Lazor, R.; Gerlich, A.P. Microstructures and mechanical properties in two X80 weld metals produced using similar heat input. *J. Mater. Process. Technol.* **2015**, *226*, 272–279. [CrossRef]

37. Beidokhti, B.; Kokabi, A.H.; Dolati, A. A comprehensive study on the microstructure of high strength low alloy pipeline welds. *J. Alloys Compd.* **2014**, *597*, 142–147. [CrossRef]

38. Beidokhti, B.; Dolati, A.; Koukabi, A.H. Effects of alloying elements and microstructure on the susceptibility of the welded HSLA steel to hydrogen-induced cracking and sulfide stress cracking. *Mater. Sci. Eng. A* **2009**, *507*, 167–173. [CrossRef]

metals

Article

TIG Dressing Effects on Weld Pores and Pore Cracking of Titanium Weldments

Hui-Jun Yi [1,*], Yong-Jun Lee [2] and Kwang-O Lee [3]

[1] Hyundai-Rotem Company, Chang-won 51407, Korea
[2] The 3rd land system team, Defense Agency for Technology and Quality, Chang-won 51472, Korea; elan4017@naver.com
[3] School of Mechanical Engineering, Pusan National University, Busan 46287, Korea; royallko@pusan.ac.kr
* Correspondence: yi.h.jun@gmail.com; Tel.: +82-10-4846-3184

Academic Editor: Giuseppe Casalino
Received: 10 August 2016; Accepted: 10 October 2016; Published: 17 October 2016

Abstract: Weld pores redistribution, the effectiveness of using tungsten inert gas (TIG) dressing to remove weld pores, and changes in the mechanical properties due to the TIG dressing of Ti-3Al-2.5V weldments were studied. Moreover, weld cracks due to pores were investigated. The results show that weld pores less than 300 μm in size are redistributed or removed via remelting due to TIG dressing. Regardless of the temperature condition, TIG dressing welding showed ductility, and there was a loss of 7% tensile strength of the weldments. Additionally, it was considered that porosity redistribution by TIG dressing was due to fluid flow during the remelting of the weld pool. Weld cracks in titanium weldment create branch cracks around pores that propagate via the intragranular fracture, and oxygen is dispersed around the pores. It is suggested that the pore locations around the LBZ (local brittle zone) and stress concentration due to the pores have significant effects on crack initiation and propagation.

Keywords: titanium welding; weld pores; cracks; TIG dressing; ductility

1. Introduction

High strength, low density, and excellent corrosion resistance are the main properties of titanium that make it attractive for a variety of applications. A variety of welding methods have been used for titanium alloys, and several studies of high-energy welding using lasers and new welding methods, such as friction stir welding, have been conducted. The welding method most widely used in industrial applications is the tungsten inert gas (TIG) welding process. The main advantages of TIG, in comparison with other welding methods, are its cost-effectiveness, workability, and ease of use. When a titanium alloy is welded using a fusion welding method such as TIG, the titanium alloy bonds easily with oxygen, nitrogen, and carbon at temperatures over 500 °C, which results in a high degree of brittleness. Therefore, the weldment needs to be protected or shielded from ambient air. The primary precaution taken is shielding the metal from any contact with air, hydrogen, carbon compounds, or other contaminants during the melting, solidification, and solid-state cooling processes associated with fusion welding. To prevent contamination, the weld joint and weld electrode must be clean, and the shielding gas (usually argon, but sometimes helium or a mixture of the two) must be free of moisture and other impurities. The torch used in the TIG process is designed to permit an inert gas to flow through it, surrounding the electrode and molten metal pool with a protective atmosphere. In addition, a trailing shield of inert gas should be used to protect the weldment while it is solidifying.

Apart from the problem of oxidation at high temperatures, the most frequently occurring problem in the welding of titanium alloys is porosity and its effects, which are the main subjects of this study [1–3]. Weld metal porosity in a titanium alloy is known to originate at the trailing edge of the weld pool, where interstitial elements (oxygen or hydrogen) are partitioned between dendrites during solidification. Partitioning occurs because of the large decrease in oxygen (or hydrogen) solubility that occurs in the transformation from liquid to solid. There are several gaseous species that could become trapped during the welding of titanium. The most likely of these is hydrogen, but other possible agents include oxygen, nitrogen, and carbon dioxide, as well as the inert shielding gases. These gases could be present as a result of any of the following: (a) desorption of the gases' elemental constituents from the parent material or welding consumables; (b) absorption into the weld pool due to inadequate shielding (through either entrapment of air in the shielding gases or a high moisture level in the shielding gases) during welding; (c) entrapment of the shielding gases; or (d) surface contamination [4–9].

Various methods, including variation of the process parameters and attention to cleanliness, have been employed to attempt to minimize porosity in titanium weldments. Removal of the hydrated layers prior to welding is critical to minimizing the potential hydrogen content of the melt pool. Consequently, the effectiveness of the method used to remove the hydrated layer and other surface contaminants may have a large influence on the weld metal porosity. In addition to the above methods for ensuring cleanliness of the titanium material prior to welding, relatively low heat input conditions and a high welding speed have been recommended to reduce the porosity of weldments for TIG and EBW (electron beam welding) [10–13]. Additionally, a new welding process was studied [14].

The effects of dispersed pores in weldments vary. Although an allowable level of porosity is specified in welding structure designs, depending on the importance of the structure, the aviation and power generation facility sectors, in which titanium alloys are widely used, regulate porosity in weldments strictly. Although the effects of porosity in weldments vary, porosity in titanium weldments is reported to result in pore cracking that initiates at the pores in the weldments. Pore cracking in titanium welds is known to occur mainly in restrained sections. Recent studies have found that alloys with hydrogen contents greater than 200 wt ppm are more prone to pore cracking and that the probability of cracking increases with increased porosity. Previous studies have attempted to explain pore cracking in Ti-6211 welds in terms of oxygen embrittlement. Therefore, the current thinking tends to associate pore cracking in titanium welds with interstitial (oxygen, hydrogen, and so forth) embrittlement [15–19].

The soundness of weldments and the presence of pores can be verified using nondestructive test methods, such as X-ray testing, after titanium alloy welding. If pores are detected in weldments, the portion of the weld in which the pores are present can be removed mechanically. This is followed by repair welding, i.e., re-welding, as required. However, this process is known to degrade the mechanical strength of weldments. In addition, complex processes, such as the complete removal of surface oxides using the chemical pickling method, must be performed prior to welding to remove the cause of the formation of pores completely. This not only reduces weld joint efficiency but also has a number of adverse effects on productivity. In this study, TIG dressing was used to remove welding pores generated in weldments more simply and effectively than has been accomplished previously using other methods. Pores in Ti-3Al-2.5V titanium alloy weldments used in power generation facilities were examined in this study, and the causes of the pore cracking phenomenon that occurs as a result of the presence of pores were analyzed. In addition, the effects of the redistribution and removal of pores formed in Ti-3Al-2.5V titanium alloy weldments via TIG dressing and the effects of TIG dressing on the strength and toughness of titanium alloy weldments at a high temperature condition were examined.

2. Materials and Methods

2.1. The Effect of TIG Dressing

To examine the effectiveness of TIG dressing in removing weld porosity in titanium alloy weldments and the effects of TIG dressing on the mechanical properties of the weldments, a specimen was manufactured using a 2.0-mm-thick Ti-3Al-2.5V alloy, as shown in Figure 1. Normally, the objective of TIG dressing is to remove welding geometrical imperfections, such as undercut by remelting the weld toe, thereby leaving the weld practically free of geometric defects. The treatment also significantly reduces the stress concentration factor of the weld toe by introducing a smooth transition. Therefore, the TIG dressing is used to improve the fatigue life of welded structures [20–22]. In this study, standard TIG welding equipment was used, without the addition of any filler material. The chemical composition of the base metal and filler metal are given in Table 1. The TIG welding was performed with 99.9% argon gas used as the shielding and purging gas. A fully equipped out-of-chamber purging device and an automatic welding machine were used to protect against oxidation during welding as shown in Figure 2. Table 2 gives the parameters of the welding and TIG dressing.

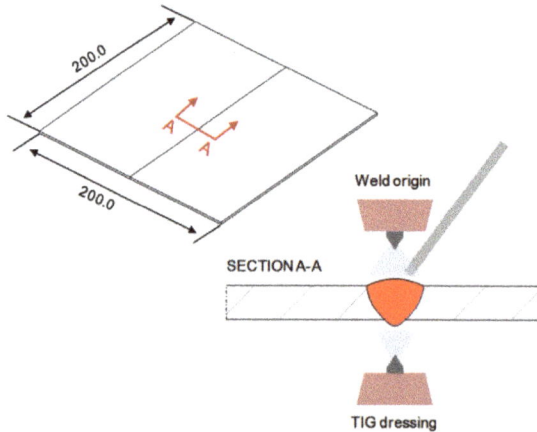

Figure 1. Illustration of welding test of Ti-3Al-2.5V and tungsten inert gas (TIG) dressing.

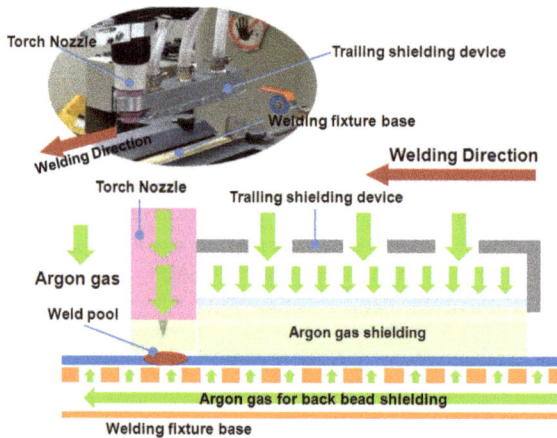

Figure 2. Illustration of out-of-chamber welding test equipment.

Table 1. Chemical compositions of base metal and filler metal (wt %).

Identification	C	Fe	Al	V	N	O	H	Ti
Base metal	0.03	0.25	3.02	2.49	0.02	0.12	0.005	Bal.
Filler metal	0.02	0.01	2.98	2.48	0.01	0.10	0.001	Bal.

Table 2. Welding and TIG dressing parameters.

Identification		Ampere (A)	Voltage (V)	Welding Speed (cm/min)	Welding Feeding Speed (cm/min)	Remark
Specimen A	WO	58	9.2	25	34	Original weldment
	R01	68	9.8	25	-	1st TIG dressing
Specimen B	WO	58	9.2	25	34	Original weldment
	R01	68	9.8	25	-	1st TIG dressing
	R02	68	9.8	25	-	2nd TIG dressing
Specimen C	WO	58	9.2	25	34	Original weldment
	R01	68	9.8	25	-	1st TIG dressing
	R02	68	9.8	25	-	2nd TIG dressing
	R03	68	9.8	25	-	3rd TIG dressing

To assess the effectiveness of the TIG dressing, a specimen was manufactured from which oxide films of the titanium alloy were not removed. After 48 h had elapsed since welding, X-ray testing was done to verify the presence of pores in the weldment and to determine pore distribution. After the first X-ray test on the specimen, the TIG dressing was conducted at a location at which a back bead in the weldment had formed, as shown in Figure 1. After the first TIG dressing was completed, X-ray testing was conducted at three times. At the time of the TIG dressing welding, the oxide film on the titanium alloy surface was not removed using a mechanical or chemical method. After the original welding and TIG dressing, X-ray inspection was performed to check for the existence of pores and to determine the pore distribution. For the X-ray test equipment, both the X-ray generator and tube used MG452 YXLON (YXLON, Hudson, NY, USA) to ensure that the required resolution was obtained. The scale bar on the radiographic film was used to confirm the pore sizes.

To assess the changes in the mechanical properties of the weldment due to the TIG dressing, tensile tests and notched tensile tests were conducted. Tensile and notch tensile specimens were prepared as per the ASTM E8M-05 standard test method as shown in Figure 3. The tensile tests were carried out in a 100 kN universal testing machine (Instron 8501, INSTRON, Norwood, MA, USA). The specimens were loaded at a rate of 1.5 kN/min, as per the ASTM standard. The tensile tests were performed at 25 °C, 300 °C, and 500 °C. At each conditions, three tensile and notch tensile specimens were used.

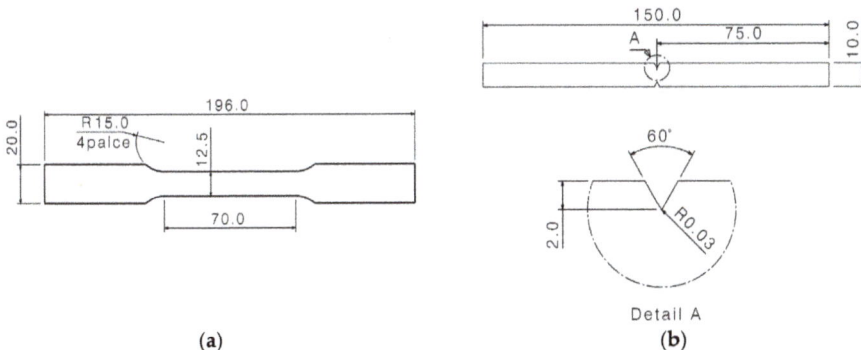

(a)

(b)

Figure 3. Illustration of tensile and notched tensile specimen: (a) tensile test; (b) notch tensile test.

2.2. Pore Craking Problems

Figure 4 shows cracks occurred Ti-3Al-2.5V (2.0-mm-thick) pipes used to move sea water in a power generation facility. Cracks also occurred in the weldments of brackets used to assemble the pipes. The cracks were observed approximately one week after the welding. The mechanical properties and chemical composition of the material used are shown in Table 1. The welding was done with the TIG welding and welding parameters are summarized in Table 3. The oxide film on the surfaces of the weldments was removed using a stainless steel brush prior to welding.

To observe the cracks in the weldments, cracked parts were removed from the structure, and the fractured cross section and microstructure of each weldment were examined using optical microscopy (OLYMPUS, Tokyo, Japan) and scanning electron microscopy (SEM, SERON, Uiwang-si, Korea). To determine the probable cause of the cracks, cross sections of cracked specimens were first observed with an optical microscope and then intentionally fractured using a tensile and bending test apparatus (Instron 8501, INSTRON, Norwood, MA, USA). The fracture surfaces were investigated with SEM to determine the fracture mode.

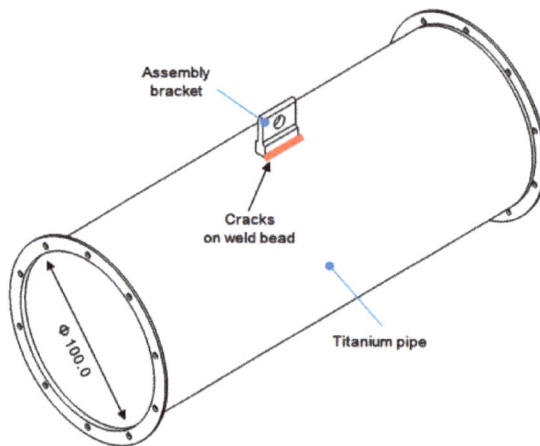

Figure 4. Illustration of cracks occurred Ti-3Al-2.5V pipe weldment.

Table 3. Welding parameters of Ti-3Al-2.5V pipe.

Ampere (A)	Voltage (V)	Welding Speed (cm/min)	Remark
75.0	10.5	25	

3. Results

3.1. Weld Porosity Redistribution Due to the TIG Dressing

Table 4 and Figure 5 show the X-ray test results and weld porosity redistribution results. The weld pore sizes were mostly within the range of 100 to 250 μm, with a few pores exceeding 250 μm in size. When the TIG dressing was conducted on the first WO weldment, the number of weld pores decreased. The change in the number of pores was detected after the first TIG dressing; the second and third TIG dressings produced no change in the number of pores or in the pore sizes. Furthermore, pores greater than 300 μm in size were not removed, nor were their sizes redistributed or reduced, by the remelting that resulted from the TIG dressing. Figure 6 shows the SEM images of the pore distribution of the WO. Large-scale SEM microscopy indicated that the inner surfaces of the pores were smooth, as shown in Figure 7. The round shapes and smooth inner surfaces suggest that the pores were formed as a result of gas evolution during the welding.

Table 4. Results of X-ray inspection, quantity, and size of pores (*d*).

Identification		$d \leq 150$ μm	150 μm$< d \leq 250$ μm	250 μm$< d \leq 500$ μm
Specimen A	WO	10	19	2
	R01	3	8	2
Specimen B	WO	12	28	2
	R01	3	12	2
	R02	3	12	2
Specimen C	WO	2	10	5
	R01	1	4	5
	R02	1	4	5
	R03	1	4	5

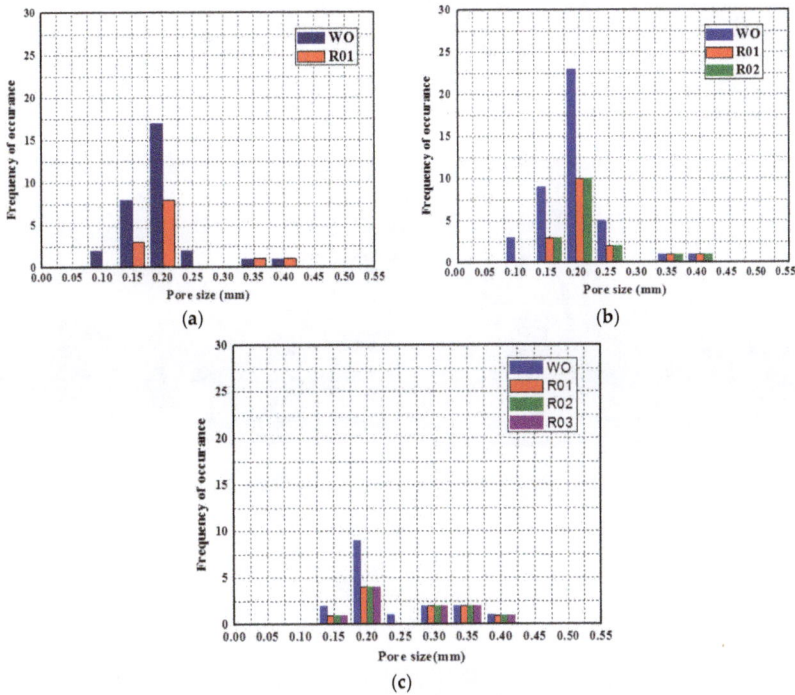

(a)

(b)

(c)

Figure 5. Effect of TIG dressing on porosity removing and distribution: (a) specimen A; (b) specimen B; (c) specimen C.

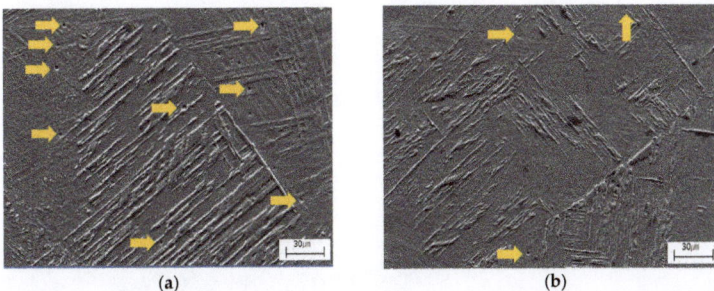

(a)

(b)

Figure 6. SEM image of pores in weld metal: Yellow arrows indicate pores: (a) WO; (b) R03.

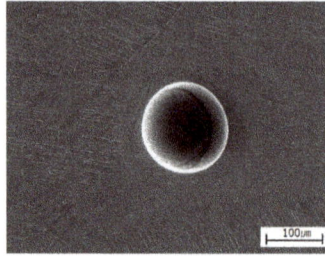

Figure 7. SEM image of pores (≥100 μm) in weld metal (WO).

3.2. Effect of the TIG Dressing on Mechanical Properties of the Weldment

The tensile test results for the WO and TIG dressings for ambient- and high-temperature conditions are shown in Figure 8. At every temperature, the ultimate tensile strength was decreased in comparison to the pretreatment strength, but not much change was observed between the first treatment and the second and third treatments. The notch toughness properties (NSR, notch strength ratio = notch tension strength/tension strength) for ambient- and high-temperature conditions are shown in Figure 9. Regardless of the temperature condition and the number of TIG dressing applications, all of the weldments have ductile properties.

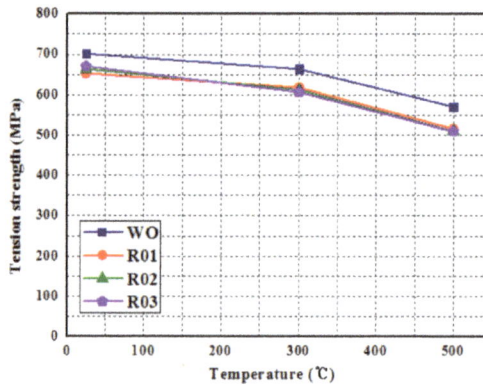

Figure 8. Results of the tensile strength test.

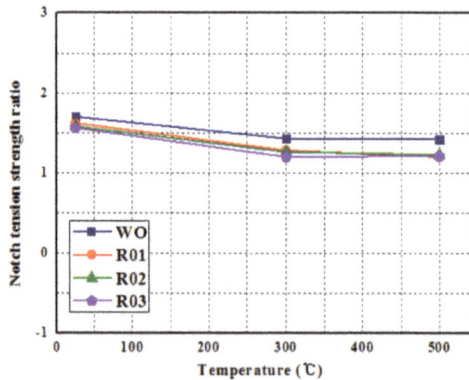

Figure 9. Results of the NSR (notch tension strength ratio).

3.3. Observation of the Microstructure of the Cracked Specimen

Metallographic examinations were conducted on the cracked specimens. Figure 10 shows the microstructure of the weldment and the heat-affected part, in which an acicular α, α', and a prior β grain boundary generated in the Ti-3Al-2.5V weldment were observed. The SEM results showed that weldment porosity was distributed over areas where cracks were formed, as shown in Figure 11. After the metallographic examinations were completed, the cracked specimens were opened and examined via SEM as shown in Figure 12. Observation of the cracks' fracture shapes revealed that cracking progressed as intragranular fracture and that small branch cracks were created during the crack propagation. That is, the propagation of the cracks formed in the Ti-3Al-2.5V alloy weldment progressed along the columnar grain boundaries through the intragranular fracture and that small branch cracks were formed as the propagation progressed.

To identify the causes of the pore formation, the chemical composition of the weld material around the pores was determined using energy dispersive spectroscopy (EDS). The results are shown in Table 5 and Figure 13. The EDS results show that the pores and the surrounding areas had higher oxygen contents than other areas, which confirms that the main cause of the porosity was oxidation. The oxide films formed in the titanium alloy surface during welding reacted with carbon in the atmosphere to form CO and CO_2, which are known to be the main cause of pore formation.

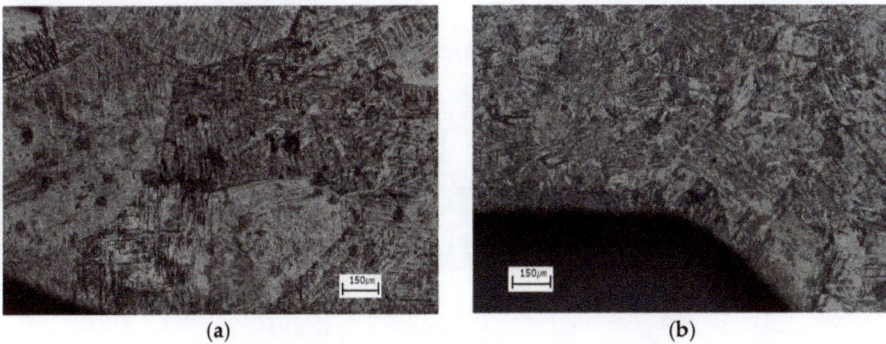

(a)   (b)

Figure 10. Optical image of Ti-3Al-2.5V pipe weldment: (**a**) weld metal; (**b**) heat affected zone.

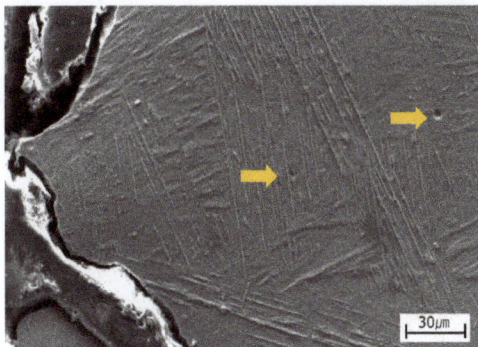

Figure 11. SEM image of cracked Ti-3Al-2.5V pipe weldment: yellow arrows indicate pores.

Figure 12. SEM image of fracture surface.

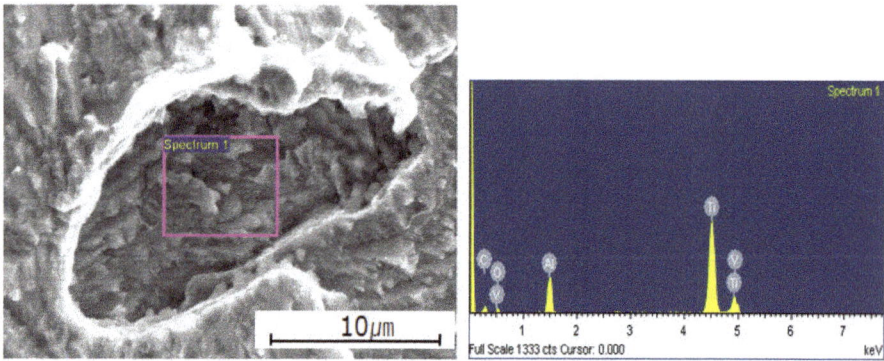

Figure 13. Energy dispersive spectroscopy (EDS) test results and SEM image of pores.

Table 5. Results of EDS test of porosity.

Identification	Weight (%)
C	7.33
O	10.77
Al	9.56
Ti	70.26
V	2.08

4. Discussion

4.1. The TIG Dressing Reheating Effects on Weld Pore Redistribution

The driving forces for fluid flow in the weld pool include the buoyancy force, the Lorentz force, the surface tension force, and the arc shear stress. In the case of the TIG welding, especially below 200 A welding parameters, the effect of arc shear stress on fluid flow is small, and the maximum velocity of buoyancy convection is far less than the forced convection driven by the Lorentz force. Therefore, the effects of arc shear stress and the buoyancy force on the pore removal mechanism by the TIG dressing can be ignored [23].

In this study, the TIG dressing was applied to a back bead region, which is different from a welded surface. As a result, weld pores that formed along the fusion line and the edge region of the weldment moved along the fluid flow of the molten pool, which moves from the center of the weldment to the edge because of the Lorentz force, as shown in Figure 14. Because of this reason, it was to be considered that the pores located near the surface area were to be removed. It was not able to verify the effect of the TIG dressing on pores size greater than 300 μm. Such a verification would require in-depth research on whether fluid flow due to the Lorentz force can influence the movement of pores greater than a certain size and whether energy exceeding the surface tension of the weld pool can be acquired when pores greater than a certain size are removed.

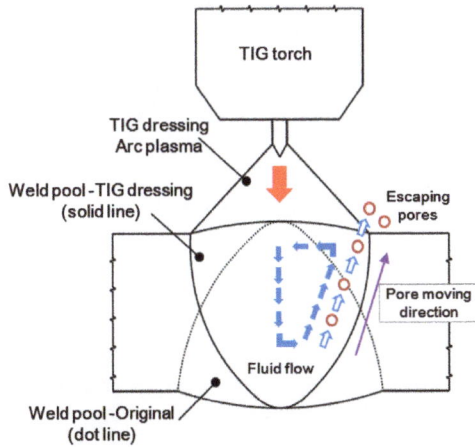

Figure 14. Schematic diagram of porosity escape.

4.2. Porosity Nucleation and Crack Mechanism

It is believed that bubble nucleation in a weld pool is heterogeneous nucleation that occurs at the boundary between the solid state and liquid state when dissolved gas exceeds its solubility. When a molten weld pool passes along the weld seam, dissolved gases from absorbed gas-forming substances act as nucleation sites for gas bubbles. The higher amount of oxygen in the pores is evidence of this mechanism. Previous studies have suggested that local embrittlement around the pores caused pore cracking. However, these previous studies did not identify any microstructural or hardness changes that could be associated with or attributed to the presence of pores. It has been suggested that embrittlement may be caused by gases present within the pores. Interstitial elements, such as oxygen, may cause local embrittlement. However, there is no metallurgical or experimental evidence of this local embrittlement around pores.

Pores that form in a titanium alloy weldment are formed at the edge of the weldment. In the weldment, the LBZ (local brittle zone) is formed along the fusion line where the parent material and welding consumables meet. The LBZ is highly brittle and is a starting point for weld fracture. In a titanium alloy weldment, weld pores form around the LBZ. Weld cracks are generated around the pores in titanium alloy weldments by complex processes that result from residual stress after welding, as pores are formed in the LBZ, which has relatively low toughness (high hardness) in comparison to the surrounding region as shown in Figure 15. The explanation for the initiation of fracture from the lowest-toughness region via the formation of a crack tip under loading is known as fracture theory. Pores are believed to act as crack tips for fracture initiation in the local brittle zone, and stress concentration due to pores geometrical effects under loading is believed to be the driving force of crack propagation.

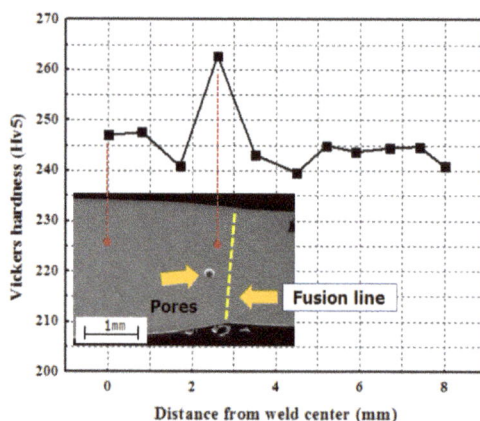

Figure 15. Hardness distribution of titanium weldment (WO).

5. Conclusions

In this study, Weld porosity redistribution due to TIG dressing and the effects of TIG dressing on the mechanical properties of titanium alloy weldments were analyzed. Additionally, the weld porosity in Ti-3Al-2.5V pipe weldments and the phenomenon of cracks due to weld porosity were observed and studied. Our findings are as follows:

1. Weld pores less than 300 μm in size were redistributed or removed via TIG dressing remelting.
2. Regardless of the tensile test temperature, the NSR rate recorded was greater than 1.0. There was a loss of 7% in the ambient- and high-temperature tensile strength of the weldments after one, two, and three TIG dressing applications, compared with the original weldment.
3. Examination of weldments in which cracking occurred showed that weld porosity was generated along the crack propagation path, that cracks were propagated through intragranular fracture, and that branch cracks were created.
4. With respect to the pore cracking mechanism, it is suggested that local features (pores forming in local brittle zones) and geometric features (stress concentration) of the pore have significant effects on crack initiation and propagation under loading conditions.

Author Contributions: Hui-Jun Yi conceived, designed and performed the experiments and wrote the papers. Yong-Jun Lee and Kwang-O Lee contributed analysis tools and reviewed papers.

Conflicts of Interest: The authors declare no conflict of interest.

References

1. Christoph, L.; Manfred, P. *Titanium and Titanium Alloys-Fundamentals and Application*, 1st ed.; Wiley-VCH: Weinheim, Germany, 2003.
2. Lutjering, G.; Williams, J.C. *Titanium*, 1st ed.; Springer: New York, NY, USA, 2007.
3. Donachie, M.J. *Titanium and Titanium Alloys*, 1st ed.; American Society for Matals: Material Park, OH, USA, 1982.
4. Williams, J.C.; Edgar, A.; Starke, J. Progress in structural materials for aerospace systems. *Acta Mater.* **2003**, *51*, 5775–5799. [CrossRef]
5. Huang, J.L.; Warnken, N.; Gebelin, J.C.; Stangwood, M.; Reed, R.C. On the mechanism of porosity formation during welding of titanium. *Acta Mater.* **2012**, *60*, 3215–3225. [CrossRef]
6. Boyer, R.R. Use of titanium in the aerospace industry. *Mater. Sci. Eng. A* **1996**, *213*, 103–114. [CrossRef]

7.   Gouret, N.; Dour, G.; Miguet, B.; Oliver, E.; Fortunier, R. Assessment of the origin of porosity in electron-beam-welded TA6V plates. *Metall. Mater. Trans. A* **2004**, *35*, 879–889. [CrossRef]
8.   Lee, P.D.; Hunt, J.D. Hydrogen porosity in directional solidified aluminium-copper alloys: In situ observation. *Acta Mater.* **1997**, *45*, 4155–4169. [CrossRef]
9.   Atwood, R.C.; Sridhar, S.; Zhang, W.; Lee, P.D. Diffusion-controlled growth of hydrogen pores in aluminium-sillcon casting: In situ obervation and modelling. *Acta Mater.* **2000**, *48*, 405–417. [CrossRef]
10. Kornilov, I.I.; Baikow, A.A. Effect of oxygen on titnaium and its alloys. *Inst. Metall.* **1973**, *10*, 2–6.
11. Matyushkin, B.A.; Gorshkov, A.I.; Murav'ev, V.I. Special features of formation and development of cracks and pores in the metal of welds of titanium alloys after the welding. *Svarochn. Proizv.* **1975**, *8*, 9–11.
12. Smith, L.S.; Gittos, M.F. *Hydride Cracking in Titanium and Its Alloys*; TWI Reasrch Report 658/1998; TWI: Cambridge, UK, 1998.
13. Bettles, C.J.; Tomus, D.D.; Gibson, M.A. The role of microsturcture in the mechanical behavior of Ti-1.6 wt % Fe alloys contaning O and N. *Mater. Sci. Eng. A* **2011**, *528*, 4899–4909. [CrossRef]
14. Fuji, A.; Horiuchi, Y.; Yamamoto, K. Friction welidng of pure titanium and pure nickel. *Sci. Tech. Weld. Join.* **2005**, *10*, 287–294. [CrossRef]
15. Atwood, R.C.; Lee, P.D. Simulation of the three-dimessional morphology of solidification porosity in an aluminium-silicon alloy. *Acta Mater.* **2003**, *51*, 5447–5466. [CrossRef]
16. Khaled, T. An inverstigation of pore cracking in titanium welds. *J. Mater. Eng. Perform.* **1992**, *3*, 419–434. [CrossRef]
17. Wu, H.; Feng, J.; He, J. Microstructure evolution and fracture behaviour for electron beam welding of Ti-6Al-4V. *Bull. Mater. Sci.* **2004**, *27*, 387–392. [CrossRef]
18. Liu, J.; Dahmen, M.; Ventzke, V.; Kashaev, N.; Poprawe, R. The effect of heat treatment on crack control and grain refinement in laser beam welded β-solidifying TiAl-based alloy. *Intermetallics* **2013**, *40*, 65–70. [CrossRef]
19. Kim, Y.W.; Kim, S.L. Effects of microstructure and C and Si additions on elevated temperature creep and fatigue of gamma TiAl alloys. *Intermetallics* **2014**, *53*, 92–101. [CrossRef]
20. Haagensen, P.J.; Maddox, S.J. *IIW Recommendations on Post Weld Improvement of Steel and Aluminum Structures*; The International Institute of Welding: Roissy, France, 2008; XIII-2200r1-07.
21. Kado, S. Influence of the Conditions in TIG Dressing on the Fatigue Strength in Welded High Tensile Strength Steels. The International Institute of Welding: Roissy, France, 1975; XIII-771-75.
22. Redchits, V. Scientific fundamentals and measures used to prevent the formation of pores in fusion welded titanium and its alloys. *Weld. Int.* **1997**, *11*, 722–728. [CrossRef]
23. Kou, S. *Welding Metallurgy*, 2nd ed.; John Wiley and Sons: New York, NY, USA, 2001.

metals

MDPI

Article

An Assessment of the Mechanical Properties and Microstructural Analysis of Dissimilar Material Welded Joint between Alloy 617 and 12Cr Steel

Hafiz Waqar Ahmad *, Jeong Ho Hwang, Ju Hwa Lee and Dong Ho Bae *

Graduate School of Mechanical Engineering, Sungkyunkwan University, Suwon 440-746, Korea; reflika@skku.edu (J.H.H.); juhwa0207@skku.edu (J.H.L.)
* Correspondence: waqar543@skku.edu (H.W.A.); bae@yurim.skku.ac.kr (D.H.B.); Tel.: +82-31-290-7443 (D.H.B.)

Academic Editor: Giuseppe Casalino
Received: 29 August 2016; Accepted: 10 October 2016; Published: 14 October 2016

Abstract: The most effective method to reduce CO_2 gas emission from the steam power plant is to improve its performance by elevating the steam temperature to more than 700 °C. For this, it is necessary to develop applicable materials at high temperatures. Ni-based Alloy 617 and 12Cr steel are used in steam power plants, due to their remarkable mechanical properties, high corrosion resistance, and creep strength. However, since Alloy 617 and 12Cr steel have different chemical compositions and thermal and mechanical properties, it is necessary to develop dissimilar material welding technologies. Moreover, in order to guarantee the reliability of dissimilar material welded structures, the assessment of mechanical and metallurgical properties, fatigue strength, fracture mechanical analysis, and welding residual stress analysis should be conducted on dissimilar material welded joints. In this study, first, multi-pass dissimilar material welding between Alloy 617 and 12Cr steel was performed under optimum welding conditions. Next, mechanical properties were assessed, including the static tensile strength, hardness distribution, and microstructural analysis of a dissimilar material welded joint. The results indicated that the yield strength and tensile strength of the dissimilar metal welded joint were higher than those of the Alloy 617 base metal, and lower than those of the 12Cr steel base metal. The hardness distribution of the 12Cr steel side was higher than that of Alloy 617 and the dissimilar material weld metal zone. It was observed that the microstructure of Alloy 617 HAZ was irregular austenite grain, while that of 12Cr steel HAZ was collapsed martensite grain, due to repeatable heat input during multi-pass welding.

Keywords: dissimilar material welding; Ni-based Alloy 617; 12Cr steel; mechanical properties; microstructural characterization

---

1. Introduction

Environmental pollution from energy generation is a topical issue worldwide. There are four ways to improve this problem: the first is to generate energy from renewable energy resources, the second is to use nuclear energy, the third is to use carbon capture and storage before pollutants are released to the atmosphere, and the fourth is to increase energy efficiency by using A-USC (advanced-ultra super critical) thermal power plants that are capable of generating energy at temperatures above 700 °C [1]. A-USC thermal power plants have improved thermal efficiency, and reduced CO_2 emissions [2,3]. However, at this elevated temperature, it is very difficult to find applicable materials that can withstand extreme environments. Ni-based super alloys and high chromium steel are suitable candidates for such extreme environments in the power generation industry due to their exceptional metallurgical stability at high temperatures, and excellent mechanical properties, such as high creep rupture strength, and good oxidation and corrosion resistance [4,5].

Alloy 617 is a nickel-chromium-cobalt-molybdenum based alloy that is most commonly used in gas turbines for combustion cans and ducts, as well as for industrial furnace components and applications. Alloy 617 is primarily known for its remarkable metallurgical stability [6]. It brings a wide variety of other outstanding properties, e.g., high temperature oxidation resistance due to added aluminum, and solid solution strengthening due to cobalt and molybdenum content [7]. It is easy to fabricate and can be easily joined through conventional welding techniques [8].

12Cr steel with the addition of Mo, Nb, N, W, and other elements possesses improved toughness, good oxidation and corrosion resistance at elevated temperatures, high tensile strength and ductility, and promising creep rupture strength. 12Cr steel is currently used for components in gas turbines, boilers, and steam power plant turbines [9,10].

A typical steam turbine rotor has three major stages: the high temperature and pressure stage, the middle temperature and pressure stage, and the low temperature and pressure stage. Since Ni-based super Alloy 617 is a very difficult material to work on, it is partially applicable to the high and low pressure stages. Use of Ni-based super Alloy 617 in lower pressure stage requires its joining with 12Cr steel, which is currently used in the low pressure stage [11]. In order to guarantee the mechanical reliability of dissimilar material weld between Ni-based Alloy 617 and 12Cr steel, it is necessary to develop welding technology between dissimilar materials and perform welding stress analysis, strength assessment, and the assessment of corrosion characteristics at the dissimilar material weld.

A significant amount of research has been carried out on Ni-based alloys as well as Cr steel alloys in recent decades [4,10,12–15]. However, it is difficult to find a systematic study on the dissimilar material welding and welded joint of Alloy 617 and 12Cr steel. In this work, a dissimilar material welding between Alloy 617 and 12Cr steel was carried out using Direct Current Straight Polarity (DCSP) tungsten inert gas (TIG) welding technology, and an assessment of the mechanical properties and a microstructural analysis of the dissimilar material weld were performed.

2. Dissimilar Material Welding between Ni-Based Alloy 617 and 12Cr Steel

2.1. Materials and Welding Procedure

In this work, DCSP TIG welding technology was used for the dissimilar material welding process. Tables 1 and 2 illustrate the chemical composition and mechanical properties of Alloy 617, 12Cr steel, and Thyssen 617, which is filler metal of a wire 1 mm in diameter. The welding conditions, such as the electrode shape, arc length, welding wave mode (CW or pulse), and welding heat input, were controlled using a real time monitoring system. The optimum welding conditions were determined by repeatedly performing preliminary welding with a variety of welding conditions at different shield gas composition and flow rates. Optimized dissimilar material welding conditions are summarized in Table 3.

Table 1. Chemical compositions of Alloy 617, Thyssen 617, and 12Cr steel.

Base/Filler Metal	Chemical Composition (% Weight)											
	Ni	Cr	Co	Mo	Al	C	Fe	Si	Ti	Cu	Mn	S
Alloy 617	44.3	22	12.5	9.0	1.2	0.07	1.5	0.5	0.3	0.2	0.5	0.008
Thyssen 617	45.7	21.5	11.0	9.0	1.0	0.05	1.0	0.1	1	-	-	-
12Cr	0.43	11.6	-	0.04	-	0.13	Bal.	0.4	-	0.1	0.58	-

Table 2. Mechanical properties of Alloy 617 and 12Cr steel.

Base Material	Y.S. (MPa)	T.S. (MPA)	Elongation	R.A. (%)	M.P. (°C)
Alloy 617	322	732	62	56	1330
12Cr	551	758	18	50	1375

Table 3. Multi-pass dissimilar material welding conditions between Alloy 617 and 12Cr steel.

Pass	Shield Gas	Current (A)	Voltage (V)	Welding Speed (cm/min)	Freq. (Hz)
1	Ar-2.5% H_2	150	10	10	0.5
2	Ar-2.5% H_2	150	13	10	0.5
3	Ar-2.5% H_2	150	16	10	0.5
4	Ar-2.5% H_2	150	16	10	0.5
5	Ar-2.5% H_2	150	16	10	0.5
6	Ar-2.5% H_2	150	16	10	0.5
7	Ar-2.5% H_2	150	16	10	0.5

In order to prevent thermal distortion, which is formidably caused by welding heat input during the welding process, both ends of the base metal were fixed with welding jigs. Figure 1 shows that the welding direction was made parallel to the rolling direction of the base metals. The groove shape of Figure 2 was machined in a U-groove to narrow the gap welding [16,17]. Thyssen 617 was used as a filler metal and its chemical composition is shown in Table 1. When welding was completed for each pass, the surface condition of the weld bead was carefully observed, and the welding condition was confirmed. After finishing each pass, for the next pass, the surface of the weld bead was brushed using a copper brush, and the temperature checked was to allow it to cool sufficiently below 70 °C. After welding, the multi-pass welds were inspected by using the ultrasonic testing method.

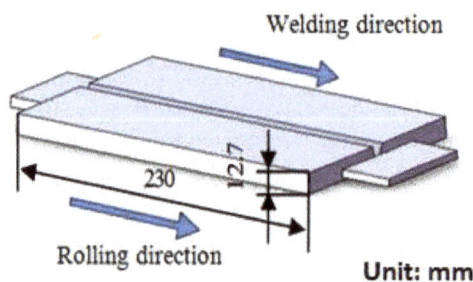

Figure 1. Welding direction for dissimilar material welding between Alloy 617 and 12Cr steel.

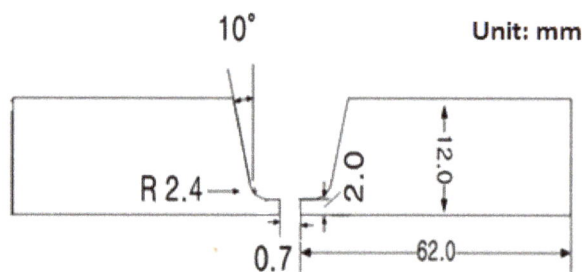

Figure 2. U-groove with narrow gap.

2.2. Results of the Dissimilar Welding

Figure 3 shows the top view and cross section of the dissimilar material weld between Alloy 617 plate and the 12Cr steel plate. Figure 4 illustrates the weld bead appearance for each pass. These figures show that there were no weld defects or oxidation phenomenon on the weld surfaces.

Even though it was assumed that out-of-plane thermal distortion would be generated by repeatable welding heat input during the multi-pass welding processes, it was prevented by the constraint of the welding jigs.

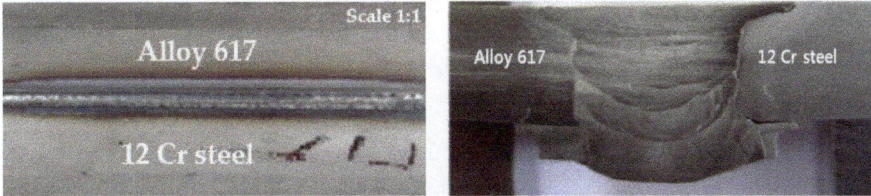

Figure 3. Dissimilar weld: top view (**left**) and cross section (**right**).

Figure 4. Weld bead appearance for each pass.

3. Assessing Mechanical Properties of Dissimilar Material Welded Joint

Specimen and Procedure

The specimen dimensions and tensile test procedure, to assess the mechanical properties of dissimilar material welded joint between Alloy 617 and 12Cr steel, was followed as recommended in the ASTM E8M standard [18]. Figure 5 shows that the weld metal, the heat affected zone (HAZ), and both base metals of Alloy 617 and 12Cr steel are included within the range of the gauge length of the specimens. Before assessing the mechanical properties of the dissimilar material welded joint, tensile tests for base metals (Alloy 617 and 12Cr steel) were performed to measure their mechanical properties. Mechanical properties of the base metals and dissimilar material welded joint were assessed using a material testing system (INSTRON 8801, Instron Korea, LLC., Seoul, Korea) as shown in Figure 6 at room temperature. In the tensile test, the loading speed was controlled by the displacement of 1 mm/min.

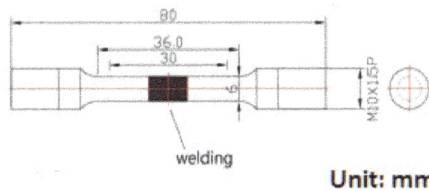

Figure 5. Configuration of tensile test specimen (ASTM E8M).

127

Figure 6. Test equipment (10 tons, Instron 8801).

The hardness distribution on the section, including base metal, HAZ, and the weld metal of dissimilar material welded joint, were measured to analyze the effect of multi-pass dissimilar material welding on metallurgical hardness change. Figure 7 shows that hardness tests were performed to compare the hardness distribution for the three positions: top, middle, and bottom of the cross section. The hydraulic micro Vickers hardness tester (Mitutoyo MVK-H2, Mitutoyo Korea Corporation, Gunpo, Korea) was used as hardness test equipment. Hardness distribution of the multi-pass dissimilar material welded joint was measured at a 200 g press-fit load for 5 s.

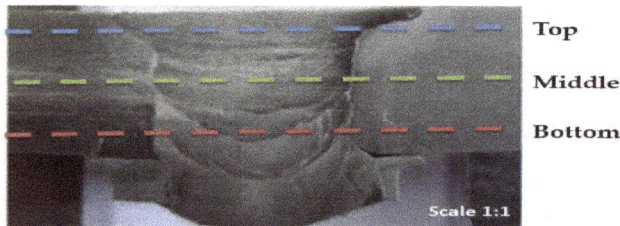

Figure 7. Three positions across the cross section of dissimilar material welded joint for hardness distribution measurements.

Optical microscopic observation, using Olympus PME 3 (Olympus Korea Co. LTD., Seoul, Korea), was carried out in order to analyze the microstructure of the dissimilar material welded joint between Alloy 617 and 12Cr steel. Specimen was fabricated from the cross section of the dissimilar material welded joint. In order to analyze the microstructures of dissimilar material welded plate, five position—(a) Alloy 617 base metal; (b) Alloy 617 HAZ; (c) weld metal; (d) 12Cr steel HAZ; and (e) 12Cr steel base metal—were optically observed as shown in Figure 8. Before analyzing the microstructures, the surface of specimen was etched according to ASTM E407 [19]. Three positions—(a), (b), and (c)—were etched for 20 s by using etchant 88 (10 mL of HCl + 20 mL of HNO_3 + 30 mL of distilled water). The other positions—(d) and (e)—were etched for 10 s by using etchant 91 (5 mL of HCl + 5 mL of HNO_3 + 1 g pf picric acid + 200 mL of ethanol). In addition, composition analysis for the five positions of the dissimilar welded joint was performed using energy dispersive X-ray spectroscopy (EDS, EDAX Inc., Mahwah, NJ, USA).

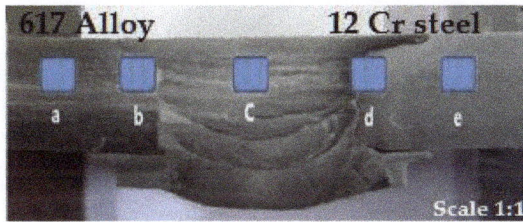

Figure 8. Positions of microstructure analysis for the dissimilar material welded joint.

4. Results and Discussion

4.1. Tensile Test Results

Figure 9a compares the tensile test results of the base metals and the dissimilar material welded joint between Alloy 617 and 12Cr steel. The magnitudes of yield and tensile strength of the dissimilar material welded joint were assessed as 490 MPa and 767 MPa, respectively. These magnitudes are higher than Alloy 617, which has a 443 MPa yield strength and a 675 MPa ultimate tensile strength, and lower than 12Cr steel, which has a 700 MPa and 817 MPa yield and tensile strength, respectively. Failure of dissimilar welded joints occurred mostly at the HAZ of 12Cr steel. Figure 9b shows one of the fractured specimens that has failed at the HAZ of 12Cr steel.

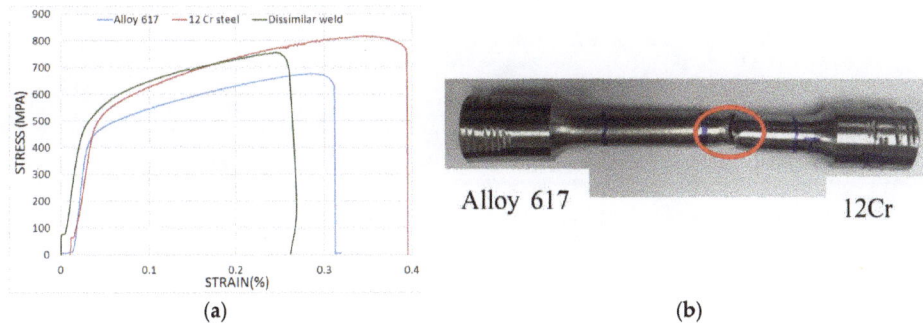

(a)

(b)

Figure 9. (a) σ-ε curves of dissimilar weld; (b) dissimilar weld fractured specimen.

Although tensile strength of 12Cr steel is higher than that of Alloy 617, but the failure of the dissimilar material weld occurred at the HAZ of 12Cr steel because the heat input during the dissimilar material welding process introduced sufficient metallurgical changes in 12Cr steel HAZ [20]. On the other hand, Ni-based Alloy 617 experienced lesser metallurgical changes compared with 12Cr steel due to heat input during the dissimilar material welding process. In fact, Ni-based Alloy 617 is primarily known for its remarkable metallurgical stability at high temperatures [6].

4.2. Hardness Distribution

Figure 10 compares the hardness distribution on the bottom, middle, and top positions of the cross section of the dissimilar material welded joint between Alloy 617 and 12Cr steel. The hardness distribution at the HAZ of 12Cr steel is higher than that of both the weld metal zone and the Alloy 617 base metal. The magnitude of the peak values were assessed as 460–490 Hv at the HAZ of 12Cr steel, and about 220–260 Hv at both the weld metal and the Alloy 617 base metal. It is notable that the hardness distribution of the weld metal zone and the Alloy 617 base metal is almost similar, and the peak values, too, do not show a major difference at the bottom, middle, or top positions in either zones. The reason for this behavior is the use of Thyssen 617 as filler material, which has a chemical composition similar to that of Alloy 617, as shown in Table 1. Another important observation is that the hardness distribution is nearly uniform on the right side of Figure 10 (weld zone and Alloy 617 zone), while irregular behavior of the hardness distribution is observed on the left side of the figure (HAZ and base metal of 12Cr steel). This behavior is a result of the fact that the microstructure and metallurgical changes in 12Cr steel and Alloy 617 are significantly different when they are affected by the welding heat input during the multi-pass welding process. In fact, it is known that the mechanical properties of Ni-based Alloy 617 are not sensitively influenced by heat [14,21]. As the bottom side experienced more heat cycles during the welding process, the hardness distribution at the bottom side was higher than the top side across the dissimilar material welded joint.

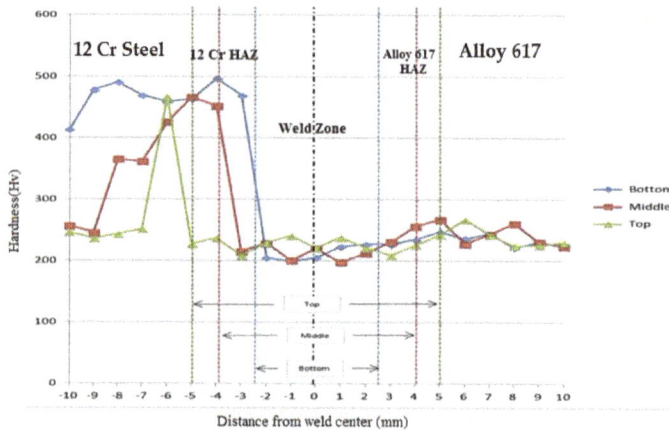

Figure 10. Micro hardness profile across the dissimilar metal welded joint.

4.3. Microstructure and Composition Analysis

Figure 11 shows the microstructures of each position of the dissimilar material welded joint. The microstructure of Alloy 617 base metal in Figure 11a shows typical austenite grain. However, HAZ of Alloy 617 in Figure 11b shows irregular austenite grain size from the effect of the welding heat input during the multi-pass welding processes. The microstructure of the weld metal in Figure 11c made by the solidification of fusion metal from the dilution of Alloy 617, Thyssen 617 filler metal, and 12Cr steel, shows dendrite grain. The HAZ of 12Cr steel in Figure 11d shows collapsed martensite grain by the effect of the welding heat input during the multi-pass welding processes, while the microstructure of the 12Cr base metal in Figure 11e shows typical martensite grain.

The composition analysis observed by energy dispersive analysis of X-rays (EDAX) for the five parts of the dissimilar material welded plate is illustrated in Table 4. The effect of the multi-pass welding process of the dissimilar materials is reflected as a slight change in composition of specimen when compared with Table 1. The major composition of the dissimilar metal weld is very similar to the Alloy 617 composition, except for Fe. Table 4 provides information about the composition change at the HAZ and the weld metal.

(a) Alloy 617   (b) Alloy 617 HAZ   (c) Dissimiler weld

(d) 12 Cr steel HAZ   (e) 12 Cr steel

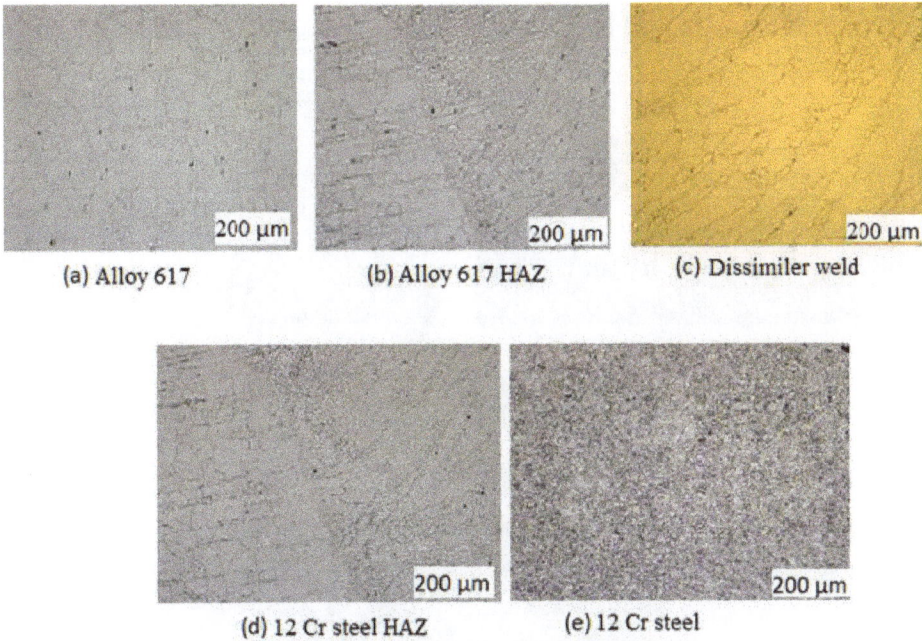

Figure 11. Microstructure observation results of dissimilar metal weld: (**a**) Alloy 617; (**b**) Alloy 617 HAZ; (**c**) Dissimilar material weld; (**d**) 12Cr steel HAZ; (**e**) 12Cr steel.

Table 4. Composition analysis of dissimilar material welded joint (wt. %).

Elements	Alloy 617 Base Metal	Alloy 617 HAZ	Weld Metal	12Cr HAZ	12Cr Base Metal
Mo	10.29	10.70	9.57	-	-
Cr	22.09	21.50	21.11	13.00	10.28
Fe	-	-	12.03	87.00	89.72
Co	13.29	13.76	10.32	-	-
Ni	54.33	54.05	46.97	-	-

Figure 12 shows the fractured tensile test specimen and microscopic observation using a scanning electron microscope (SEM) on the fractured surface of the weld, which occurred on the HAZ of 12Cr steel. As mentioned above, the microstructure of the HAZ of 12Cr steel showed collapsed martensite grain by the effect of welding heat input during the multi-pass welding processes. Additionally, as illustrated in Table 4, the major compositions (Cr and Fe) were slightly changed in the HAZ of 12Cr steel. Even though the microstructure in the HAZ of 12Cr steel shows both collapsed martensite grain and the complicate grain size, as shown in Figure 12b,c, some dimples were observed by the fractography on the fractured surface of the HAZ. Therefore, it was supposed that the failure mode in the HAZ of 12Cr steel was a brittle fracture combined ductile characteristics.

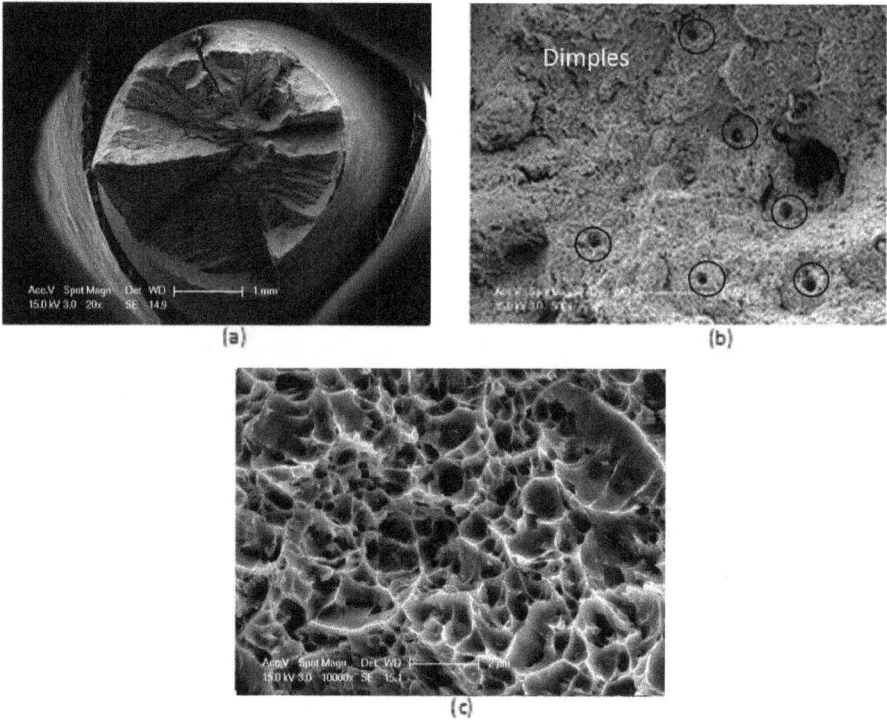

Figure 12. Fractography of the dissimilar metal weld at the 12Cr steel side: (**a**) at 1 mm resolution; (**b**) at 50 μm resolution; (**c**) at 2 μm resolution.

5. Conclusions

This paper performed and analyzed a dissimilar material welding between Ni-based Alloy 617 and 12Cr steel by DCSP TIG welding technology. In order to guarantee the mechanical reliability of the dissimilar material weld, the mechanical properties, including the tensile strength, the hardness distribution, and changes in microstructure were assessed. The conclusions are as follows:

1. Dissimilar material welding technology using DCSP TIG welding between Ni-based Alloy 617 and 12Cr steel was developed. Optimized major conditions for the DCSP TIG welding were determined as shield gas (Ar-2.5% H_2 mixed gas), 150 Amp, and 10–16 V.

2. The magnitudes of yield strength and tensile strength of the multi-pass dissimilar material welded joint were assessed as 490 MPa and 767 MPa, respectively. Dissimilar material welded joints mostly failed at the HAZ of 12Cr steel. The mechanical properties of the dissimilar material weld, including yield and tensile strength, were higher than those of the Alloy 617 base metal, and less than those of the 12Cr base metal.

3. The hardness distribution at the HAZ of 12Cr steel is higher than that of the weld metal zone and the Alloy 617 base metal. The magnitudes of the peak values were assessed as 460–490 Hv for the HAZ of 12Cr steel, and about 220–260 Hv for both the weld metal and the Alloy 617 base metal. The hardness distributions for the weld metal zone and Alloy 617 HAZ and the base metal did not show a significant difference.

4. The microstructures of the dissimilar material welded joint, including the Alloy 617 base metal and the HAZ, the weld metal, and the 12Cr steel HAZ and 12Cr base metal, were metallurgically changed via welding heat input during the multi-pass welding process. The microstructures of

Ni-based Alloy 617 base plate and the HAZ of Alloy 617 were analyzed as a typical austenite grain and an irregular austenite grain. However, 12Cr steel HAZ and the 12Cr base metal were analyzed as collapsed martensite and martensite grain, respectively. The microstructure of the weld metal was analyzed as dendrite grain.

Acknowledgments: This work was supported by the Reliability Evaluation Lab of the Mechanical Engineering Department, Sungkyunkwan University, Suwon, Korea.

Author Contributions: H. Waqar Ahmad conceived and designed the experiments; H. Waqar Ahmad and J.H. Hwang performed the experiments and analyzed the data under the supervision of D.H. Bae; the microstructure analysis was performed in Sungkyunkwan University, Korea; H. Waqar Ahmad wrote the paper.

Conflicts of Interest: The authors declare no conflict of interest.

Abbreviations

The following abbreviations are used in this manuscript:

DMW	Dissimilar Metal Welding
EDAX	Energy Dispersive Analysis of X-rays
HAZ	Heat Affected Zone
DCSP	Direct Current Straight Polarity
TIG	Tungsten Inert Gas

References

1. Mitigating Climate Change Through Renewables. Available online: http://www.irena.org/remap/REmap-FactSheet-8-Climate%20Change.pdf (accessed on 13 October 2016).
2. Weitzel, P.S.; Tanzosh, J.M.; Boring, B. *Advanced Ultra-Supercritical Power Plant (700 to 760C) Design for Indian Coal*; Power Generation Group, Inc.: Barberton, OH, USA, 2012.
3. Higher Efficiency Power Generation Reduces Emissions. Available online: https://www.scribd.com/document/144610465/Beer-Emissions (accessed on 13 October 2016).
4. Maile, K. Qualification of Ni-Based Alloys for Advanced Ultra Supercritical Plants. *Proced. Eng.* **2013**, *55*, 214–220. [CrossRef]
5. Xie, X.; Wu, Y.; Chi, C.; Zhang, M. Superalloys for Advanced Ultra-Super-Critical Fossil Power Plant Application. In *Superalloys*; InTech: Rijeka, Croatia, 2015; pp. 51–76.
6. Development of Materials for Use in A-USC Boilers. Available online: https://www.mhi.co.jp/technology/review/pdf/e524/e524027.pdf (accessed on 12 October 2016).
7. Guo, Y.; Wang, B.; Hou, S. Aging Precipitation Behavior and Mechanical Properties of Inconel 617 Superalloy. *Acta Met. Sin.* **2013**, *26*, 307–312. [CrossRef]
8. Nickel Alloy 617, Inconel® 617. Available online: http://continentalsteel.com/nickel-alloys/grades/inconel-617/ (accessed on 12 October 2016).
9. Klueh, R.L.; Harries, D.R. Development of High (7%–12%) 2 Chromium Martensitic Steels. In *High-Chromium Ferritic and Martensitic Steels for Nuclear Applications*; ASTM: West Conshohocken, PA, USA, 2001; pp. 5–27.
10. Taban, E.; Kaluc, E.; Atici, T.; Kaplan, E. 9%–12%Cr Steel: Properties and Weldability Aspects, The Situation in Turkish Industry. In Proceedings of the 2nd Internaional Conference on Welding Technologies and Exhibiton, Ankara, Turkey, 23–25 May 2012; pp. 203–212.
11. Latest Technologies and Future Prospects for a New Steam Turbine. Available online: https://www.mhi-global.com/company/technology/review/pdf/e522/e522039.pdf (accessed on 12 October 2016).
12. Degallaix, G.; Vogt, J.B.; Foct, J. Low Cycle Fatigue of a 12Cr Martensitic Stainless Steel: The Role of Microstructure. In *Low Cycle Fatigue and Elasto-Plastic Behaviour of Materials*; Springer: Dordrecht, The Netherlands, 1987; pp. 95–100.
13. Abe, F. Research and Development of Heat-Resistant Materials for Advanced USC Power Plants with Steam Temperatures of 700 °C and Above. *Engineering* **2015**, *1*, 211–224. [CrossRef]
14. Mankins, W.L.; Hosier, J.C.; Bassford, T.H. Microstructure and Phase of INCONEL Alloy 617 Stability. *Metall. Trans.* **1974**, *5*, 2579–2590. [CrossRef]

15. Microstructure and Strength Characteristics of Alloy 617 Welds. Available online: https://searchworks.stanford.edu/view/11171654 (accessed on 13 October 2016).
16. Corlett, B.J.; Lucas, J.; Smith, J.S. Sensors for Narrow-Gap Welding. *IEEE Proc. A Sci. Meas. Technol.* **1991**, *138*, 213. [CrossRef]
17. Park, K.D.; Ksmpe, J.Y. A Study on Welding Characteristics of Environment for the Stuctural Inconel. In Proceedings of the KSMPE Conference, Daegu, Korea, November 2004; pp. 216–220.
18. Standard Test Method for Tension Testing of Metallic Materials. Available online: https://www.astm.org/Standards/E8.htm (accessed on 12 October 2016).
19. Standard Practice for Micro-Etching Metals and Alloys. Available online: https://www.scribd.com/document/259609551/ASTM-E407-07-Standard-Practice-for-Microetching-Metals-and-Alloys (accesed on 12 October 2016).
20. Du Toit, M.; van Rooyen, G.T.; Smith, D. Heat-Affected Zone Sensitization and Stress Corrosion Cracking in 12% Chromium Type 1.4003 Ferritic Stainless Steel. *Corros. Sci.* **2007**, *63*, 395–404. [CrossRef]
21. Microstructure and Strength Characteristics of Alloy 617 Weld. Available online: https://inldigitallibrary.inl.gov/sti/3310959.pdf (accessed on 12 October 2016).

![metals logo] *metals*

MDPI

Article

The Effects of Pulse Parameters on Weld Geometry and Microstructure of a Pulsed Laser Welding Ni-Base Alloy Thin Sheet with Filler Wire

Dongsheng Chai, Dongdong Wu, Guangyi Ma *, Siyu Zhou, Zhuji Jin and Dongjiang Wu *

Key Laboratory for Precision and Nontraditional Machining Technology of Ministry of Education, Dalian University of Technology, Liaoning 116024, China; chai_dongsheng@mail.dlut.edu.cn (D.C.); wudongdong@mail.dlut.edu.cn (D.W.); zhousiyu@mail.dlut.edu.cn (S.Z.); kimsg@dlut.edu.cn (Z.J.)
* Correspondence: gyma@dlut.edu.cn (G.M.); djwudut@dlut.edu.cn (D.W.); Tel.: +86-411-847-076-25 (D.W.)

Academic Editor: Giuseppe Casalino
Received: 31 August 2016; Accepted: 28 September 2016; Published: 8 October 2016

Abstract: Due to its excellent resistance to corrosive environments and its superior mechanical properties, the Ni-based Hastelloy C-276 alloy was chosen as the material of the stator and rotor cans of a nuclear main pump. In the present work, the Hastelloy C-276 thin sheet 0.5 mm in thickness was welded with filler wire by a pulsed laser. The results indicated that the weld pool geometry and microstructure were significantly affected by the duty ratio, which was determined by the pulse duration and repetition rate under a certain heat input. The fusion zone area was mainly affected by the duty ratio, and the relationship was given by a quadratic polynomial equation. The increase in the duty ratio coarsened the grain size, but did not obviously affect microhardness. The weld geometry and base metal dilution rate was manipulated by controlling pulsed parameters without causing significant change to the performance of the weld. However, it should be noted that, with a larger duty ratio, the partial molten zone is a potential weakness of the weld.

Keywords: pulsed laser welding; filler wire; Hastelloy C-276 thin sheet; weld geometry; microstructure

---

1. Introduction

The Ni-based Hastelloy C-276 alloy is widely used in chemical processing and the nuclear industry, and as marine engineering components such as pumps, valve parts, and spray nozzles, due to its excellent resistance to corrosive environments and its superior mechanical properties [1,2]. The stator and rotor cans of a nuclear main pump in a third-generation nuclear power plant are made of Hastelloy C-276 thin sheets via welding. The unique service conditions of the stator and rotor cans demand high quality of the weld of Hastelloy C-276 thin sheets.

Arc welding, the most common welding method, has been used to weld Hastelloy C-276 by many researchers. Cleslak et al. [3] indicated that intermetallic secondary solidification constituents, a combination of p and μ phases, were found to be associated with weld metal hot cracks in Hastelloy C-276. Li et al. [4] investigated the effects of plate thickness and the annealing process on the microstructure and properties of the Hastelloy C-276 welding line by GTAW, and the grain coarsened with the increase in plate thickness. When filler metal was used, many secondary phase particles were observed after the welding process. The grain size increased in both fusion and the heat affected zone (HAZ), and the tensile strength of the weld decreased. The corrosive resistance was better than that of the autogenous weld bead. The 0.4-mm-thick Hastelloy C-276 thin sheet was welded by GTAW without filler metal, and both the heat affected zone (HAZ) and a secondary phase were detected. Manikandan et al. [5–7] reported that, by current pulsing, the microstructure and mechanical behavior of GTAW of Hastelloy C-276 could be improved. Pulsed Current Gas Tungsten Arc (PCGTA) weldments were found to be the best in terms of (i) freedom from microsegregation, (ii) strength, and (iii) freedom from unwanted secondary phases.

Ahmad et al. [8] studied the microstructure and hardness of the electron beam welded zone of 3-mm-thick Hastelloy C-276. The molten zone (MZ) was found to be of a fine lamellar type, and the hardness was 35% higher compared to as-received alloy, while a hardness reduction of about 5%–8% was observed in the HAZ. Van der Eijk et al. [9] welded NiTi to Hastelloy C-276 with and without filler wire. It was found that the mixed zone of the weld contained a number of brittle phases, and there was a tendency of the NiTi to absorb elements from the Hastelloy C-276.

Compared with conventional fusion welding methods, laser welding has advantages such as a low heat input, a narrow HAZ, low distortion, and ease of automation [10]. Wu and Ma et al. [11–13] performed a series of studies about the laser welding of 0.5-mm-thick Hastelloy C-276 thin sheets without filler wire, and the MZ was found to be of much finer grains, and the element segregation was found. However, the trend of the brittle phase's formation was weakened, no HAZ was found, and the mechanical properties were comparable to the as-received alloy. Ventrella et al. [14] welded Hastelloy C-276 thin foil with a 100-micron thickness. The results indicated that, by using a precise control of the pulse energy and the dilution rate, sound welds could be obtained.

Reports about arc welding Hastelloy C-276 thin sheets with thicknesses of 0.5 mm or less are rare. Studies have proved that pulsed laser welding is an appropriate technique for welding such thin sheets. Laser welding without filler wire has high demands for preparation and clamping. The defect of weld sag is a serious problem when the gap width is larger than 10% of the workpiece thickness [15]. In the present work, a 0.5-mm-thick Hastelloy C-276 thin sheet was welded by a pulsed Nd:YAG laser with 0.5-mm filler wire. In order to highlight the characteristics of the effects of the pulsed parameters, a fixed total heat input was used. The influence of pulsed parameters on weld bead geometry and microstructure was investigated.

2. Experimental Setup

The experimental setup is shown in Figure 1. A millisecond pulsed Nd:YAG laser system (GSI LUMONICS, JK701H, Rugby, UK) with a 1064-nm wavelength and multimode beam was used. The collimated beam diameter was 23.5 mm. The focal length was 80 mm, and focal beam diameter was approximately 0.6 mm. The incident direction of pulsed laser was perpendicular to the substrate surface. The argon shroud gas was delivered through a nozzle with a flowrate of 12 L/min, and the nozzle had a diameter of 6 mm and was set 30 mm away from the weld pool. The experiment was conducted on a commercially available with dimensions of 100 mm × 40 mm × 0.5 mm. The yield strength and the ultimate tensile strength of the as received Hastelloy C-276 thin sheet (Haynes International, Kokomo, IN, USA) are 391 MPa and 857 MPa at room temperature, respectively [16]. The filler metal was 0.5-mm-diameter ERNiCrMo-4 filler wire. The chemical compositions are shown in Table 1.

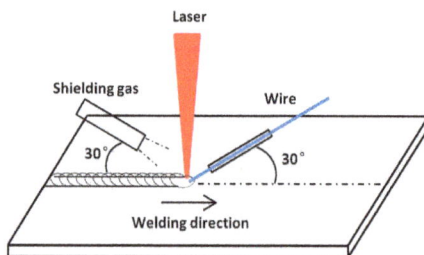

Figure 1. Experiment setup.

Table 1. Chemical composition of Hastelloy C-276 and ERNiCrMo-4 (wt. %).

Sample	Ni	Fe	Cr	Mo	W	Co	Mn	C	Si	P	S	V
Hastelloy C-276	Bal.	5.14	16.00	15.58	3.45	1.26	0.53	0.001	0.02	0.006	0.003	0.01
ERNiCrMo-4	Bal.	5.30	16.00	15.20	3.30	0.11	0.41	0.009	0.03	0.003	0.001	0.01

The bead-on-plate tests were used to analyze the weld shapes. The welding direction was vertical to the rolling direction. The substrate surfaces were grinded with 600 grit silicon carbide paper and then cleaned by ethanol to remove the oxide and oil. The weld parameters are shown in Table 2.

Table 2. Weld parameters.

Average Power (P: W)	Welding Speed (mm/min)	Feeding Speed (mm/min)	Focus (mm)	Pulse Duration (τ: ms)	Pulse Frequency (f: Hz)
75	350	350	−1	3, 4, 5, 6, 7, 8, 9	60
				6	40, 50, 60, 70, 80, 90, 100

The average laser power (P) and the peak power (P_p) are defined as follows:

$$P = E_p \cdot f; \tag{1}$$

$$P_p = E_p / \tau, \tag{2}$$

where E_p (J) is the pulse energy, f (Hz) is the pulse frequency, and τ (ms) is the pulse duration. The pulse duration and pulse frequency, as two independent variables, are chosen to analyze the effects of the pulsed laser on the welding process. The average laser power is fixed in the experiments. The pulse energy decreases with the increase in the pulse frequency, causing a decrease in the peak power when the pulse duration is kept constant. The pulse energy and frequency are kept constant when the pulse duration is increased, so the peak power drops with the increase in the pulse duration.

The pulse duty ratio (α) is defined as the ratio of pulse duration to pulse period and can be expressed as

$$\alpha = f \cdot \tau / 1000 \tag{3}$$

The pulse duty ratio represents the ratio of the heating time in one pulse cycle. A larger pulse duty ratio means a longer heating time and a shorter cooling time in the welding process.

Figure 2 shows the typical cross-section of the weld. The weld width (L_u, L_l), the fusion zone area (A ($A = A_u + A_s + A_l$), A_s), the reinforcement height (H_u, H_l), and the reinforcement height-to-width ratio (r_u, r_l) were used to evaluate the weld shape. A smaller reinforcement height-to-width ratio means a better spreadability of the filler metal [17].

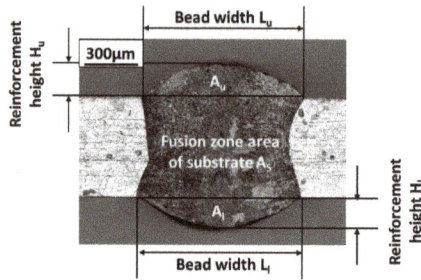

Figure 2. Cross-section of weld bead of pulsed laser welding with filler wire.

3. Results and Discussion

When the pulse duration was shorter than 4 ms or the frequency was lower than 50 Hz, the welds showed lack of penetration defects, as shown in Figure 3a,c. When the pulse duration was longer than 8 ms or the frequency was higher than 90 Hz, a very irregular outline and even undercut sometimes occurred on the upper surface of the weld, as shown in Figure 3b,d. The irregular upper surfaces indicate that the weld pool was unstable during the welding process. These welds are not included in further discussion.

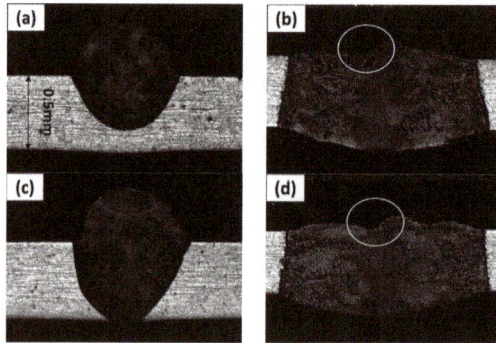

Figure 3. Defects of the welds: (**a**) $\tau = 3$ ms; (**b**) $\tau = 9$ ms; (**c**) $f = 40$ Hz; (**d**) $f = 100$ Hz.

3.1. Weld Bead Geometry

Figure 4 shows the cross-sections of the weld with different pulse duration. The laser power was 75 W, the welding speed and the wire feeding speed were both 350 mm/min, and the pulse frequency was 60 Hz. The weld beads showed a smooth outline, except when pulse duration = 4 ms. Most of the filler metal was distributed on the upper side. The weld width was too small for the spread of the filler metal, and an unsmooth surface was generated, as shown in Figure 4a.

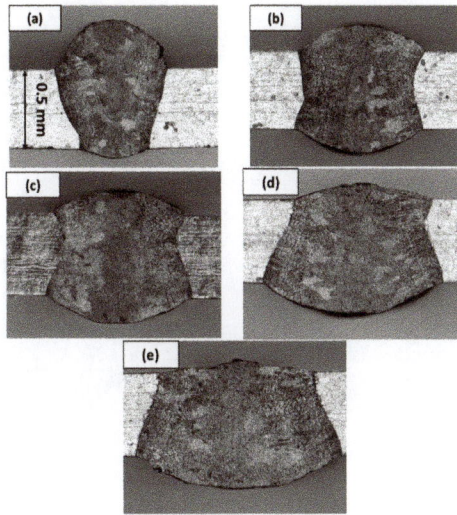

Figure 4. Cross-sections of weld bead under different pulse duration: (**a**) $\tau = 4$ ms; (**b**) $\tau = 5$ ms; (**c**) $\tau = 6$ ms; (**d**) $\tau = 7$ ms; (**e**) $\tau = 8$ ms.

Figure 5 shows the cross-sections of the welds with different pulse frequencies. The laser power was 75 W, the welding speed and wire feeding speed were both 350 mm/min, and the pulse duration was 6 ms. The welds showed smooth outline and were free from defects. Different from the case when $\tau = 4$ ms, the unsmooth upper surface appeared when the weld was quite wide, as shown in Figure 5e. This phenomenon may be related to surface tension effects in a relatively wide weld pool to the thickness of the substrate [18].

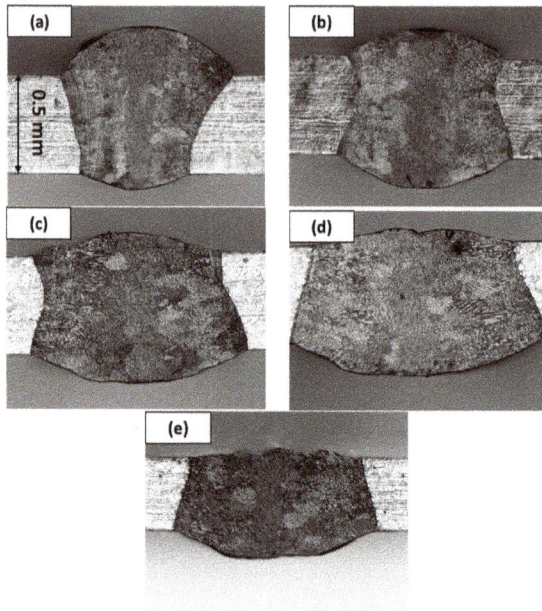

Figure 5. Cross-section of weld bead under different pulse frequency: (**a**) $f = 50$ Hz; (**b**) $f = 60$ Hz; (**c**) $f = 70$ Hz; (**d**) $f = 80$ Hz; (**e**) $f = 90$ Hz.

Figures 6a and 7a show the variations of the weld fusion zone area with pulse duration and pulse frequency, respectively. The fusion area of the weld increased linearly with the increase in the pulse duration or the pulse frequency. The welding speed and the wire feeding speed were constants, so the increase in the fusion area led to an increase in the fusion quantity of the base metal.

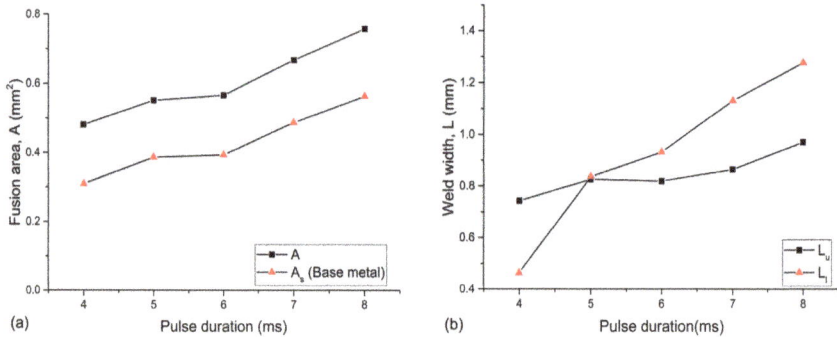

Figure 6. *Cont.*

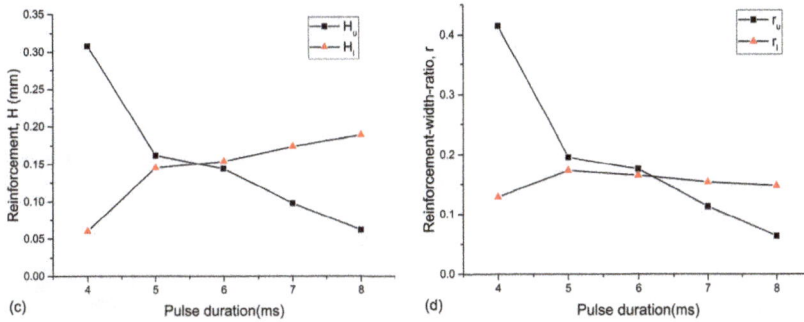

Figure 6. Variations of weld bead parameters with pulse duration: (**a**) fusion zone area; (**b**) weld width; (**c**) reinforcement height; (**d**) reinforcement height-to-width ratio.

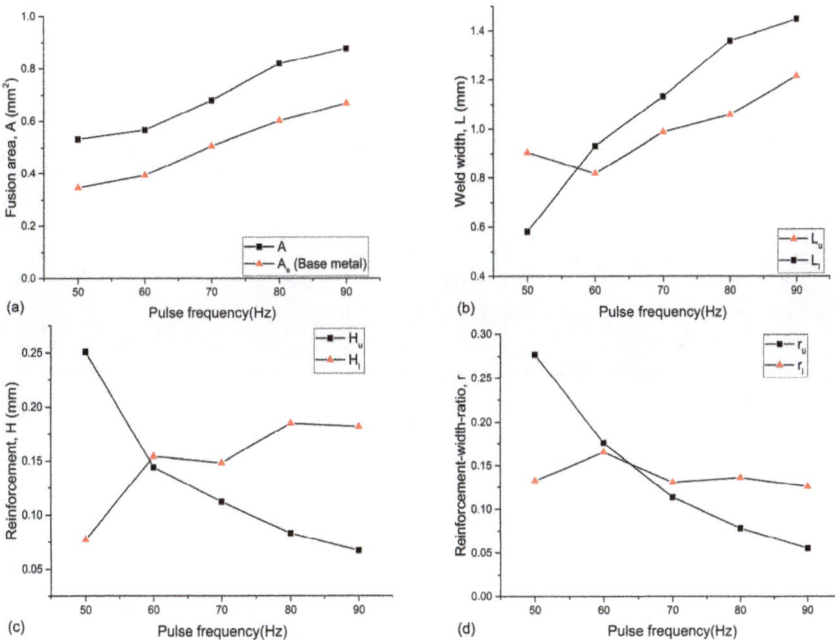

Figure 7. Variations of weld bead parameters with pulse frequency: (**a**) fusion zone area; (**b**) weld width; (**c**) reinforcement height; (**d**) reinforcement height-to-width ratio.

Figures 6b and 7b show the variations of the weld width with pulse duration and frequency. The width of the lower surface was more sensitive to the pulse parameters. The weld width to the pulse duration or frequency curve of the lower surface had a higher slope than that of the upper surface. The lower surface was wider than the upper surface when the pulse duration was longer than 5 ms or the frequency was higher than 60 Hz. When the frequency was 50 Hz, most of the filler metal solidified on the upper surface of the substrate, and the spreadability of the filler metal was good when the pulse duration was 6 ms, so the width of the upper surface was larger than that of 60 Hz, as shown in Figure 7b.

The variations in the reinforcement height with pulse duration and frequency are shown in Figures 6c and 7c. With a longer duration or higher frequency, the molten pool was wider and had a longer existence time, which was beneficial to the downward flow and the spreading of the filler metal. The reinforcement height of the upper surface reduced significantly at first then changed to linearly with a lower speed. The welding speed and the wire feeding rate were kept constant, meaning that the fusion area of the reinforcements ($A_u + A_l$) was a constant. Therefore, the variation of the reinforcement height of the lower surface had an opposite trend.

Figures 6d and 7d show that the reinforcement-to-width ratio of the upper surface decreases with the increase in pulse duration or pulse frequency. The upper reinforcement height decreased, and the upper surface widened with the increased pulse duration or pulse frequency. The reinforcement-to-width ratio of root reinforcement increased at first and then reduced slowly with the increase in the pulse duration or pulse frequency, and the inflection point of the r curve was near the point where the width of the upper surface was equal to that of the lower surface. Both the height and width of the root reinforcement increased with the increase in pulse duration and frequency, but the width increased with higher speed. Therefore, the reinforcement-to-width ratio decreased. The reinforcement-to-width ratio can be used to describe the spreadability of the filler metal. Define $\theta = \arctan(2H/L)$ as the contact angle of the reinforcement. Generally, a lower reinforcement-to-width ratio means a better spreadability of the filler metal and a lower contact angle between the reinforcement and the base plate, and it is beneficial for the service time of the welded joint when being used with alternate load or in an erosion-corrosion environment.

The pulse duty ratio is defined as the ratio of pulse duration-to-pulse period. The higher the duty ratio, the longer the heating time is in the welding process, compared with the cooling time. The pulse intensity decreases and the duty ratio increases with the increase in the pulse duration when the average power and pulse frequency are fixed. The decreased pulse intensity reduces the highest temperature of the molten pool. However, the increased heating time enhances heat accumulation and increases the volume of the molten pool. The extended existence time and enlarged volume of the molten pool enhance the downward flow and the spreading of the filler metal. Thus, the weld widens and the root reinforcement gets larger with the increase in the pulse duration.

The pulse frequency has similar effects on weld geometry to those of pulse duration. The increase in the pulse frequency does not change the heating time, and the pulse energy and intensity reduce with the increase in frequency. However, the cooling time reduces with the increase in the frequency; thus, the duty ratio is increased.

The duty ratios of the welding parameters in Table 2 were calculated with Equation (3). The variations in fusion zone area (A) with laser pulse duty ratio (α) are shown in Figure 8. It shows that, by increasing the pulse duration or the frequency, the variations in the fusion zone area to duty ratio change in the same way.

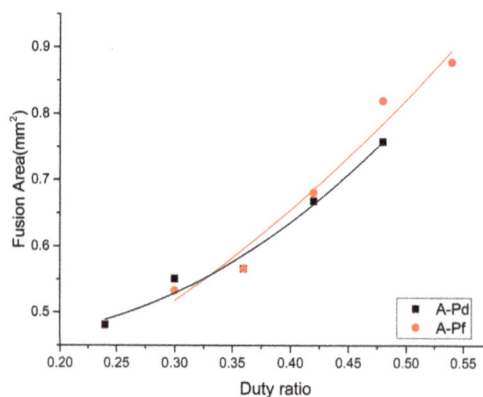

Figure 8. Variations of fusion zone area with laser pulse duty ratio.

According to the analyses above, the least-square polynomial fit was used to build the relation between the fusion zone area and duty ratio. The relationship between A and α is shown in Equation (4), and the prediction curve is given in Figure 9.

$$A = 0.48 - 0.59\alpha + 2.48\alpha^2 \quad (0.25 < \alpha < 0.55). \tag{4}$$

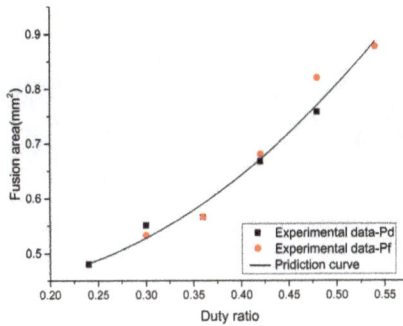

Figure 9. Relationship between fusion zone area and duty ratio.

This indicates that, for the pulsed laser welding of a 0.5 mm Hastelloy C-276 thin sheet with filler wire at a fixed total heat input, the volume of the molten pool during the welding process is dominated by the duty ratio. As mentioned above, the variation in the fusion area shows the variation in the fusion quantity of the base metal. Thus, the dilution of the base metal can be obtained by this relationship. This will be useful when control of the element composition in the fusion zone is needed. The geometry of the weld bead, especially the distribution of the filler metal, is closely related to the duty ratio. Therefore, the weld geometry can be approximately predicted and manipulated by adjusting the duty ratio.

3.2. Microstructure

When the total heat input is fixed, the pulse parameters determine the weld geometry by influencing the thermal process in the welding process. The pulse parameters have significant effects on the microstructure of the weld.

The microstructure of the weld with pulse durations (4, 6, and 8 ms) are shown in Figure 10a–c, and the microstructure of the weld with the pulse frequency is shown in Figure 10d–f. The microstructure of the base metal consists of equiaxed grains, and annealing twins are observed. From the fusion line to the weld joint center, the microstructure of the fusion zone changed from a small quantity of planar crystal, columnar dendrites to equiaxed dendrites. For the rapid cooling rate in pulsed laser welding, the growth of the planar crystal was restrained, as shown in Figure 10a. With the increase in the pulse duration and the pulse frequency, the components of columnar dendrites were increased, and the microstructure showed a coarsening trend.

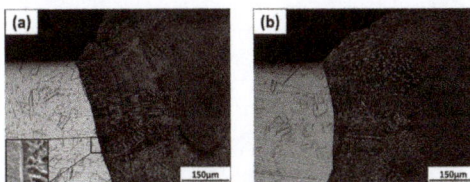

Figure 10. *Cont.*

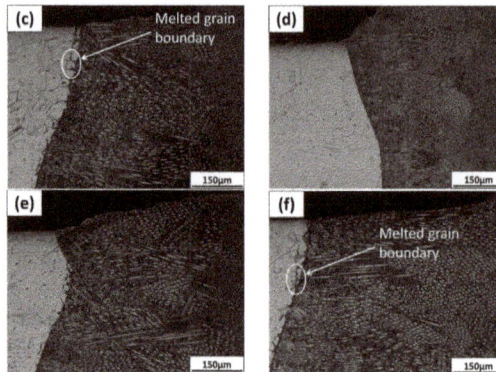

Figure 10. Variations of microstructure of the weld with pulse duration and frequency: (**a**) τ = 4 ms; (**b**) τ = 6 ms; (**c**) τ = 8 ms; (**d**) f = 50 Hz; (**e**) f = 70 Hz; (**f**) f = 90 Hz.

With a high duty ratio, although no typical characteristic of the heat affect zone, such as an increase in the grain size of the base metal near the fusion line, was found, the base metal near the fusion line showed a tendency of grain boundary liquation. According to the study by Lippold et al. [19], due to the epitaxial growth of the fusion zone grains, grain boundaries in the base metal near the fusion line are contiguous with the solidification grain boundaries in the fusion zone. In the fusion zone metal solidification process, some of the low-melting-point solute may segregate toward the grain boundary. The solute or impurity elements can be transported down the boundary pipeline into the base metal grain boundaries near the fusion line. The solute or impurity elements will decrease the melting point of the grain boundaries, resulting in grain boundary liquation. The grain boundary liquation area, which has also been called the partially melted zone, can lead to liquation cracking, loss of ductility, and hydrogen cracking [20]. It may be a weakness of the weld.

Figures 11 and 12 show the morphology of the columnar dendrites and equiaxed dendrites in the fusion zone with different pulse durations and pulse frequencies. The magnified locations are shown in Figures 11b and 12b. The primary dendrite arm spacing and second dendrite arm spacing increase with the pulse duty ratio. The average equiaxed dendrites grain size in the fusion zone center increases from about 4 µm to 7 µm when the pulse duration increases from 4 ms to 8 ms, and increases from about 5 µm to 9 µm when the pulse frequency increases from 50 Hz to 90 Hz. The microstructure of the fusion zone is dominated by temperature gradient G and growth rate R. Generally, the growth rate R is related to the welding speed V, $R \cdot \cos(\alpha - \beta) = V \cdot \cos\alpha$ [19], α is the angle between the welding direction and the normal of the molten pool boundary, and β is the angle between the welding direction and the growth direction of the dendrite at that point. The welding speed is kept constant, so there is no large variation of R when the duty ratio is changed. The peak power reduces with the increase in the duty ratio, so the temperature of the molten pool center reduces. At the same time, the volume of the molten pool increases. Thus, the temperature gradient G of the molten pool reduces with the increase in the duty ratio. The decreasing value of $G \cdot R$ results in the coarsening of the microstructure of the weld.

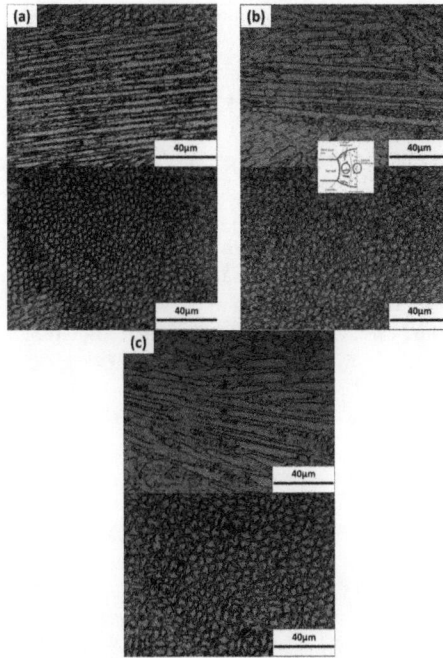

Figure 11. Variations of grain size with pulse duration: (**a**) τ = 4 ms; (**b**) τ = 6 ms; (**c**) τ = 8 ms.

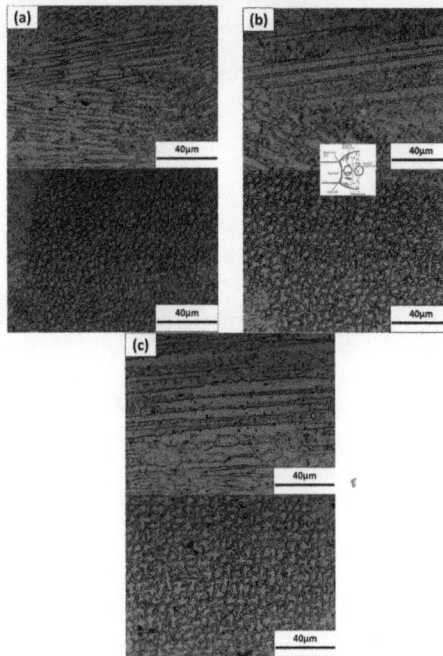

Figure 12. Variations of grain size with pulse frequency: (**a**) f = 50 Hz; (**b**) f = 70 Hz; (**c**) f = 90 Hz.

In order to reveal the effects of the pulse parameters on the microhardness distributions of the weld joint, hardness tests were done on the cross-section of the joints, as shown in Figure 13. The microhardness profiles for the welds with 4, 6, and 8 ms are shown in Figure 13a. The microhardness profiles for the welds with 50, 70, and 90 Hz are shown in Figure 13b. No decrease in hardness was found in the weld; the hardness of the fusion zone had the same value as the base metal. The grain refinement increases the hardness of the weld. At the same time, the welding process may weaken the solution strengthening of the base metal and result in a tendency to decrease in hardness. These two effects indicated that no decrease in hardness was found in the weld. Moreover, no difference was found in the hardness profiles of the welds. The average hardness of fusion zone and base metal both were approximately 250 HV0.1. These results show that, although the increase in duty ratio will cause the coarsening of the fusion zone microstructure, there will be no significant hardness loss in the fusion zone. The grain boundary liquation area is quite narrow, and it is impossible to distinguish it on a polished surface during hardness tests, so its microhardness data is not available.

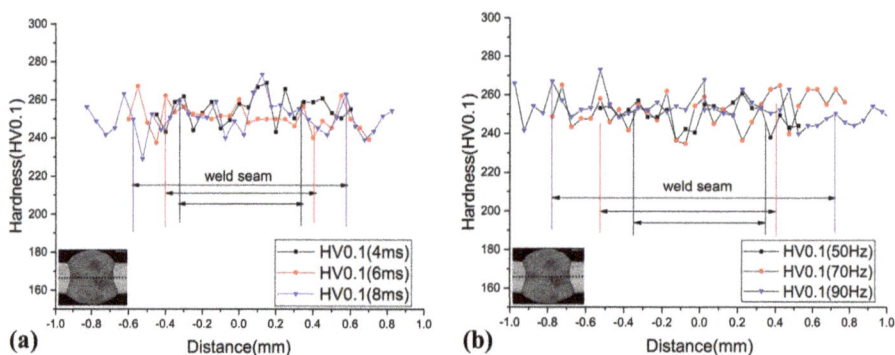

Figure 13. Variations of microhardness of the weld with pulse duration and frequency: (**a**) Microhardness profiles for the welds with different pulse durations; (**b**) Microhardness profiles for the welds with different pulse frequencies.

4. Conclusions

A 0.5-mm-thick Hastelloy C-276 thin sheet was welded via pulsed laser with filler wire. The effects of the pulse parameters on the weld bead geometry and microstructure were investigated when the total heat input was fixed.

The pulse duration and pulse frequency have similar effects on the weld bead geometry. The duty ratio dominated by pulse duration and frequency can be used to predict and control the weld bead geometry.

The distribution of the filler metal between the upper and lower surface influences the weld width. The lower surface is wider than the upper surface when the lower reinforcement is larger than the upper reinforcement. With the increase in duty ratio, the upper surface reinforcement height-to-width ratio decreases, while the lower surface reinforcement height-to-width ratio increases at first and then decreases slowly. The increase in the duty ratio is beneficial for the spreadability of the filler metal and reduces the contact angle of the reinforcement.

The relationship between the fusion zone area and the duty ratio is given by $A = 0.48 - 0.59\alpha + 2.48\alpha^2$ ($0.25 < \alpha < 0.55$). The dilution rate of the base metal can be obtained by this equation, and this will be useful when control of the element composition in the fusion zone is needed.

The results show that, although the increase in duty ratio will cause the coarsening of the fusion zone microstructure, there is no significant hardness loss in the fusion zone. With a larger duty ratio, the base metal near the fusion line shows a tendency towards grain boundary liquation, and the grain boundary liquation area may reduce the performance of the weld bead, so an overly large duty ratio needs to be avoided.

The melting state of the molten pool and filler wire between each pulse is unknown. Further study will focus on direct observation of the molten pool development and the melting mechanism of the filler wire with different pulse parameters.

Acknowledgments: The authors would like to acknowledge the financial support from the National Key Basic Research Program of China (No. 2015CB057305), the Science Fund for Creative Research Groups (No. 51321004), and the National Nature Science Foundation of China (No. 51402037).

Author Contributions: Dongsheng Chai and Dongdong Wu conceived the experiments; Guangyi Ma and Siyu Zhou contributed by analyzing data and revising the writing; Zhuji Jin and Dongjiang Wu supervised the work and further analyzed the data. Dongsheng Chai wrote the article.

Conflicts of Interest: The authors declare no conflict of interest.

References

1. Zhang, Q.; Tang, R.; Yin, K.; Luo, X.; Zhang, L. Corrosion behavior of Hastelloy C-276 in supercritical water. *Corros. Sci.* **2009**, *51*, 2092–2097. [CrossRef]

2. Hashim, M.; Babu, K.E.S.R.; Duraiselvam, M.; Natu, H. Improvement of wear resistance of Hastelloy C-276 Through laser surface melting. *Mater. Des.* **2013**, *46*, 546–551. [CrossRef]

3. Cieslak, M.J.; Headley, T.J.; Romig, A.D. The welding metallurgy of Hastelloy alloys C-4, C-22, and C-276. *Metall. Trans. A* **1986**, *17*, 2035–2047. [CrossRef]

4. Yue-sheng, L.X.; Xia-hui, L.Y. Study on microstructure and properties of welded joint of extra thin Ni-based alloy plate. *J. Plast. Eng.* **2011**, *18*, 91–97.

5. Manikandan, M.; Arivazhagan, N.; Rao, M.N.; Reddy, G.M. Microstructure and mechanical properties of alloy C-276 weldments fabricated by continuous and pulsed current gas tungsten arc welding techniques. *J. Manuf. Process.* **2014**, *16*, 563–572. [CrossRef]

6. Manikandan, M.; Arivazhagan, N.; Rao, M.N.; Reddy, G.M. Improvement of microstructure and mechanical behavior of gas tungsten arc weldments of alloy C-276 by current pulsing. *Acta Metall. Sin.* **2015**, *28*, 208–215. [CrossRef]

7. Manikandan, M.; Sasikumar, P.; Arul Murugan, B.; Sathishkumar, M.; Arivazhagan, N. Microsegregation studies on pulsed current gas tungsten arc welding of alloy C-276. *Int. J. Sci. Eng. Res.* **2015**, *6*, 33–38.

8. Ahmad, M.; Akhter, J.I.; Akhtar, M.; Iqbal, M.; Ahmed, E.; Choudhry, M.A. Microstructure and hardness studies of the electron beam welded zone of Hastelloy C276. *J. Alloy. Compd.* **2005**, *390*, 88–93. [CrossRef]

9. Van der Eijk, C.; Fostervoll, H.; Sallom, Z.K.; Akselsen, O.M. Plasma welding of NiTi to NiTi, stainless steel and Hastelloy C-276. In Proceedings of ASM Materials Solutions 2003 Conference, Pittsburgh, PA, USA, 13–15 October 2003.

10. Moradi, M.; Ghoreishi, M. Influences of laser welding parameters on the geometric profile of Ni-base superalloy Rene 80 weld-bead. *Int. J. Adv. Manuf. Technol.* **2011**, *55*, 205–215. [CrossRef]

11. Ma, G.; Wu, D.; Chai, D.; Guo, Y.; Guo, D. Near-free-defect laser welding of AP1000 nuclear reactor coolant pump can. *J. Mech. Eng.* **2015**, *51*, 1–8. [CrossRef]

12. Wu, D.J.; Ma, G.Y.; Niu, F.Y.; Guo, D.M. Pulsed laser welding of Hastelloy C-276: High-temperature mechanical properties and microstructure. *Mater. Manuf. Process.* **2013**, *28*, 524–528. [CrossRef]

13. Ma, G.; Wu, D.; Niu, F.; Zou, H. Microstructure evolution and mechanical property of pulsed laser welded Ni-based superalloy. *Opt. Laser. Eng.* **2015**, *72*, 39–46. [CrossRef]

14. Ventrella, V.A.; Berretta, J.R.; de Rossi, W. Pulsed Nd: YAG laser welding of Ni-alloy Hastelloy C-276 foils. *Phys. Proced.* **2012**, *39*, 569–576. [CrossRef]

15. Hao, K.; Li, G.; Gao, M.; Zeng, X. Weld formation mechanism of fiber laser oscillating welding of austenitic stainless steel. *J. Mater. Process. Technol.* **2015**, *225*, 77–83. [CrossRef]

16. Ma, G.Y.; Wu, D.J.; Guo, Y.Q.; Gao, Z.M.; Guo, D.M. Tensile Properties of Weld Joint on Thin Hastelloy C-276 Sheet of Pulsed Laser Welding. *Rare Metal Mater. Eng.* **2013**, *42*, 1241–1245.

17. Qin, G.L.; Lei, Z.; Lin, S.Y. Effects of Nd: YAG laser + pulsed MAG arc hybrid welding parameters on its weld shape. *Sci. Technol. Weld. Join.* **2007**, *12*, 78–86. [CrossRef]

18. Kim, J.; Kim, S.; Kim, K.; Jung, W.; Youn, D.; Lee, J.; Ki, H. Effect of beam size in laser welding of ultra-thin stainless steel foils. *J. Mater. Process. Technol.* **2016**, *233*, 125–134. [CrossRef]
19. Lippold, J.C.; Baeslack, W.A.; Varol, I. Heat-affected zone liquation cracking in austenitic and duplex stainless steels. *Weld. J.* **1992**, *71*, 1–14.
20. Kou, S. *Welding Metallurgy*, 2nd ed.; John Wiley & sons: Hoboken, NJ, USA, 2002.

metals

MDPI

Article

Influences of Laser Spot Welding on Magnetic Property of a Sintered NdFeB Magnet

Baohua Chang [1,*]**, Dong Du** [1]**, Chenhui Yi** [1]**, Bin Xing** [1] **and Yihong Li** [2]

[1]   State Key Laboratory of Tribology, Department of Mechanical Engineering, Tsinghua University,
     Beijing 100084, China; dudong@tsinghua.edu.cn (D.D.); yich06@163.com (C.Y.); xingb13@126.com (B.X.)
[2]   Taiyuan Tuolituo Technology Co. Ltd., Taiyuan 030032, China; liyihongtongli@aliyun.com
*   Correspondence: bhchang@tsinghua.edu.cn; Tel.: +86-10-6278-1182

Academic Editor: Giuseppe Casalino
Received: 11 May 2016; Accepted: 22 August 2016; Published: 26 August 2016

Abstract: Laser welding has been considered as a promising method to join sintered NdFeB permanent magnets thanks to its high precision and productivity. However, the influences of laser welding on the magnetic property of NdFeB are still not clear. In the present paper, the effects of laser power on the remanence (B_r) were experimentally investigated in laser spot welding of a NdFeB magnet (N48H). Results show that the B_r decreased with the increase of laser power. For the same welding parameters, the B_r of magnets, that were magnetized before welding, were much lower than that of magnets that were magnetized after welding. The decrease in B_r of magnets after laser welding resulted from the changes in microstructures and, in turn, the deterioration of magnetic properties in the nugget and the heat affected zone (HAZ) in a laser weld. It is recommended that the dimensions of nuggets and HAZ in laser welds of a NdFeB permanent magnet should be as small as possible, and the magnets should be welded before being magnetized in order to achieve a better magnetic performance in practical engineering applications.

Keywords: laser welding; NdFeB magnet; magnetic property; process parameters

1. Introduction

NdFeB rare earth permanent magnets have been widely used in many industrial fields, such as aeronautics, astronautics, automotive, appliance, computers, and communications, thanks to their excellent magnetic property [1–3]. Sintered NdFeB permanent magnets are brittle and with poor mechanical properties. Therefore, in practical applications, the magnets are often joined with other materials using adhesive bonding or mechanical joining methods, which, however, have lower productivity and are hard to be applied on miniature structures [4–6].

Due to the high quality, high efficiency, and unrelenting repeatability, lasers have been widely applied in many materials processing processes, such as welding [7–13], cutting [14–16], additive manufacturing [17–20], and so on, and they have been used to process many types of materials, including metals, ceramics, glass, and polymers. Owing to the high power density, the laser generally results in welds with high precision and a small heat affected zone (HAZ) and, therefore, is considered as a good candidate to join the tiny sintered NdFeB components. However, studies carried out so far on the laser welding of sintered NdFeB magnet are still very limited. Microstructures and mechanical behavior have been studied for laser welds of NdFeB magnets [21] and laser welds of dissimilar materials of a NdFeB magnet and mild steel [22]. Nevertheless, it is still not clear how the laser welding will affect the magnetic property and how the optimal magnetic performance can be obtained. As we know, the magnetic property and performance are critical for the NdFeB permanent magnetic functional material in engineering applications, which makes it stringent to understand the change in the magnetic response of NdFeB when subject to laser welding processing.

In this paper, the influences of laser power on the magnetic property in terms of remanence are studied experimentally in micro-laser spot welding of an NdFeB permanent magnet, and the causes of the influences have been analyzed.

2. Materials and Methods

Sintered powder NdFeB permanent magnets N48H, without coating on surfaces, were used in this study. The compositions of the material are listed in Table 1. The Nd-Fe-B alloy particles with diameters of 3–5 microns were pressed and then sintered at 1060 °C for 2 h under vacuum conditions, followed by two tempering processes, which are 900 °C for 2 h, and 500 °C for 2 h, respectively. The microstructures of the magnet are shown in Figure 1, from which it can be seen that the main magnetic phase ($Nd_2Fe_{14}B$) of the magnet is gray and distributed non-uniformly, the white parts are the Nd-rich phase existing along grain boundaries or at intersections of grain boundaries of the main phase, and there exist certain amounts of voids which formed in the powder metallurgy process. The magnets were cut into specimens with dimensions of $7.5 \times 3.5 \times 0.7$ mm.

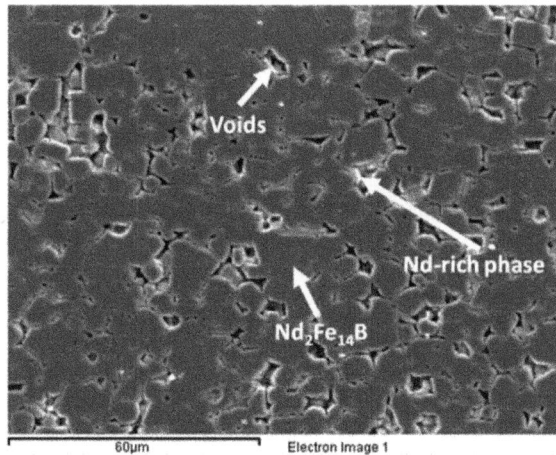

Figure 1. Microstructure of a NdFeB permanent magnet.

Table 1. Chemical compositions of the sintered NdFeB permanent magnet N48H (wt %).

Elements	Nd	Fe	B	Pr	Dy	Co	Cu	Nb
Contents	20.63	66.75	1.00	6.88	2.99	1.50	0.15	0.10

An IPG Photonics YLS-2000 (2 kW) (IPG Photonics, Oxford, MS, USA) type continuous waveform (cw) fiber laser was used in the welding trials, which had a wavelength of 1064 nm, and maximum power output of 2 kW. A schematic of welding setup is shown in Figure 2. Laser spot welds were made on the specimens at different laser powers, with a defocusing distance of +1 mm. The surfaces of the magnets were ground prior to welding to remove oxide layer, and 99.99% high-purity argon was used to protect the spot welds from oxidation, with a flow rate of Ar of 10 L/min. After welding, the spot welds were mounted, ground, polished, and then chemically etched with 4% nitric acid alcohol. Optical microscopy (Olympus, Tokyo, Japan) was employed to observe the shapes and dimensions of weld spots, and scanning electronic microscopy (SEM) (Hitachi, Tokyo, Japan) was used to analyze the microstructures of spot welds.

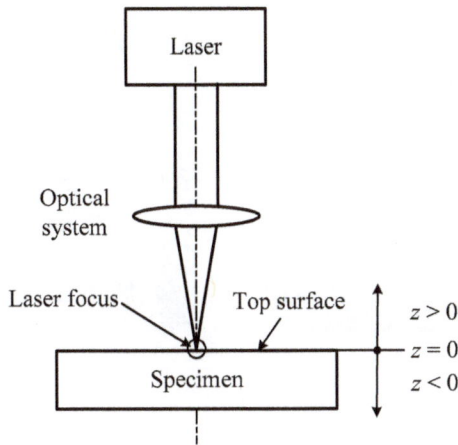

Figure 2. Schematic setup of laser spot welding of sintered NdFeB.

The remanence, Br, is one of the main parameters used to characterize the magnetic property of materials. In this study, the magnetic fluxes generated in coils by magnets were measured using a TA102E-1 type maxwellmeter (MKY, Beijing, China), in order to quantify the variation in remanence after laser welding. Two sequences were used in the present study. In the first sequence, the magnets were firstly magnetized, then spot welded by laser, and finally the remanences were measured; in the second sequence, the magnets were firstly welded by laser, then magnetized, and finally the measurements of remanences were carried out.

3. Results

3.1. Effects of Laser Power on Weld Dimensions

During laser welding of the sintered NdFeB, the materials with temperatures higher than the melting point (1180 °C) of the main phase $Nd_2Fe_{14}B$ will completely melted and form weld pools, and upon cooling, will finally form a nugget in a weld. The materials with temperatures ranging between 655 °C (melting point of Nd-rich phase) and 1180 °C will only partially melted, i.e., the Nd-rich phase will melt, while the main phase will not. This part of the material forms the heat affected zones (HAZ) in a weld spot. For materials with temperatures lower than 655 °C, melting will not take place, and the base metal of magnet is not affected as far as the microstructure is concerned. The constitution of a laser spot weld of the sintered NdFeB is shown in Figure 3a, in which two geometric parameters, i.e., surface diameter D_s and penetration depth P_n, are defined to characterize the dimensions of the spot weld. From Figure 3a, two different regions can be noticed in the nugget and, based on previous study [21], it has been found that they represent the columnar grain zone along fusion line (the interface between nugget and HAZ), and the equiaxed grain zone in the middle part of the nugget, respectively. The dashed lines in Figure 3a represent the boundaries between the nugget, HAZ, and base metal. Figure 3b more clearly shows the three distinct regions in a laser weld. The boundary between the base metal and the HAZ can be identified by whether the Nd-rich phase is melted or not.

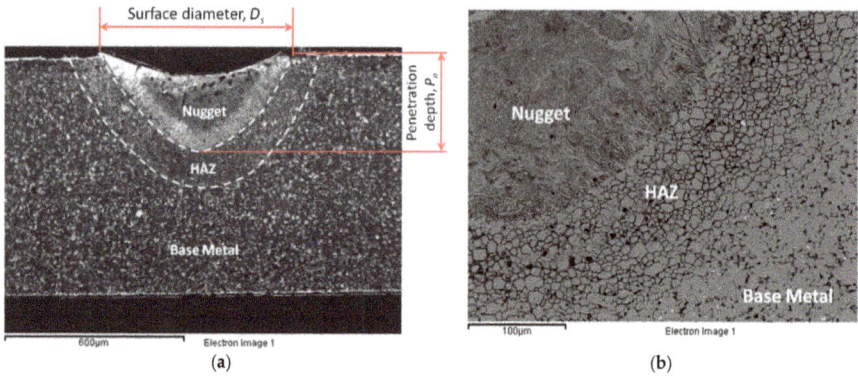

Figure 3. Constitution of a laser spot weld of the sintered NdFeB: (**a**) Definition of characteristic dimensions; (**b**) Three typical regions in a laser weld.

Dimensions of the weld spots produced with different laser powers ranging from 80 W to 180 W are given in Figure 4, from which it can be seen that the penetration depths increase monotonically for increased laser power, and a full penetration (0.7 mm) spot weld is obtained with a laser power of 180 W. The surface diameters of weld spots increase when increase the laser power from 80 W to 160 W, while a slight decrease in surface diameter can be noted with a further increase in the laser power from 160 W to 180 W. This can be attributed to the transition from partial penetration to full penetration of welds, which promotes the heat transfer in depth direction while reducing it in the radial direction. This phenomenon has also been reported in continuous laser welding trials with aluminum alloys [23].

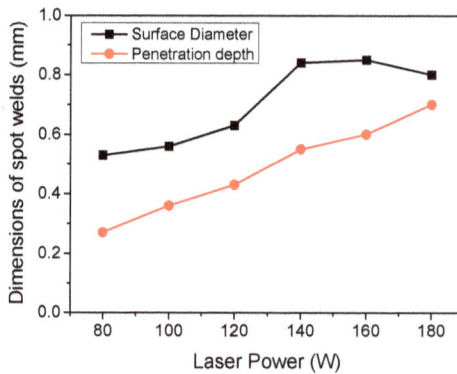

Figure 4. Dimensions of spot welds fabricated with different laser powers (weld time: 0.05 s, defocusing distance: +1 mm).

3.2. Effects of Laser Power on Magnetic Property

Figure 5 presents the effect of laser power on the magnetic flux. It can be seen that for the magnets welded before being magnetized, the magnetic flux decreased slightly for increased laser power. The minimum magnetic flux of 142.3 mWb was obtained at a laser power of 180 W, which was 6.5% lower than that of the magnets not welded, which was 152.2 mWb. For the magnets welded after being magnetized, the magnetic flux also decreased with the increasing of laser power, and the decrease was much more significant than the magnets welded before being magnetized. It can be seen that a 45.0% reduction in magnetic flux was caused by laser welding with a laser power of 180 W, resulting in a magnetic flux of 83.7 mWb.

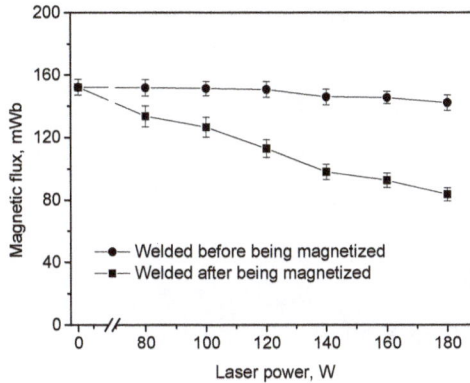

Figure 5. Effect of laser power on magnetic flux.

4. Discussion

4.1. Characteristics of Microstructures of a Laser Spot Weld

The cross-sections of laser spot welds of the magnet were observed by SEM. Similar microstructure features were found for welds made with different laser powers, despite the different dimensions of nuggets and heat affected zones, as indicated in the preceding section. As a representative, the microstructure of a partial penetrated spot weld (shown in Figure 6a) is presented here. Figure 6b demonstrates the details of section A of the nugget shown in Figure 6a. It can be seen from Figure 6b that the nugget of a NdFeB spot weld is composed mainly by ultra-fine grains of isotropic $Nd_2Fe_{14}B$, with disordered orientations. The different grain orientations of different regions are the result of epitaxial growth of columnar grains from fusion line, in addition to the growth of equiaxed grains in all directions during solidification [21]. According to the Stoner-Wohlfarth model [3], the remanence (Br) of isotropic NdFeB permanent magnets with a single easy magnetization axis is one half of the saturation magnetization (Bs) of the magnet. In contrast, the Br of anisotropic NdFeB permanent magnets with single easy magnetization axis is approximately equal to Bs. Obviously, the change in microstructure of nugget region from anisotropic to isotropic due to the melting and solidification during laser welding will deteriorate the magnetic property of the magnet.

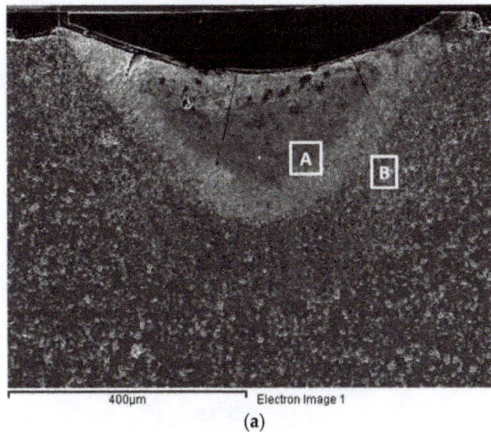

(a)

Figure 6. *Cont.*

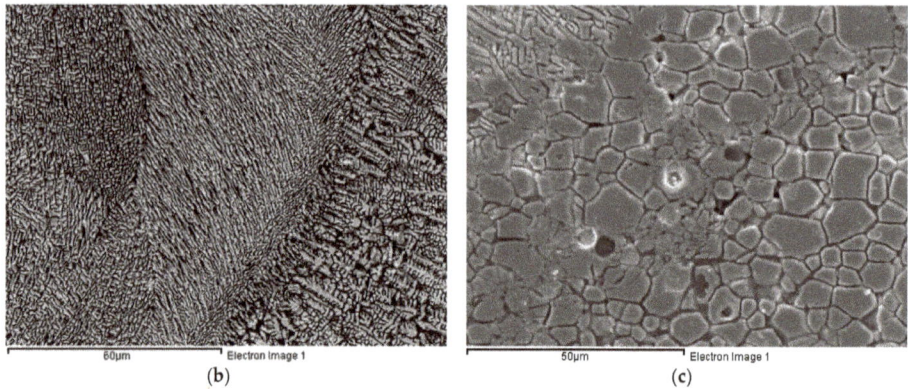

(b)   (c)

Figure 6. The cross-section of laser spot weld (**a**); and the microstructures of nugget (**b**) and HAZ (**c**) in a laser spot weld.

Figure 6c presents the details of the heat affected zone (section B in Figure 6a). It can be seen that the melting of Nd-rich phase at grain boundaries will result in micro gaps and consequently weaken the bonding between grains of main phase $Nd_2Fe_{14}B$. Some main phase grains may be broken into small pieces, which may rotate when the melted Nd-rich phase flows locally. In this way, the orientation of the easy magnetization axes of these grains will deviate from their original orientation, which may result in the decrease in the degree of orientation and, in turn, the deterioration of the magnetic property.

4.2. Characteristics of Thermal Demagnetization of NdFeB

According to the experimental studies on the magnetic domains and magnetic property of NdFeB [24,25], the thermal demagnetization behavior of a NdFeB permanent magnet can be described with the curve shown in Figure 7. The magnetic property is basically unchanged for temperatures lower than the maximum working temperature (about 120 °C); for temperatures of 120 °C $< T <$ 240 °C, the magnetic property decreases slightly; for $T >$ 240 °C, the magnetic property decreases dramatically, and when the temperature is greater than about 300 °C (Curie temperature of NdFeB), the magnetic property disappears completely.

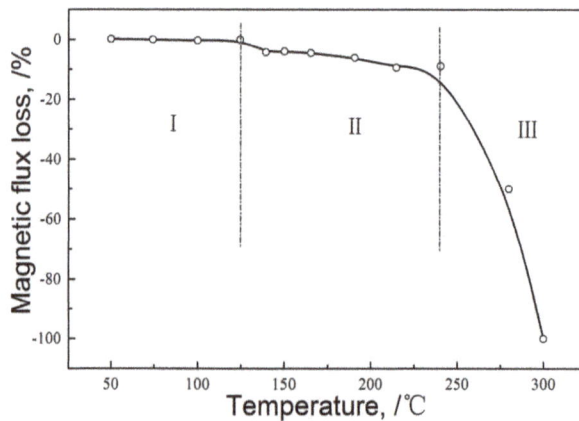

Figure 7. Thermal demagnetization curve of NdFeB permanent magnets [24,25].

4.3. Divisions of Regions in a Laser Spot Weld of NdFeB

Based on the discussion above, a spot weld of the NdFeB magnet can be divided into five regions, as shown in Figure 8.

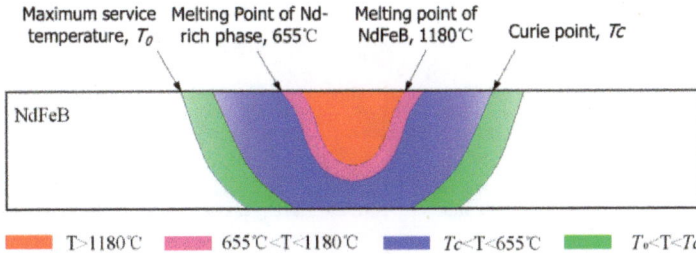

Figure 8. Schematic of temperature distribution and different regions in a spot weld of NdFeB permanent magnet.

The first region is nugget region that experiences temperatures above the melting point of magnet (1180 °C), which has lower magnetic property than the base metal due to its isotropic ultra-fine solidification structures, and the magnetic property of this region is not recoverable by magnetization.

The second region is heat affected zone, which experiences temperatures from the melting point of Nd-rich phase (655 °C) to that of base metal (1180 °C). It has lower magnetic property than base metal because of the melting of Nd-rich phase and in turn the weakened bonding between main phase grains. Similar to the nugget region, the magnetic property of this region is also not recoverable by magnetization.

The third region experiences temperatures ranging from the Curie point of the magnet (about 300 °C) to the melting point of the Nd-rich phase (655 °C) during the welding process. The magnetic property of this region will be reduced to zero if the magnet is magnetized, and can be fully recovered by re-magnetization after welding because the microstructure is not affected by welding.

The fourth region has temperatures ranging from the maximum working temperature (about 120 °C) to the Curie point of the magnet (about 300 °C) in laser welding. The magnetic property of this region will be partially reduced if the magnet is magnetized, and can be fully recovered by re-magnetization after welding.

The last region is the part with temperatures lower than 120 °C during welding, both microstructure and magnetic property of this region are not influenced by welding.

For magnets laser welded after being magnetized, the magnetic property of regions with temperatures higher than the maximum working temperature 120 °C (regions 1–4 as discussed above) is affected, which lead to a notable decrease in magnetic flux as shown in Figure 5. In contrast, for those magnets laser welded before being magnetized, although the magnetic properties of regions 1–4 may also be affected by welding, the magnetic properties of region 3 and 4 can be fully recovered. As a result, the deterioration of magnetic property is much less than those welded after magnetization. In order to obtain a better magnetic performance, it is recommended that the NdFeB permanent magnets should be welded before being magnetization in practical engineering applications, while the dimensions of nuggets and HAZ should be as small as possible.

5. Conclusions

(1) The magnetic property (in terms of Br) of NdFeB decreases with the increase of laser power. For the same welding parameters, the magnetic property of magnets that were magnetized before laser welding is much lower than that of magnets that were magnetized after laser welding.

(2)   The decrease in the magnetic property of magnets after laser welding results from the changes in microstructures and, in turn, the deterioration of magnetic properties in the nugget and the heat affected zone (HAZ) in a laser weld.

(3)   In order to obtain better magnetic performance, it is recommended that the NdFeB permanent magnets should be welded before being magnetized in practical engineering applications, while the dimensions of nuggets and HAZ should be as small as possible.

Acknowledgments: This research was supported by a Marie Curie International Incoming Fellowship with the 7th European Community Framework Programme (Grant No. 919487), whose financial support is gratefully appreciated.

Author Contributions: Baohua Chang and Dong Du conceived and designed the experiments; Chenhui Yi and Bin Xing performed the experiments and analyzed the data; Yihong Li contributed NdFeB materials used in the study; Baohua Chang wrote the paper.

Conflicts of Interest: The authors declare no conflict of interest.

References

1.   Croat, J.J.; Herbst, J.F.; Lee, R.W.; Pinkerton, F.E. Pr-Fe and Nd-Fe-based materials: A new class of high-performance permanent magnets. *J. Appl. Phys.* **1984**, *55*, 2078–2082. [CrossRef]

2.   Shield, J.E.; Zhou, J.; Aich, S.; Ravindran, V.K.; Skomski, R.; Sellmyer, D.J. Magnetic reversal in three-dimensional exchange-spring permanent magnets. *J. Appl. Phys.* **2006**, *99*, 08B508. [CrossRef]

3.   Zhou, S.Z.; Dong, Q.F. *Super Permanent Magnet: Rare Earth Iron Series Permanent Magnet*, 2nd ed.; Metallurgical Industry Press: Beijing, China, 2004.

4.   Je, S.S.; Rivas, F.; Diaz, R.E.; Kwon, J.; Kim, J.; Bakkaloglu, B.; Kiaei, S.; Chae, J. A compact and low-cost MEMS loudspeaker for digital hearing aids. *IEEE Trans. Biomed. Circuits Syst.* **2009**, *3*, 348–358. [CrossRef] [PubMed]

5.   Kim, H.J.; Kim, D.H.; Koh, C.S.; Shin, P.S. Application of polar anisotropic NdFeB ring-type permanent magnet to brushless DC motor. *IEEE Trans. Magn.* **2007**, *43*, 2522–2524. [CrossRef]

6.   Mosca, E.; Marchetti, A.; Lampugnani, U. Laser welding of PM Materials. *Powder Metall. Int.* **1983**, *15*, 115–118.

7.   Chang, B.H.; Allen, C.; Blackburn, J.; Hilton, P.; Du, D. Fluid Flow Characteristics and Porosity Behavior in Full Penetration Laser Welding of a Titanium Alloy. *Metall. Mater. Trans. B* **2015**, *46*, 906–918. [CrossRef]

8.   Casalino, G.; Campanelli, S.L.; Ludovico, A.D. Laser-arc hybrid welding of wrought to selective laser molten stainless steel. *Int. J. Adv. Manuf. Technol.* **2013**, *68*, 209–216. [CrossRef]

9.   Casalino, G.; Mortello, M. Modeling and experimental analysis of fiber laser offset welding of Al-Ti butt joints. *Int. J. Adv. Manuf. Technol.* **2016**, *83*, 89–98. [CrossRef]

10.   Kuryntsev, S.V.; Gilmutdinov, A.K. The effect of laser beam wobbling mode in welding process for structural steels. *Int. J. Adv. Manuf. Technol.* **2015**, *81*, 1683–1691. [CrossRef]

11.   Casalino, G.; Mortello, M.; Leo, P.; Benyounis, K.Y.; Olabi, A.G. Study on arc and laser powers in the hybrid welding of AA5754 Al-alloy. *Mater. Des.* **2014**, *61*, 191–198. [CrossRef]

12.   Kuryntsev, S.V.; Gilmutdinov, A.K. Heat treatment of welded joints of steel 0.3C-1Cr-1Si produced by high-power fiber lasers. *Opt. Laser Technol.* **2015**, *74*, 125–131. [CrossRef]

13.   Wu, S.J.; Gao, H.M.; Zhang, Z.Y. A preliminary test of a novel molten metal filler welding process. *Int. J. Adv. Manuf. Technol.* **2015**, *80*, 647–655. [CrossRef]

14.   Hilton, P.A.; Lloyd, D.; Tyrer, J.R. Use of a diffractive optic for high power laser cutting. *J. Laser Appl.* **2016**, *28*, 012014. [CrossRef]

15.   Kaakkunen, J.J.J.; Laakso, P.; Kujanpaa, V. Adaptive multibeam laser cutting of thin steel sheets with fiber laser using spatial light modulator. *J. Laser Appl.* **2014**, *26*, 032008. [CrossRef]

16.   Hilton, P.; Khan, A.; Walters, C. The laser alternative. *Nucl. Eng. Int.* **2010**, *55*, 18–20.

17.   Xing, B.; Chang, B.H.; Yang, S.; Du, D. A study on the cracking behaviour in laser metal deposition of IC10 directionally solidified nickel-based superalloy. *Mater. Res. Innov.* **2015**, *19*, 281–285. [CrossRef]

18.   Sexton, L.; Lavin, S.; Byrne, G.; Kennedy, A. Laser cladding of aerospace materials. *J. Mater. Proc. Technol.* **2002**, *122*, 63–68. [CrossRef]

19. Casalino, G.; Campanelli, S.L.; Contuzzi, N.; Ludovico, A.D. Experimental investigation and statistical optimisation of the selective laser melting process of a maraging steel. *Opt. Laser Technol.* **2015**, *65*, 151–158. [CrossRef]
20. Campanelli, S.L.; Angelastro, A.; Signorile, C.G.; Casalino, G. Investigation on direct laser powder deposition of 18 Ni (300) marage steel using mathematical model and experimental characterization. *Int. J. Adv. Manuf. Technol.* **2016**. [CrossRef]
21. Chang, B.H.; Yi, C.H.; Du, D.; Zhang, H.; Li, Y.H. Characteristics of microstructures in laser spot welds of a sintered NdFeB permanent magnet for different welding modes. *J. Tsinghua Univ. Sci. Technol.* **2014**, *54*, 1138–1142.
22. Chang, B.H.; Bai, S.J.; Du, D.; Zhang, H.; Zhou, Y. Studies on the micro-laser spot welding of an NdFeB permanent magnet with a low carbon steel. *J. Mater. Proc. Technol.* **2010**, *210*, 885–891. [CrossRef]
23. Chang, B.H.; Blackburn, J.; Allen, C.; Hilton, P. Studies on the spatter behaviour when welding AA5083 with a Yb-fibre laser. *Int. J. Adv. Manuf. Technol.* **2016**, *84*, 1769–1776. [CrossRef]
24. Luo, Y.; Zhang, N.; Su, X.J. Thermal analysis and domain observation of Nd-Fe-B magnet. *Acta Metall. Sin.* **1987**, *23*, 136–140.
25. Zhang, N.; Luo, Y. Thermal analysis of Nd-Fe-B magnet. *J. Iron Steel Res.* **1986**, *6*, 91–98.

metals

MDPI

Article

The Effects of Laser Welding Direction on Joint Quality for Non-Uniform Part-to-Part Gaps

Rocku Oh [1], Duck Young Kim [1,*] and Darek Ceglarek [2]

[1] Smart Factory Laboratory, Ulsan National Institute of Science and Technology, UNIST-gil 50, Ulsan 44919, Korea; org817@unist.ac.kr

[2] Warwick Manufacturing Group, University of Warwick, Coventry CV4 7AL, UK; d.j.ceglarek@warwick.ac.uk

* Correspondence: dykim@unist.ac.kr; Tel.: +82-52-217-2713

Academic Editor: Giuseppe Casalino
Received: 25 May 2016; Accepted: 3 August 2016; Published: 6 August 2016

Abstract: Controlling part-to-part gaps is a crucial task in the laser welding of galvanized steel sheets for ensuring the quality of the assembly joint. However, part-to-part gaps are frequently non-uniform. Hence, elevations and depressions from the perspective of the heading direction of the laser beam always exist throughout the gap, creating ascending, descending, and flat travelling paths for laser welding. In this study, assuming non-uniform part-to-part gaps, the effects of welding direction on the quality of the joint of galvanized steel sheets—SGARC440 (lower part) and SGAFC590DP (upper part)—were examined using 2-kW fiber and 6.6-kW disk laser welding systems. The experimental analysis of coupon tests confirmed that there is no statistically significant correlation between the direction of welding and weld pool quality if the gap exceeds the tolerable range. However, when the gap is controlled within the tolerable range, the welding direction can be considered as an important process control variable to enhance the quality of the joint.

Keywords: laser welding; part-to-part gap; welding direction; process control

---

1. Introduction

From the perspective of automotive body-in-white assemblies, laser welding has many desirable features such as high joining speed, excellent repeatability, and non-contact single-sided access, resulting in a greater degree of freedom in car body design. Nevertheless, laser welding is yet to be widely and successfully used, especially for joining of complex galvanized steel parts in lap-joint configurations [1], for which conventional joining methods such as resistance spot welding are generally employed.

Important design parameters for laser welding have been investigated in many empirical studies. For example, Benyounis et al. [2] identified the importance of laser power, focal position, and welding speed on the assembly joint's quality such as heat input and weld bead geometry (i.e., penetration depth, widths of welded zone, and heat-affected zone). Wu et al. [3] also confirmed that there is a statistically significant correlation between a joint's quality (i.e., welding penetration and the width of a weld seam) and laser power and welding speed.

In general, individual part variations caused by the deformation of metal sheets result in unexpected gaps between the upper (top) and lower (bottom) parts, as shown in Figure 1. Hence, tight part-to-part gap control is required to ensure the assembly joint quality of the galvanized steel sheets; these sheets are characterized by the lower evaporation point of Zn (906 °C) than the melting point of Fe (1538 °C). The vaporized zinc gas hampers the formation of stabilized keyholes and often causes serious weld defects such as porosity, spatter, intermetallic brittle phases, and discontinuities formed by zinc vapor entrapment in the welding joints [4].

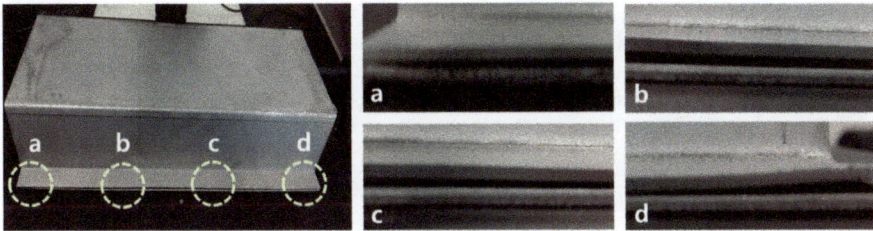

Figure 1. Part-to-part gap variation and angles of elevation and depression in a simplified side-member part assembly.

Several ad hoc methods have been developed to control part-to-part gaps; these methods include laser dimpling [5], shim insertion between parts [6], usage of porous powder metal, pre-drilling [7], and synchronous rolling technique [8]. Laser dimpling allows us to create small dimples on bottom parts by using a relatively low-powered laser beam prior to the corresponding laser-welding process. Laser dimpling is a commonly used practical method to realize the minimum required gap because small dimples are created by the same laser system that is used for laser welding [5]. Table 1 summarizes the recent research efforts to identify the major welding parameters that affect the quality of laser lap welding of galvanized steel.

Table 1. The major process parameters in the laser lap welding of galvanized steel from literature.

Major Process Parameters	Laser Welding Quality	References
Laser power, focal position, welding speed	Heat input and weld bead geometry (i.e., penetration depth, widths of welded zone, and heat-affected zone)	Benyounis et al. [2], Wu et al. [3]
Part-to-part gap	Weld depth, weld width, and concavity	Zhao et al. [9]
Laser power, welding speed, focal position, and shielding gases	Static tensile strength	Mei et al. [10]
Shielding gases	Tensile strength and widths of heat-affected zone	Chen et al. [11], Yang et al. [12]
Clamp pressure	Lap shear strength and weld seam width	Acherjee et al. [13], Anawa et al. [14]

What makes laser welding more complex is that part-to-part gaps are very often non-uniform, as shown in Figure 1. Owing to the non-uniformity, elevation and depression angles always exist throughout the gap, creating ascending, descending, and flat travelling paths for laser welding from the perspective of the direction of the laser beam.

This study examined the effect of the non-uniformity of the part-to-part gap on the weldment quality of a joint during the laser welding of galvanized steel sheets. We conducted an experimental analysis of 84 coupon tests under the in-tolerance condition of part-to-part gap (0.3 mm) and 66 coupon tests under the out-of-tolerance condition (0.5 mm). Laser lap joining of two different galvanized steel sheets, namely, SGARC440 (lower part) and SGAFC590DP (upper part) were performed by using 2-kW fiber and 6.6-kW disk laser welding systems.

The quality characteristic of a weldment was evaluated based on the top and bottom *s*-values and concavities of the weld pool, and their correlation with the weld direction was examined based on the analysis of variance (ANOVA).

The purpose of the experiment was to examine whether the laser welding direction is as an important process control variable for non-uniform part-to-part gaps, especially in the case of the remote laser welding system with a scanning mirror head used as the end-effector (e.g., Comau's Smart Laser™). Note that a robotic remote laser welding system can easily change not only the welding position, but also the direction of welding, simply by controlling the tilting mirrors in the scan head, as shown in the Figure 2. This control procedure for repositioning and redirection does not affect the required cycle time of the weld process.

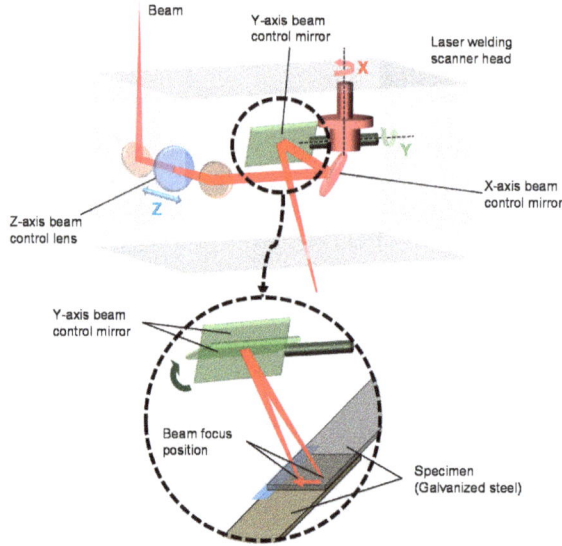

Figure 2. A schematic illustration of a laser welding scanner head that can easily change the direction of welding.

2. The Experiments

2.1. Laser Welding Systems

In order to provide reliable empirical evidence, we conducted coupon tests by using two different laser welding systems, namely, a 2-kW fiber laser welding system and a 6.6-kW disk laser welding system. The former system is a 2.5 axis gantry-based automated welding system that delivers a laser beam from IPG YLS-2000-AC fiber laser source (IPG Photonics, Oxford, MA, USA) with a maximum output discharge of 2 kW in the TEM01 mode of laser radiation. The 6.6-kW disk laser welding system is a five-axis KUKA robot-based remote laser welding system that delivers a laser beam from TRUMPF TruDisk 6602 disk laser (TRUMPF, Schramberg, Germany). Table 2 lists the technical parameters of both systems.

Table 2. Technical parameters of the laser welding systems.

Parameters	Unit	Fiber Laser YLS-2000AC	Disk Laser TruDisk6602
Max. laser power	kW	2.0	6.6
Beam quality	mm × mrad	6.0	8.0
Fiber diameter	μm	600	200
Emission wavelength	nm	1070	1030
Focal length	mm	278	533

2.2. Experimental Materials

We conducted laser welding experiments using sheets made of galvanized steels SGARC440 (lower part: 1.8 mm thickness) and SGAFC590DP (upper part: 1.4 mm thickness); these materials are currently used in the side-member parts of a car model. The amount of zinc coating on the lower and upper parts and their chemical compositions are summarized in Table 3. The mechanical properties of the tested materials are listed in Table 4.

Table 3. Chemical composition (weight %) of the test materials.

Tested Materials	Dimension (mm) (length × width × thickness)	Zinc Coating (g/m^2)	C (%)	Si (%)	Mn (%)	P (%)	S (%)
SGARC440 (Lower part)	130 × 30 × 1.8	45.5	0.12	0.5	1.01	0.021	0.004
SGAFC590DP (Upper part)	130 × 30 × 1.4	45.4	0.09	0.26	1.79	0.03	0.003

Table 4. Mechanical properties of the test materials.

Tested Material	Tensile Test		
	Yield Strength (N/m^2)	Max-Tensile Strength (N/m^2)	Elongation (%)
SGARC440 (Lower part)	327.5	451.1	38
SGAFC590DP (Upper part)	413.8	625.7	28

2.3. Non-Uniform Part-to-Part Gap

In general, for the successful laser welding of galvanized steel, it is necessary first to control the gap usually to be within 10% of the thickness of the upper part on which the laser beam is incident and additionally to allow for the minimum gap between the parts in order to provide a channel to vent out the vaporized zinc [13]. In general, in the case of automotive body parts, gaps of a maximum of 0.3 mm and minimum of 0.05 mm are required for laser welding of galvanized steel sheets.

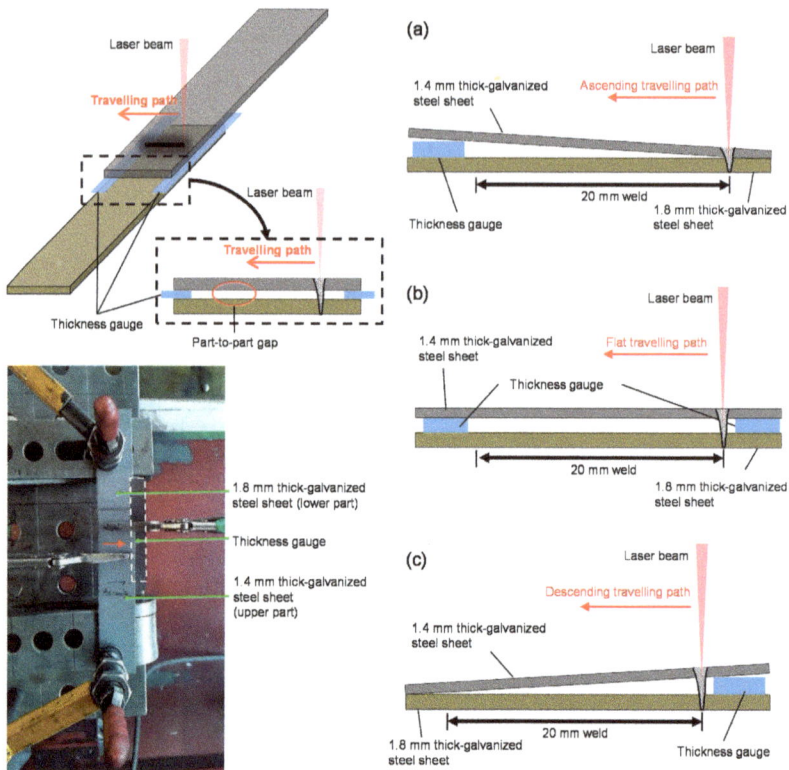

Figure 3. Weld joint configuration clamped at the two corners and the three types of travelling paths: (**a**) ascending; (**b**) flat; (**c**) descending.

As shown in Figure 1, these gaps are very often non-uniform because of individual part variations in the galvanized steel sheets. In order to simplify these random part-to-part gaps for the experiment, we linearized the travelling paths of the laser into only three linear types: ascending, descending, and flat, depending on the angles of elevation and depression from the perspective of the heading direction of the laser beam. In other words, we created part-to-part gaps and ascending (Type A), flat (Type B), and descending (Type C) travelling paths of welding by inserting a conventional metal thickness gauge (thickness: 0.3 mm, 0.5 mm) between the upper and the lower parts that were to be joined.

Figure 3 shows the schematic representation of experimental setups. Note that all the specimens and thickness gauges were washed using an alcohol-based cleaner to remove any dust and oil layers, and the specimen was tightly clamped at two corners to minimize any unexpected part-to-part gaps.

2.4. Experimental Design

The surface appearance and cross sectional macrostructure significantly affects the weld quality, which is tensile shear strength, as shown in the previous works by Sinha et al. [4] and Wei et al. [15]. Chen et al. [16] also investigated relations between geometry of weld seam and tensile strength using conventional destructive techniques for measuring the geometry of the weld seam and tensile strength. Ceglarek et al. [17] used the measures as the key joint quality indicators to monitor laser welding process and simultaneous joint quality evaluation. Especially on concavity, Westerbaan et al. [18] observed higher amount of concavity reduced the tensile strengths and fatigue resistance. Based on the related works, the concavities as well as the s-value but should be considered as quality measures. We defined the weld pool quality by considering the top and the bottom s-values and concavities, as follows:

$$\text{Weld pool quality} = (\text{top}\,s-\text{value} + \text{bottom}\,s-\text{value})/2 - (|\text{top concavity}| + |\text{bottom concavity}|) \quad (1)$$

We performed coupon tests with the 2-kW fiber and the 6.6-kW disk laser welding systems to investigate statistically whether the type of travelling path is an important process parameter for determining the joint's quality, especially in terms of the weld pool quality (see Figure 4). The weld specimens were cut at the center of the welded seam utilizing wire-cut EDM which is SODICK's SL400G. The s-values, top and bottom concavity were measured by using a LEICA DMS300 microscopic system and its embedded software, Leica Application Suite EZ (LAS EZ).

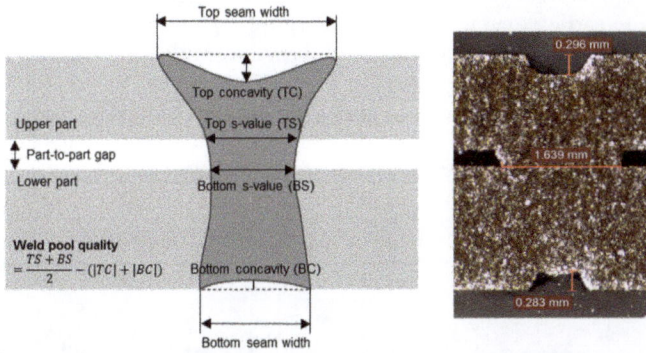

Figure 4. Definition of weld pool quality and an example of the cross sectional view at the middle of the specimen.

To analyze the result further, we conducted additional experiments considering a new joining quality measure, namely, the tensile strength of the weld. The tensile tests of welds were conducted using the testing instruments, INSTRON 5982 (100 kN capacity, INSTRON, Norwood, MA, USA). The tested specimen and setup was exactly the same with the experiments #1 and #2 (see Table 5).

Table 5. Experimental design.

Experiments	Experimental Factors		Part-to-Part Gap (mm)	Welding Speed (mm/min)
	Laser Power (W)	Type of Travelling Path		
#1-1 3 levels × 2 levels with 2 replicates (2 kW fiber)	1600 1800 2000	Ascending Descending	0.3 (in-tolerance)	800 1000 1250
#1-2 3 levels × 2 levels with 5 replicates (6.6 kW disk)	4000 5000 6000	Ascending Descending	0.3 (in-tolerance)	4000 5000 6000
#2-1 3 levels with 10 replicates (2 kW fiber)	2000	Ascending Flat Descending	0.5 (out-of-tolerance)	900
#2-2 3 levels × 3 levels with 4 replicates (6.6 kW disk)	4000 5000 6000	Ascending Flat Descending	0.5 (out-of-tolerance)	3000 4000 4000

Through the sufficient trials and pre-experiments, we identified the appropriate magnitude of laser power and its corresponding welding speed for the experiments in order to minimize noise, thereby maintaining stable weld quality. The tilting of the laser beam because of the non-uniformity of the part-to-part gap was assumed to be negligible. Each replicate was blocked in order to eliminate the effect of nuisance factors.

3. The Effects of Laser Welding Direction

3.1. Weld Pool Quality

The result of experiment #1 is summarized in Tables 6 and 7. Note that the two values below the photo of each cross section describe the average s-value (mm) and the weld pool quality (mm) respectively. ANOVA for experiment #1-1 (Table 8) indicated that the travelling path has no statistically significant effect on the weld pool quality at the 0.05 level of significance under the condition of in-tolerance part-to-part gap and low laser power. However, it is likely that the descending path outperforms the ascending path, as illustrated in the main effect plot in Figure 5a. ANOVA for experiment #1-2 (Table 9) and the main effect plot (Figure 5b) show that changes in the travelling path during laser welding, i.e., changes in the welding direction do not influence the weld pool quality, even when a relatively high power laser beam is used under the condition of in-tolerance part-to-part gap.

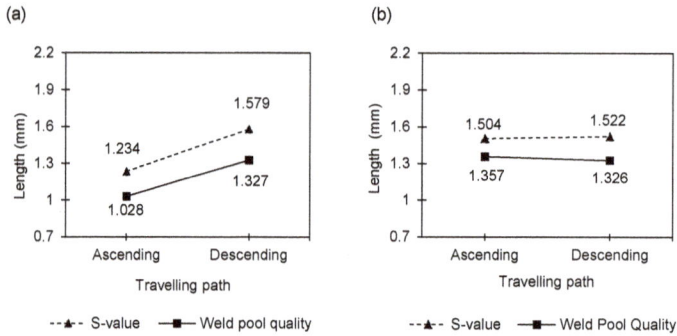

Figure 5. Main effect plots of the average s-value and weld pool quality in terms of the travelling path for experiment (**a**) #1-1 (laser power: 1.6, 1.8, and 2 kW, part-to-part gap: 0.3 mm) and (**b**) #1-2 (laser power: 4, 5, and 6 kW, part-to-part gap: 0.3 mm).

Table 6. The laser welding experimental data for experiment #1-1.

Laser Power	Average s-Value/Weld Pool Quality (mm)	
	Ascending	**Descending**
1.6 kW	1.344/0.979 1.336/1.219	1.506/1.342 1.691/1.301
1.8 kW	1.218/1.096 1.134/0.901	1.458/1.218 1.845/1.544
2 kW	1.292/1.120 1.082/0.854	1.519/1.383 1.456/1.178

Table 7. The laser welding experimental data for experiment #1-2.

Laser Power	Average s-Value/Weld Pool Quality (mm)	
	Ascending	**Descending**
4 kW	1.473/1.411 1.528/1.429 1.542/1.360 1.474/1.373 1.541/1.471	1.445/1.238 1.532/1.389 1.519/1.411 1.365/1.127 1.424/1.243
5 kW	1.621/1.399 1.625/1.510 1.343/1.099 1.632/1.476 1.492/1.407	1.485/1.345 1.701/1.544 1.576/1.387 1.564/1.400 1.567/1.414
6 kW	1.387/1.224 1.449/1.317 1.465/1.339 1.433/1.210 1.551/1.338	1.567/1.312 1.499/1.269 1.487/1.170 1.587/1.355 1.510/1.294

Dimensional variation caused by the deformation of metal sheets often results in a large part-to-part gap (larger than 0.3 mm); this impedes the maintenance of laser lap welding quality. Experiment #2 was conducted to investigate the effect of welding direction under the condition of out-of-tolerance part-to-part gap, and its result is summarized in Tables 10 and 11.

Table 8. ANOVA table for experiment #1-1 (2-kW fiber laser welding system).

Source	Degree of Freedom	Sum of Squares	Mean Square	F-Ratio	P-Value
Blocks	1	0.00165	0.00165	0.05	0.828
Laser power	2	0.01256	0.00628	0.2	0.824
Travelling path	1	0.26895	0.26895	8.6	0.033
Laser power × Travelling path	2	0.01281	0.00641	0.2	0.821
Error	5	0.15628	0.03126		
Total	11	0.45224			

Table 9. ANOVA table for experiment #1-2 (6.6-kW disk laser welding system).

Source	Degree of Freedom	Sum of Squares	Mean Square	F-Ratio	P-Value
Block	4	0.0479	0.0120	1.34	0.291
Laser power	2	0.0666	0.0333	3.71	0.043
Travelling path	1	0.0072	0.0072	0.81	0.380
Laser Power × Travelling path	2	0.0373	0.0187	2.08	0.151
Error	20	0.1793	0.0090		
Total	29	0.3383			

Table 10. The laser welding experimental data of experiment #2-1.

Laser Power	Average s-Value/Weld Pool Quality (mm)					
	Flat		Ascending		Descending	
	1.999/1.373	1.887/1.409	1.532/0.976	1.880/1.405	1.719/0.943	1.501/1.107
	2.120/1.179	2.107/1.422	1.700/1.329	1.685/1.265	1.600/1.324	1.755/1.310
2 kW	2.260/1.732	1.900/1.373	2.015/1.567	1.560/1.217	1.802/1.290	1.383/1.059
	2.143/1.272	2.202/1.248	1.431/1.047	1.521/1.075	2.073/1.173	1.809/1.479
	2.215/1.136	2.260/1.195	1.753/1.388	1.573/1.219	1.460/1.209	1.743/1.387

As in the case of the result of experiment #1, ANOVA for experiment #2-1 (Table 12) also indicates that the travelling path has no statistically significant effect on the weld pool quality at the 0.05 level of significance under the condition of out-of-tolerance part-to-part gap and low laser power. However, the results of experiment #2-2 (Table 13) show that the travelling path may cause variations in the weld pool quality when the part-to-part gap is large and the laser power is relatively high. However, the main effect plot in Figure 6 showed that the root cause of this variation was the quality difference between the flat travelling path and the other paths, rather than any difference between the ascending and descending paths.

Table 11. The laser welding experimental data of experiment #2-2.

Laser Power	Average *s*-Value/Weld Pool Quality (mm)					
	Flat		Ascending		Descending	
4 kW	1.731/1.046	1.796/1.079	1.692/1.158	1.717/1.221	1.865/1.030	1.847/1.017
	1.697/0.990	1.668/0.945	1.620/1.144	1.715/1.136	1.681/1.077	1.692/1.257
5 kW	1.517/0.662	1.493/0.563	1.806/1.225	1.751/1.322	1.701/1.393	1.509/1.169
	1.545/0.779	1.491/0.734	1.869/1.444	1.660/1.215	1.720/1.392	1.845/1.400
6 kW	1.509/0.584	1.634/0.715	1.637/1.151	1.738/1.182	1.757/1.220	1.701/1.110
	1.649/0.776	1.618/0.423	1.765/1.285	1.781/1.203	1.670/1.298	1.853/1.313

(a)   (b)

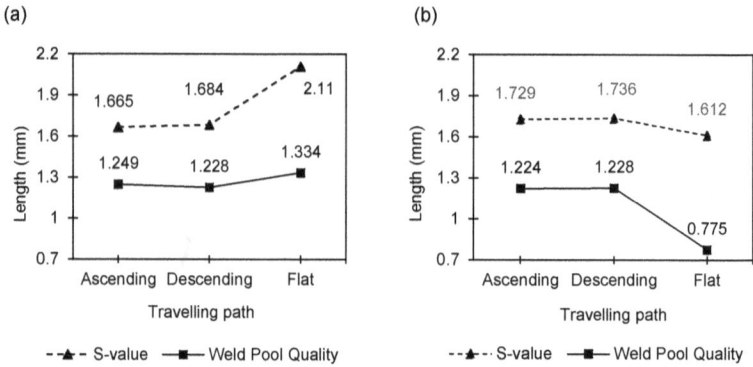

Figure 6. Main effect plots of the average s-value and weld pool quality in terms of the travelling path for experiment (**a**) #2-1 (laser power: 2 kW, part-to-part gap: 0.5 mm) and (**b**) #2-2 (laser power: 4, 5, and 6 kW, part-to-part gap: 0.5 mm).

Table 12. ANOVA table of the s-value for experiment #2-1 (2-kW fiber laser welding system).

Source	Degree of Freedom	Sum of Squares	Mean Square	F-Ratio	p-Value
Travelling path	2	0.0630	0.0315	1.06	0.361
Error	27	0.8026	0.0297		
Total	29	0.8656			

Table 13. ANOVA table for experiment #2-2 (6-kW disk laser welding system).

Source	Degree of Freedom	Sum of Squares	Mean Square	F-Ratio	p-Value
Blocks	3	0.0437	0.0146	1.62	0.212
Laser power	2	0.0507	0.0254	2.81	0.08
Travelling path	2	1.6103	0.8051	89.33	0.00
Laser power × Travelling path	4	0.4616	0.1154	12.8	0.00
Error	24	0.2163	0.0090		
Total	35	2.3826			

Note that the weld pool quality of experiment #2-1 significantly differs from that of experiment #2-2 in the case of the flat travelling path. This is because the welding process is relatively longer because of the low laser power, and hence, there is sufficient time to create a keyhole through the top and the bottom parts. Hence, an acceptable s-value is attained despite the large part-to-part gap. This large gap, however, usually creates large top concavity, which has a negative effect on the weld pool quality.

3.2. Tensile Strength

We conducted additional experiments and evaluated the welding quality by tensile strength. The tensile tests of welds were conducted using the testing instruments, INSTRON 5982 (100 kN capacity). Table 14 shows maximum tensile strengths for different laser powers.

Table 14. The laser welding experimental data of experiment #3-1 (laser power: 2 kW, part-to-part gap: 0.3 mm) and experiment #3-2 (laser power: 4, 4.5, 5.5, and 6 kW, part-to-part gap: 0.3 mm)

Laser Power	Experiment #3-1 Maximum Tensile Strength (MPa)		Laser Power	Experiment #3-2 Maximum Tensile Strength (MPa)	
	Ascending	Descending		Ascending	Descending
	143.150	138.044		130.195	154.848
	157.330	161.804	4 kW	132.497	174.842
	157.330	142.091		137.479	141.421
	167.639	175.656		143.150	146.920
	125.989	151.240	4.5 kW	128.442	139.133
	177.610	113.757		126.111	170.180
2 kW	124.059	155.150		162.380	133.611
	139.451	166.041	5.5 kW	138.745	142.645
	156.543	167.844		122.292	157.112
				152.604	145.174
			6 kW	163.173	145.174
				163.173	122.911

Results of experiment #3-1 (Table 15) indicated that the travelling path under the condition of in-tolerance part-to-part gap and low laser power had statistically significant effects on the maximum tensile shear strength at the 0.05 level of significance. Furthermore, the descending travelling path yielded better results than the ascending path, as shown in Figure 7a. The results of experiment #3-2 (Table 16) showed that changes in the travelling path during laser welding did not influence the tensile strength in the case of a relatively high power laser source under the condition of in-tolerance part-to-part gap. However, from the main effect plot shown in Figure 7b, we observed the tendency of joining quality: the descending travelling path outperformed the ascending one. There is a similar tendency when we are dealing with the weld pool quality. Based on the results, weld pool quality can be an indirect measure of the tensile shear strength.

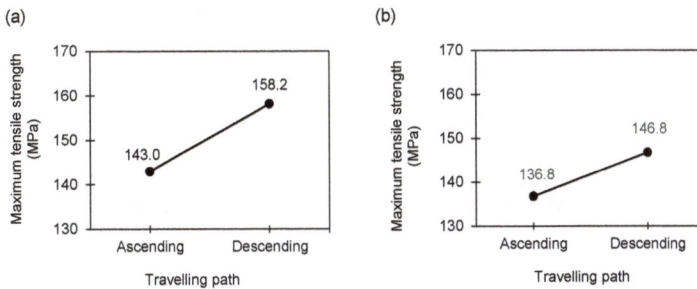

Figure 7. Main effect plots of maximum tensile strength in terms of the travelling paths for experiment (a) #3-1 (laser power: 2 kW, part-to-part gap: 0.3 mm) and (b) #3-2 (laser power: 4, 4.5, 5.5, and 6 kW, part-to-part gap: 0.3 mm).

Table 15. ANOVA table of the tensile strength for experiment #3-1 (2-kW fiber laser welding system).

Source	Degree of Freedom	Sum of Squares	Mean Square	F-Ratio	p-Value
Travelling path	1	1037	1037	5.44	0.033
Error	16	3053	191		
Total	17	4090			

Table 16. ANOVA table of the tensile strength for experiment #3-2 (6.6-kW disk laser welding system).

Source	Degree of Freedom	Sum of Square	Mean Square	*F*-Ratio	*p*-Value
Laser power	3	419	139.7	0.62	0.612
Travelling path	1	552.9	552.9	2.45	0.137
Laser power × Travelling path	3	1200.7	400.2	1.78	0.192
Error	16	3603.5	225.2		
Total	23	5776.1			

4. Discussion

Table 17 summarizes the results of the four sets of experiments. The results did not provide a statistically significant evidence to correlate the direction of welding with the weld pool quality. Nevertheless, we observed that the descending travelling path yields a slightly better joining quality than the ascending path in the case of the relatively low power laser beam under the condition of in-tolerance part-to-part gap, as illustrated in the main effect plot shown in Figure 5.

Table 17. Summary of the four experiments.

Experiments	Part-to-Part Gap (mm)	Significance of Welding Direction
#1-1 (2-kW fiber) 3 levels × 2 levels with 2 replicates	0.3 (in-tolerance)	Descending ⩾ Ascending
#1-2 (6.6-kW disk) 3 levels × 2 levels with 5 replicates	0.3 (in-tolerance)	X
#2-1 (2-kW fiber) 3 levels with 10 replicates	0.5 (out-of-tolerance)	X
#2-2 (6.6-kW disk) 3 levels × 3 levels with 4 replicates	0.5 (out-of-tolerance)	X

From the experimental results, we inferred that the descending path is usually better than the ascending path if the peak part-to-part gap does not exceed the tolerable range. In the case of ascending travelling path, the gap at the starting point of welding is not sufficient to create and sustain a stable keyhole owing to insufficient degassing. In contrast, in the case of the descending path, the gap at the starting point is acceptable as in the case of the flat travelling path. This allows vaporized zinc gas to escape effectively through the gap even before forming a stable cavity or keyhole. Once the keyhole formed and the full penetration through the top and the bottom parts was realized, the keyhole itself acted as a channel for venting out the zinc vapor in spite of the small gap at the finishing point of the descending travelling path. The images of the lateral and the top surfaces of the weld joints shown in Figure 8 present the result of this phenomenon.

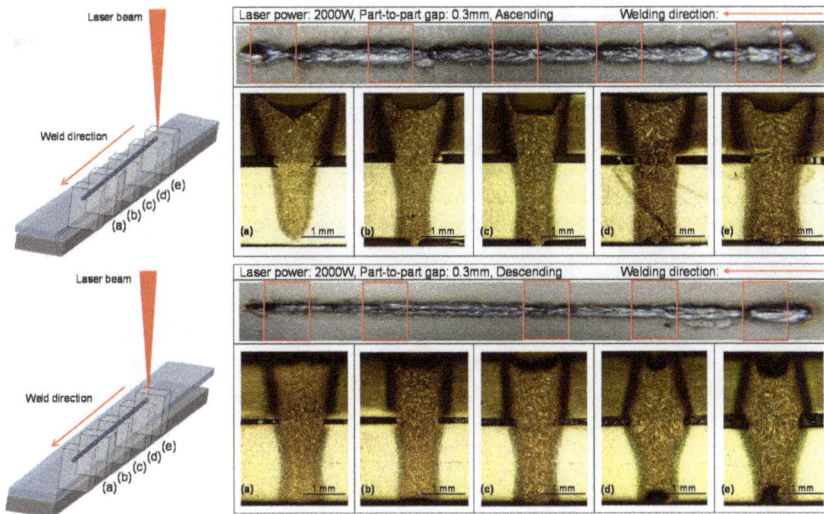

Figure 8. The lateral images of weld joints in the cases of ascending (**top**) and descending (**bottom**) travelling paths.

In summary, the experimental results provided some evidence that the laser welding direction can be considered as an important process control variable to enhance the quality of a joint so long as the part-to-part gap is controlled within the tolerable range. This finding, however, calls for further study considering other experimental parameters such as different materials and the amount of zinc. Furthermore, effectively identifying the part deformation that will generate different types of travelling paths during laser welding is a challenge.

5. Conclusions

The effects of welding direction on the quality of joints were investigated. The main findings are summarized as follows:

- Individual part variation often causes non-uniform part-to-part gaps.
- If the part-to-part gap exceeds the tolerable range, the direction of welding does not affect the weld pool quality significantly.
- If the part-to-part gap exceeds the tolerable range, laser power adjustment is more sensitive to the weld pool quality than welding direction change.
- If the part-to-part gap is controlled within the tolerable range, then the direction of welding can be considered as an important process control variable to enhance the quality of the joint.

These findings motivate further research to determine the status of part-to-part gaps by in-process weld signal monitoring. By using the status information, the magnitudes of process parameters such as laser power, welding speed, and the direction of welding can be adjusted for the next welding operations in the same batch of parts to be joined, where individual part variations tend to have similar patterns.

Acknowledgments: Special thanks to go to Sungwoo Hitech Co. Ltd. for supporting the experiments with the 6.6-kW disk laser welding system. The research reported in this paper is financially supported by the international collaborative R&D program of the Korea Institute for Advancement of Technology (Grant No. EUFP-M0000224) which is linked with the EU FP7 project of the European Commission (Grant No. FP7 Project 285051).

Author Contributions: D.-Y. Kim coordinated the overall work of the paper and conceived the experiments; R. Oh developed the detailed design of experiments and analyzed the results; D. Ceglarek discussed the experimental results.

Conflicts of Interest: The authors declare no conflict of interest.

References

1. Katundi, D.; Tosun-Bayraktar, A.; Bayraktar, E.; Toueix, D. Corrosion behaviour of the welded steel sheets used in automotive industry. *J. Achiev. Mater. Mfg. Eng.* **2010**, *38*, 146–153.
2. Benyounis, K.; Olabi, A.; Hashmi, M. Effect of laser welding parameters on the heat input and weld-bead profile. *J. Mater. Prcess. Technol.* **2005**, *164*, 978–985. [CrossRef]
3. Wu, Q.; Gong, J.; Chen, G.; Xu, L. Research on laser welding of vehicle body. *Opt. Laser Technol.* **2008**, *40*, 420–426. [CrossRef]
4. Sinha, A.; Kim, D.; Ceglarek, D. Correlation analysis of the variation of weld seam and tensile strength in laser welding of galvanized steel. *Opt. Laser Eng.* **2013**, *51*, 1143–1152. [CrossRef]
5. Colombo, D.; Colosimo, B.; Previtali, B. Comparison of methods for data analysis in the remote monitoring of remote laser welding. *Opt. Laser Eng.* **2013**, *51*, 34–46. [CrossRef]
6. Rito, N.; Ohta, M.; Yamada, T.; Gotoh, J.; Kitagawa, T. Laser Welding Method. U.S. Patent No. 4,745,257, May 1988.
7. Pennington, E. Laser Welding of Galvanized Steel. U.S. Patent No. 4,642,446, February 1987.
8. Ozaki, H.; Kutsuna, M.; Nakagawa, S.; Miyamoto, K. Laser roll welding of dissimilar metal joint of zinc coated steel to aluminum alloy. *J. Laser Appl.* **2010**, *22*, 1–6. [CrossRef]
9. Zhao, Y.; Zhang, Y.; Hu, W.; Lai, X. Optimization of laser welding thin-gage galvanized steel via response surface methodology. *Opt. Laser Eng.* **2012**, *50*, 1267–1273. [CrossRef]
10. Mei, L.; Chen, G.; Jin, X.; Zhang, Y.; Wu, Q. Research on laser welding of high-strength galvanized automobile steel sheets. *Opt. Laser Eng.* **2009**, *47*, 1117–1124. [CrossRef]
11. Chen, H.; Pinkerton, J.; Li, L.; Liu, Z.; Mistry, T. Gap-free fibre laser welding of Zn-coated steel on Al alloy for light-weight automotive applications. *Mater. Des.* **2011**, *32*, 495–504. [CrossRef]
12. Yang, S.; Carlson, B.; Kovacevic, R. Laser welding of high-strength galvanized steels in a gap-free lap joint configuration under different shielding conditions. *Weld. J.* **2011**, *90*, 8s–18s.
13. Acherjee, B.; Misra, D.; Bose, D.; Venkadeshwaran, K. Prediction of weld strength and seam width for laser transmission welding of thermoplastic using response surface methodology. *Opt. Laser Technol.* **2009**, *41*, 956–967. [CrossRef]
14. Anawa, M.; Olabi, A. Optimization of tensile strength of ferritic/austenitic laser-welded components. *Opt. Laser Eng.* **2008**, *46*, 571–577. [CrossRef]
15. Wei, S.T.; Liu, R.D.; Liu, D.; Lin, L.; Lu, X.F. Effects of welding parameters on fibre laser lap weldability of galvanised DP1000 steel. *Sci. Technol. Weld. Join.* **2015**, *20*, 433–442. [CrossRef]
16. Chen, L.; Xu, W.W.; Gong, S.L. The effect of weld size on joint mechanical properties of 5A90Al-Li alloy. *Adv. Mater. Res.* **2011**. [CrossRef]
17. Ceglarek, D.; Colledani, M.; Va'ncza, J.; Kim, D.; Marine, C.; Kogel-Hollacher, M.; Mistry, A.; Bolognese, L. Rapid deployment of remote laser welding processes in automotive assembly systems. *CIRP Ann. Manuf. Technol.* **2015**, *64*, 389–394. [CrossRef]
18. Westerbaan, D.; Parkes, D.; Nayak, S.S.; Chen, D.L.; Biro, E.; Goodwin, F.; Zhou, Y. Effects of concavity on tensile and fatigue properties in fibre laser welding of automotive steels. *Sci. Technol. Weld. Join.* **2014**, *19*, 60–68. [CrossRef]

![metals logo] *metals*

MDPI

Article

Gas Metal Arc Welding Using Novel CaO-Added Mg Alloy Filler Wire

Minjung Kang, Youngnam Ahn and Cheolhee Kim *

Joining R & D Group, Korea Institute of Industrial Technology, 156 Gaetbeol-ro (Songdo-dong), Yeonsu-Gu, Incheon 21999, Korea; kmj1415@kitech.re.kr (M.K.); welidng@kitech.re.kr (Y.A.)
* Correspondence: chkim@kitech.re.kr; Tel.: +82-32-850-0222

Academic Editor: Giuseppe Casalino
Received: 25 February 2016; Accepted: 23 June 2016; Published: 8 July 2016

Abstract: Novel "ECO Mg" alloys, i.e., CaO-added Mg alloys, which exhibit oxidation resistance during melting and casting processes, even without the use of beryllium or toxic protection gases such as SF_6, have recently been introduced. Research on ECO Mg alloys is still continuing, and their application as welding filler metals was investigated in this study. Mechanical and metallurgical aspects of the weldments were analysed after welding, and welding behaviours such as fume generation and droplet transfer were observed during welding. The tensile strength of welds was slightly increased by adding CaO to the filler metal, which resulted from the decreased grain size in the weld metal. When welding Mg alloys, fumes have been unavoidable so far because of the low boiling temperature of Mg. Fume reduction was successfully demonstrated with a wire composed of the novel ECO Mg filler. In addition, stable droplet transfer was observed and spatter suppression could be expected by using CaO-added Mg filler wire.

Keywords: magnesium; fusion welding; CaO-added filler; fumes; microstructure

---

1. Introduction

Weight reduction is becoming an increasingly important issue in the automotive industry, to adhere to global CO_2 emission and fuel consumption regulations. Magnesium alloys are a promising lightweight alternative to aluminium alloys and high-strength steel in the automotive industry. Although Mg alloys have many advantages such as high specific strength, low density, and good casting ability, they are inherently reactive during processing of the molten Mg alloy; therefore, sulphur hexafluoride (SF_6) protection gas must be used or beryllium must be added to the Mg alloy to prevent ignition. Recently, so-called "ECO Mg" alloys have been invented by introducing CaO to the conventional Mg alloys. Previous studies have reported that Mg oxidation and ignition could be controlled and minimized in various ECO Mg alloys during melting [1–4], and improved mechanical properties and metallurgical characteristics have been reported [5–7].

Welding and joining are essential processes to manufacture automotive parts, and several welding processes have been proposed for Mg alloys. However, welding Mg alloys is inherently difficult because of their material characteristics such as low boiling temperatures, a high thermal expansion coefficient, and high thermal conductivity. Friction stir welding, a solid-state process, has been suggested to overcome the disadvantages of Mg alloys [8–11]. Also, numerous research studies for fusion welding processes that are more familiar and flexible have been recently published to optimize the fusion welding processes. These include gas tungsten arc welding [12,13], gas metal arc welding (GMAW) [14], and laser welding [15,16]. Recent studies on friction stir welding of ECO Mg alloy revealed that the mechanical and metallurgical characteristics of friction stir welds were also improved by the enhanced mechanical properties of ECO Mg alloy [17,18]. Other interesting results come from the laser welding of ECO Mg alloy. In addition to improved mechanical and metallurgical characteristics, laser-induced plasma can be controlled using ECO Mg alloy, which can prevent plasma interference with laser irradiation and can consequently lead to increased welding speed to realize fully penetrated welds [19].

In GMAW, the filler metal electrode has two important roles: as an electrode and as the source of deposited metal in the welds. The filler wire, which is normally employed as the positive pole, is exposed to the high-temperature welding arc, and electrons are condensed into the filler wire. Thus, the molten drop hanging on the end of the filler wire is easily overheated. Mg in the filler wire can result in droplet evaporation due to its low boiling point. However, most previous studies of droplet evaporation and fume generation have focused on Mg-containing Al filler wires [20–23] and only spatter generation and porosity in Mg filler wire were discussed [14]. On the other hand, as the source of deposited metal in welds, ECO Mg filler wire dissolved in the weld metal can improve mechanical and metallurgical characteristics as confirmed in autogenous laser welding of ECO Mg alloy [19].

In this study, ECO Mg alloy was employed as a filler metal for GMAW, and the weldability was studied. First, various mechanical and metallurgical characteristics of welds are explained, and then welding phenomena such as fume generation and droplet transfer monitored during GMAW will be discussed.

2. Experimental Setup

In this study, the base material was a commercial AZ 31 alloy sheet with a measured tensile strength of 278 MPa whose measured chemical composition is given in Table 1. The sheet was 150 mm long, 120 mm wide, and 1.5 mm thick. Before welding, the oxide film on the specimen was removed with a stainless brush in the area intended for welding.

Table 1. Chemical composition of base material (wt. %).

Al	Zn	Mn	Si	Fe	Cr	Mg
3.098	0.981	0.304	0.037	0.013	0.013	Bal.

The filler materials were a commercial AZ 31 alloy (AZ31-A) and CaO-added AZ 31 alloy (AZ31-B). Both fillers were 1.2 mm in diameter, and their chemical compositions measured by ICP are given in Table 2.

Table 2. Chemical compositions of filler wires (wt. %).

Filler Wire	Al	Zn	Mn	Si	Ca	Mg
AZ31-A	2.76	0.92	0.40	0.019	-	Bal.
AZ31-B	3.10	0.63	0.19	1.18	0.89	Bal.

In the welding experiments, bead-on-plate (BOP) welding was conducted and the fillers were fed into the weld pool with a lead angle of 20° from perpendicular, as shown in Figure 1a. Figure 2 shows the arc welding system with a six-axis articulate robot, a wire feeding system, and a welding power source. Argon shielding gas was supplied by a welding torch with a flow rate of 20 L/min. Mg alloys have a low boiling temperature, and a tremendous amount of spatters are generated when welding in the standard mode [14]. To avoid spatter generation and minimize heat input during welding, a Fronius CMT (cold-metal-transfer) 3200 was used as the power source, operating in the CMT mode, which is a type of short circuit transfer mode welding. The welding current, welding voltage, wire feed speed, and welding speed selected were 80 A, 10.2 V, 6.4 m/min, and 0.6 m/min, respectively.

After welding, the metallurgical and mechanical characteristics of the welds were examined. The specimens were polished and etched for 50 s in a solution of 2 mL hydrochloric acid and 100 mL ethanol in order to observe the microstructure of welds using a light microscope. Static tensile tests with a test speed of 5 mm/min were conducted on the specimen shown in Figure 3a according to ISO 6892-1:2009. The specimens were prepared with and without the weld reinforcements in Figure 3b,c. Mechanical machining was used to remove the weld reinforcement. The micro-Vickers hardness was measured at 0.3 mm intervals in the welds under a load of 0.49 N (50 gf) and a holding time of 10 s.

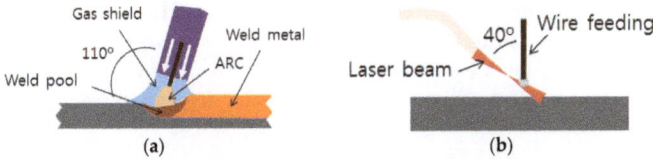

Figure 1. Setup for experiments: (**a**) Gas metal arc welding; (**b**) Laser melting.

Figure 2. Picture of the arc welding system used.

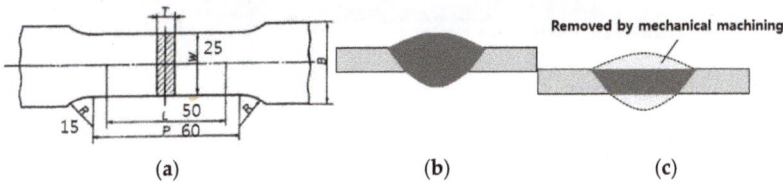

Figure 3. Schematic diagrams of prepared tensile test specimen (all dimensions in mm). (**a**) X-Y plane; (**b**) X-Z plane section with reinforcement; (**c**) X-Z plane section without reinforcement.

In the fume generation experiments, GMAW and laser melting experiments were conducted. The setup for GMAW was the same as for welding, and two welding modes—the CMT and pulse welding modes—were employed. In the CMT mode welding, the welding conditions were identical to the previous welding experiments. In the pulse mode welding, the average welding current, welding voltage, wire feed speed, and welding speed were 82 A, 22.4 V, 8.7 m/min, and 0.5 m/min, respectively.

During the laser melting experiment, as shown in Figure 1b, the filler wires were fed perpendicularly to the specimen with a wire feed speed of 3 m/min. The laser beam with a power of 4 kW was delivered using a 200-μm-diameter optical fibre (LLK-D 02, Trumpf Laser- und Systemtechnik GmbH, Ditzingen, Germany) from a Yb:YAG disk laser source (Trumpf HLD 4002, Trumpf Laser- und Systemtechnik GmbH, Ditzingen, Germany) and irradiated the end of the filler wire at an inclination angle of 40° and a defocusing distance of 3 mm.

High-speed photography using a Photron FASTCAM Ultima APX camera was employed with a capture speed of 2000 frames per second to record fume generation behaviour. A neutral density filter (ND400) and metal-halide back lighting were used to obtain clear images.

3. Welding Characteristics

Figure 4 shows the bead appearance and cross-section for welds with each of the filler wires. Sound bead appearance and cross-section were achieved for both fillers under the given welding conditions. Tensile tests were carried out for specimens with and without weld reinforcement. Joint efficiency was defined as the ratio of average tensile strength of the weldment to that of the base material, and is given in Figure 5. The specimens with weld reinforcement were fractured at the heat-affected zone, whereas those without weld reinforcement were fractured at the weld metal as shown in Figure 6. The fracture location did not vary with the filler wire. However, the tensile strengths of weldments welded with the CaO-containing AZ31-B wire were slightly higher than those welded with the conventional AZ31-A wire.

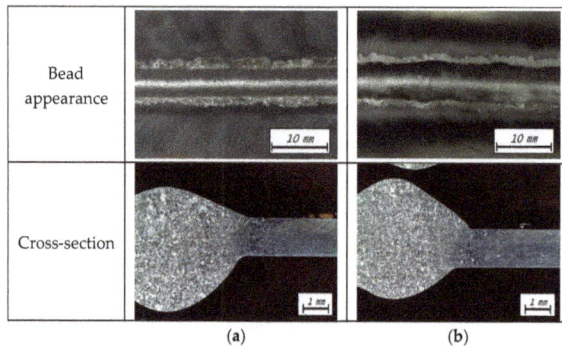

Figure 4. Bead shapes for bead-on-plate (BOP) welding: (**a**) AZ31-A; (**b**) AZ31-B.

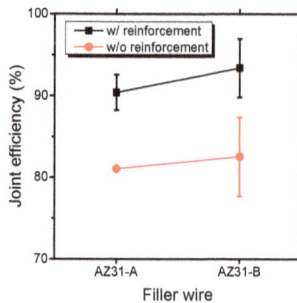

Figure 5. Joint efficiency for BOP welding.

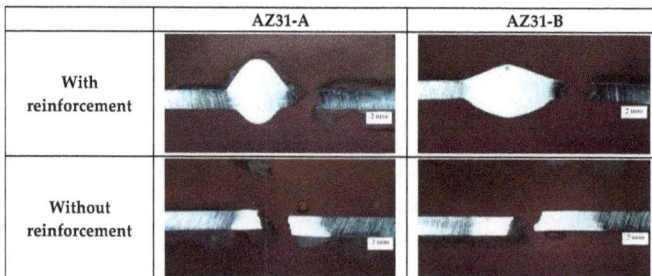

Figure 6. Fractured specimen with and without reinforcement by tensile test.

Figures 7 and 8 show the hardness profiles and microstructures, respectively, of the weldments. The increase in the hardness of the weld metal with the AZ31-B wire is clearly observed, while the hardness of the weld metal with the AZ31-A wire is even lower than that of the base metal. A previous study explained that the increase in hardness results from the Al_2Ca intermetallic compound in the CaO-containing Mg alloy [2]. The microstructure of the base metal and the heat-affected zone were identical because only the filler metal varied. In the heat-affected zone, abnormal grain growth was observed due to the heat input during welding. Grain size in the weldment—even those containing CaO—is larger than that of the base material fabricated by hot rolling, which leads to dynamic recrystallisation due to intense plastic deformation [24]. The grains found for the weldment with the AZ31-B wire were finer than those for the AZ31-A wire. The previous research (Reference [5]) demonstrated that grain refining was driven by the secondary phase, Al_2Ca. In Figure 8c,d, the grain size in the weld metal was measured by using ISO 643:2003. The measured values are 59.9 μm and 20.4 μm for the weld metals with AZ31-A and AZ31-B, respectively, which can explain the difference in the hardness profile in the weld metal.

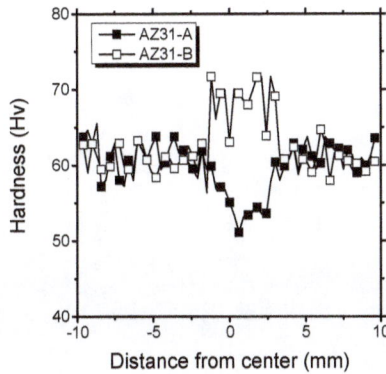

Figure 7. Hardness profile for BOP welding.

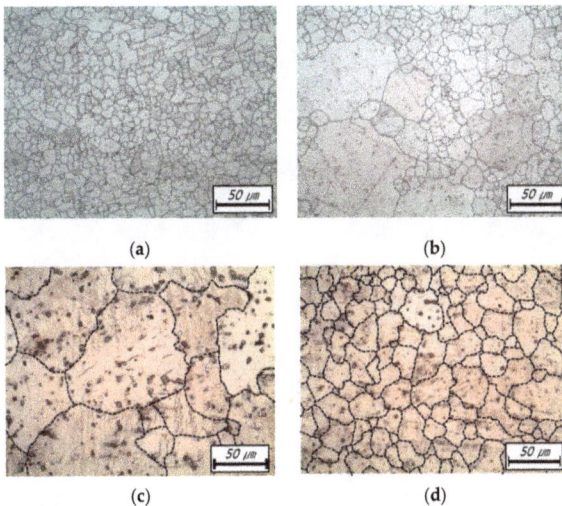

Figure 8. Microstructures of (**a**) base metal (BM); (**b**) heat-affected zone (HAZ); and weld metal (WM) (**c**) AZ31-A; (**d**) AZ31-B.

4. Fume Generation and Droplet Transfer

Figure 9 compares high-speed images taken during the CMT mode welding. Comparing both filler wires shows that fume generation was reduced when the AZ31-B filler wire was used. In the case of the CaO-containing Mg alloy, a thin CaO-rich barrier layer was formed at the surface; this layer can successfully suppress rapid oxidation and burning during the melting process [4]. During welding, the filler wire was melted and a droplet was formed at the end of the wire. The burning resistance of the AZ31-B filler wire could result in fume suppression, and this is more clearly confirmed during the laser melting experiment, as shown in Figure 10. The welding fumes can be generated from either the molten droplet or the weld pool [25], but no fumes were generated from the weld pool in the laser melting experiment.

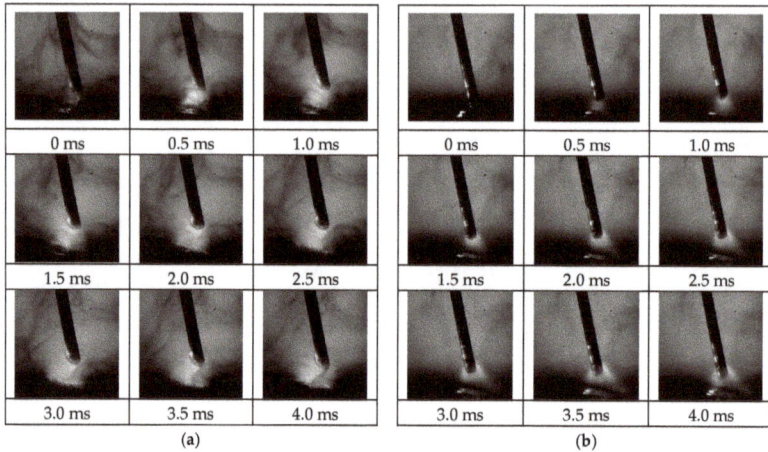

Figure 9. High-speed images during the cold-metal-transfer mode welding: (a) AZ31-A; (b) AZ31-B.

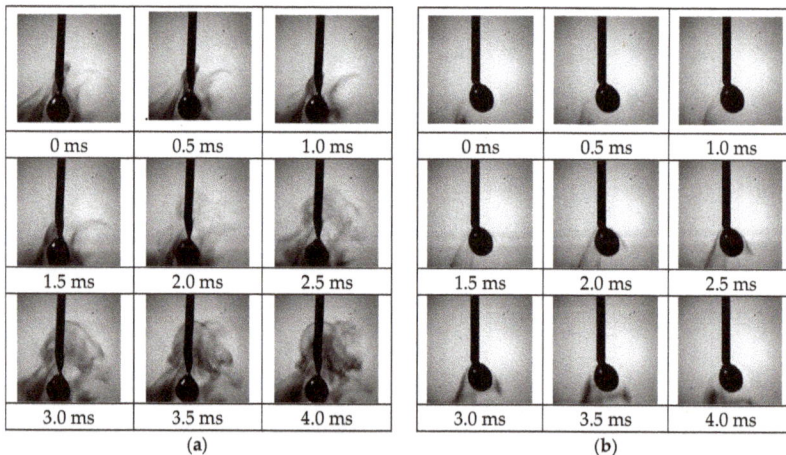

Figure 10. High-speed images during laser melting: (a) AZ31-A; (b) AZ31-B.

Pulse mode welding was conducted to implement relatively high-current welding. Figure 11 shows the high-speed images obtained during pulse mode welding. During the pulse mode welding, fume suppression by the AZ31-B wire was not clearly observed because of the high welding current.

However, the taper was formed at the end of the wire and the pinch force to detach the droplet increased, which can enhance spray transfer [26]. Therefore, the droplet transfer mode changed from globular transfer to spray transfer mode in which droplet size was reduced and the droplets were more rapidly transferred into the weld pool. In the conventional GMAW of Mg alloys, a large droplet hanging on the wire can be expelled by sudden arc expansion which causes severe spattering [14]. Small droplet size and fast transfer time would be helpful to suppress spatter generation in welding Mg alloys.

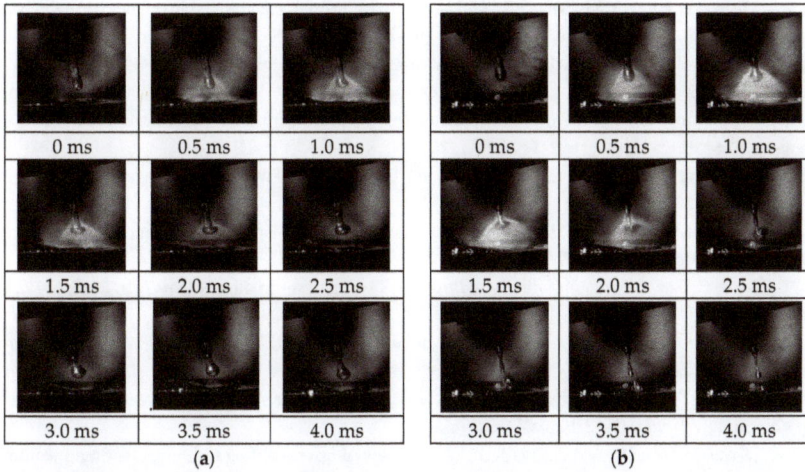

Figure 11. High-speed images during the pulse mode welding: (a) AZ31-A; (b) AZ31-B.

5. Conclusions

In this study, welding characteristics, fume generation, and droplet transfer behaviour of the CaO-added AZ31 filler wire were compared with the conventional AZ31 filler wire. The conclusions are as follows:

(1) The tensile strength of weldments was slightly increased by using the CaO-added AZ31 filler wire. By adding CaO, the grain size in the weld metal decreased from 59.9 μm to 20.4 μm, and the average hardness in the weld metal increased from 55.3 Hv to 68.9 Hv.
(2) Welding fumes were successfully suppressed during the CMT mode welding. The fume suppression was due to the burning resistance of CaO-added Mg alloy, and this has been confirmed by using the laser melting test.
(3) In the pulse mode welding, spray transfer mode welding was achieved, which enabled stable droplet transfer. Also, the suppression of spatter was expected because of the small droplet size and fast transfer time.

Author Contributions: M. Kang and Y. Ahn performed the experiments; M. Kang and C. Kim wrote the paper.

Conflicts of Interest: The authors declare no conflicts of interest.

References

1. Lee, J.K.; Yoon, Y.O.; Kim, S.K. Development of environment-friendly CaO added AZ31 Mg alloy. *Solid State Phenom.* **2007**, *124*, 1481–1484. [CrossRef]
2. Ha, S.-H.; Lee, J.-K.; Kim, S.K. Effect of CaO on oxidation resistance and microstructure of pure Mg. *Mater. Trans.* **2008**, *49*, 1081–1083. [CrossRef]

3. Lee, J.-K.; Kim, S.K. Effect of CaO composition on oxidation and burning behaviors of AM50 Mg alloy. *Trans. Nonferrous Met. Soc. China* **2011**, *21*, 23–27. [CrossRef]

4. Lee, D.B. High temperature oxidation of AZ31 + 0.3 wt. % Ca and AZ31 + 0.3 wt. % CaO magnesium alloys. *Corros. Sci.* **2013**, *70*, 243–251. [CrossRef]

5. Kim, S.K. Design and development of high-performance Eco-Mg alloys. In *Magnesium Alloys-Design, Processing and Properties*; Czerwinski, F., Ed.; InTech: Rijeka, Croatia, 2011; pp. 431–468.

6. Nam, T.H.; Kim, S.H.; Kim, J.G.; Kim, S.K. Corrosion resistance of extruded Mg-3Al-1Zn alloy manufactured by adding CaO for the replacement of the protective gases. *Mater. Corros.* **2014**, *65*, 577–581. [CrossRef]

7. Nam, N.; Bian, M.; Forsyth, M.; Seter, M.; Tan, M.; Shin, K. Effect of calcium oxide on the corrosion behaviour of AZ91 magnesium alloy. *Corros. Sci.* **2012**, *64*, 263–271. [CrossRef]

8. Esparza, J.; Davis, W.; Trillo, E.; Murr, L. Friction-stir welding of magnesium alloy AZ31B. *J. Mater. Sci. Lett.* **2002**, *21*, 917–920. [CrossRef]

9. Park, S.H.C.; Sato, Y.S.; Kokawa, H. Effect of micro-texture on fracture location in friction stir weld of Mg alloy AZ61 during tensile test. *Scr. Mater.* **2003**, *49*, 161–166. [CrossRef]

10. Carlone, P.; Palazzo, G. Characterization of TIG and FSW weldings in cast ZE41A magnesium alloy. *J. Mater. Process. Technol.* **2015**, *215*, 87–94. [CrossRef]

11. Carlone, P.; Astarita, A.; Rubino, F.; Pasquino, N. Microstructural Aspects in FSW and TIG Welding of Cast ZE41A Magnesium Alloy. *Metall. Mate. Trans. B* **2016**, *47*, 1340–1346. [CrossRef]

12. Liu, L.; Dong, C. Gas tungsten-arc filler welding of AZ31 magnesium alloy. *Mater. Lett.* **2006**, *60*, 2194–2197. [CrossRef]

13. Liu, L.-M.; Cai, D.-H.; Zhang, Z.-D. Gas tungsten arc welding of magnesium alloy using activated flux-coated wire. *Scr. Mater.* **2007**, *57*, 695–698. [CrossRef]

14. Wagner, D.; Yang, Y.; Kou, S. Spatter and porosity in gas-metal arc welding of magnesium alloys: Mechanisms and elimination. *Weld. J.* **2013**, *92*, 347–362.

15. Cao, X.; Jahazi, M.; Immarigeon, J.; Wallace, W. A review of laser welding techniques for magnesium alloys. *J. Mater. Process. Technol.* **2006**, *171*, 188–204. [CrossRef]

16. Liu, L.; Wang, J.; Song, G. Hybrid laser—TIG welding, laser beam welding and gas tungsten arc welding of AZ31B magnesium alloy. *Mater. Sci. Eng. A* **2004**, *381*, 129–133.

17. Choi, D.H.; Ahn, B.W.; Kim, S.K.; Yeon, Y.M.; Kim, Y.J.; Park, S.-K.; Jung, S.B. Microstructure evaluation of friction stir welded AZ 91 with CaO Mg alloy. *Mater. Trans.* **2011**, *52*, 802–805. [CrossRef]

18. Choi, D.-H.; Kim, S.-K.; Jung, S.-B. The microstructures and mechanical properties of friction stir welded AZ31 with CaO Mg alloys. *J. Alloy. Compd.* **2013**, *554*, 162–168. [CrossRef]

19. Kang, M.; Kim, C. Effect of CaO contents on Yb:YAG disk laser weldability of AZ31 Mg alloy. *Mater. Sci. Forum* **2015**, *804*, 31–34. [CrossRef]

20. Wang, J.; Nishimura, H.; Katayma, S.; Mizutani, M. Evaporation phenomena of magnesium from droplet at welding wire tip in pulsed MIG arc welding of aluminium alloys. *Sci. Technol. Weld. Join.* **2011**, *16*, 418–425. [CrossRef]

21. Kim, C.-H.; Ahn, Y.-N.; Lee, K.-B. Droplet transfer during conventional gas metal arc and plasma-gas metal arc hybrid welding with Al 5183 filler metal. *Curr. Appl. Phys.* **2012**, *12*, 178–183. [CrossRef]

22. Semenov, I.; Krivtsun, I.; Demchenko, V.; Semenov, A.; Reisgen, U.; Mokrov, O.; Zabirov, A. Modelling of binary alloy (Al-Mg) anode evaporation in arc welding. *Model. Simul. Mater. Sci. Eng.* **2012**, *20*. [CrossRef]

23. Reisgen, U.; Mokrov, O.; Zabirov, A.; Krivtsun, I.; Demchenko, V.; Lisnyi, O.; Semenov, I. Task of volumetrical evaporation and behaviour of droplets in pulsed MIG welding of AlMg alloys. *Weld. World* **2013**, *57*, 507–514. [CrossRef]

24. Fatemi-Varzaneh, S.; Zarei-Hanzaki, A.; Beladi, H. Dynamic recrystallization in AZ31 magnesium alloy. *Mater. Sci. Eng. A* **2007**, *456*, 52–57. [CrossRef]

25. Chae, H.; Kim, C.; Kim, J.; Rhee, S. Fume generation behaviors in short circuit mode during gas metal arc welding and flux cored arc welding. *Mater. Trans.* **2006**, *47*, 1859–1863. [CrossRef]

26. Kim, Y.; Eagar, T. Analysis of metal transfer in gas metal arc welding. *Weld. J.* **1993**, *72*, 269–278.

![metals logo] *metals*

MDPI

Article

Effects of Reflow Time on the Interfacial Microstructure and Shear Behavior of the SAC/FeNi-Cu Joint

Yunxia Chen [1,*], Xulei Wu [2], Xiaojing Wang [2] and Hai Huang [2]

[1] School of Mechanical and Electrical Engineering, Shanghai Dianji University, Shanghai 201306, China
[2] School of Material and Science Engineering, Jiangsu University of Science and Technology, Zhenjiang 212003, China; m18852890017@163.com (X.W.); xjwang.0@163.com (X.W.); 18252580705@163.com (H.H.)
* Correspondence: cyx1978@yeah.net; Tel.: +86-21-3822-6089; Fax: +86-21-3822-3186

Academic Editor: Giuseppe Casalino
Received: 21 March 2016; Accepted: 28 April 2016; Published: 11 May 2016

Abstract: Effects of reflow time on the interfacial microstructure and shear strength of the SAC/FeNi-Cu connections were investigated. It was found that the amount of Cu_6Sn_5 within the solder did not have a noticeable increase after a long time period of reflowing, indicating that the electro-deposited FeNi layer blocked the Cu atoms effectively into the solder area during a long period under liquid-conditions. The ball shear test results showed that the SAC/FeNi-Cu joint had a comparable strength to the SAC/Cu joint after reflowing, and the strength drop after reflowing for 210 s was less than that of the SAC/Cu joint.

Keywords: FeNi; UBM; IMCs

1. Introduction

Solder bump connection is an important interconnect in flip chip packages, where the solder bumps deposited on metal terminals on the chip are connected to the metal pads on the substrate [1]. The metal terminals consist of successive layers of metal, under bump metal (UBM), which provides a strong mechanical and electrical connection. It is often regarded as a metallurgical process because of the formation of intermetallic compounds (IMCs) [2,3]. Cu film is widely used as a UBM or soldering pad because of its outstanding wetting property, conductivity and cheapness. However, the Cu film is consumed too quickly in the liquid reaction with solder bump alloy during reflow and the solid thermal aging process. Therefore, a barrier film such as Ni-P layer is utilized to protect the excessive loss of Cu film [4–6]. However, the existence of Kirkendall voids around the solder/Ni interface and the P-rich layer formations can degrade the reliability of the connection dramatically. To some extent, the thinner IMC layer has been realized to be an effective approach to improve the reliability of the interconnections.

FeNi alloy has been widely applied in precision instruments and used as lead-frame materials due to its excellent low-expansion property. Early studies on the Fe-Ni alloy are mainly about the brazing or solid bonding processes. Recently, a few publications have reported on the interfacial reactions between Sn based solders and Fe-Ni alloys, discussing its potential application as UBM layer for Sn-based solders [7–16]. It is reported that the FeNi layer has an acceptable wettability for SnAgCu (SAC) solder with or without an adequate pre-treatment in soldering [7], exhibiting a slower interfacial reaction rate compared to the traditional UBMs [8]. Moreover, the shear resistance behavior of an SAC solder joint can be slightly improved by using the FeNi alloy rather than the Cu substrate [9,10]. Moreover, it is found that only $FeSn_2$ phase with minor Ni solubility formed between the FeNi substrate and Sn solder during liquid reactions at 270 °C [11–14]. Subsequently, the IMC formed at the Sn/FeNi interface was shown to be very sensitive to the concentration of Ni in FeNi alloys [15], and could transformed from (Ni,Fe)$_3$Sn$_4$ into FeSn$_2$ phase when the Fe content in the Fe-Ni layer was between 10% to 12.5% [16]. The Sn-Bi and Sn-Cu eutectic solders reacted with the Fe-Ni substrate. Yen *et al.* revealed that only the FeSn$_2$ phase was formed at the interfacial reactions [17]. In this study, the interfacial stability during reflow for different time periods was implemented. The microstructure and the shear properties of SAC/FeNi-Cu connection were investigated to discuss the potential application of FeNi layer as UBM layer for a Sn-based solder.

2. Experiments

The FeNi film surface was electro-deposited on a copper substrate with the atomic ratio of Fe:Ni near to 24:76 by adjusting the composition of the plating solution. The single lap shear joints of Cu/SAC/Cu (eutectic SAC solder) and Cu-FeNi/SAC/FeNi-Cu were fabricated using a solder ball of SAC alloy about 1 mm in diameter. The reflow process was executed on a BGA rework station at 260 °C for different time: 90, 150, 210 and 270 s respectively. Lap shear tests were carried out on a micromechanical testing system at a displacement control of 0.4 mm/min. The interfacial microstructures and the fracture surfaces after shear were observed by Scanning Electron Microscopy (SEM, Leica Cambridge, Cambridge, UK) and Energy Dispersive X-ray Spectroscopy (EDS, Shimadzu, Kyoto, Japan).

3. Results and Discussion

3.1. Interfacial Microstructure After Reflow

Figure 1 shows the surface morphology of FeNi film as deposition, a mirror like appearance. None is detected on gross or microscopic examination. Further EDS results, namely the semi-quantitative analysis, shows that element Fe in the prepared film is 25 at. %, see Figure 1b.

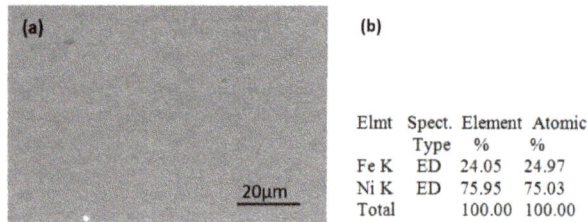

Elmt	Spect. Type	Element %	Atomic %
Fe K	ED	24.05	24.97
Ni K	ED	75.95	75.03
Total		100.00	100.00

Figure 1. (a) Surface morphology and (b) Energy Dispersive X-ray Spectroscopy (EDS) result of as-deposited FeNi plating.

Figure 2 displays the cross-sectional images of the interfaces between SAC solder and substrates. A scallop IMC layer can be seen in Figure 2a along the SAC/Cu interface, mainly showing the Cu$_6$Sn$_5$ phase indicated by EDS analysis. The peak thickness of the scallop morphology IMC was above 5 μm. When the SAC solder reacted with the deposited FeNi film under the same conditions, by contrast, a flat and thin IMC in sub-micrometer thickness was formed at the interface of SAC and the FeNi layer, see Figure 2b. This planar IMC was very thin, about 180 nm, indicating that the consumption rate of FeNi film was enormously slow compared with the Cu substrate during the reflow process.

Based on the results regarding SAC/FeNi-Cu and SAC/FeNi from Guo and Hwang [7,14], this very thin IMC was determined to be FeSn$_2$ phase with a tetragonal crystal structure determined by Zhu at the same 260 °C reflowing conditions [9]. It is completely in accordance with our previous results on the liquid reaction between Sn based solder and FeNi substrate [18], where Sn reacts with Fe preferentially over Ni.

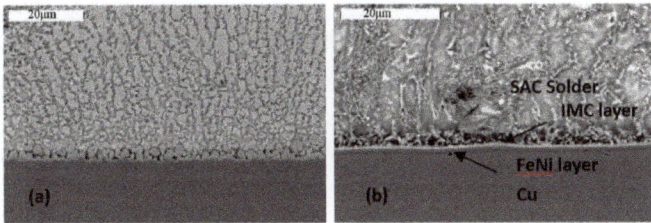

Figure 2. Intermetallic compounds (IMCs) between SnAgCu (SAC) solder and substrates after reflowing for 90 s: (**a**) Cu and (**b**) FeNi-Cu.

3.2. Interfacial Microstructures After Reflowing for Different Time

During the reflowing processes at 260 °C, it was found that the total IMC thickness of Cu_6Sn_5 between SAC solder and substrate Cu grew thicker with the reflow time prolonged from 90 to 270 s (see Figure 3). At such a high temperature, Cu atoms easily diffused into SAC solder, and firstly reacted with Sn at the SAC/Cu interface to form Cu_6Sn_5 IMC. Along with this, Cu atoms also diffused into the inner SAC solder and formed irregular Cu_6Sn_5 phase. During the process, no other IMC phases (*i.e.*, Cu_3Sn emerging during the solid state aging) showed in the reflow time-increasing procedure for the formation needing longer incubation time [19]. The scallop morphology of Cu_6Sn_5 also remained.

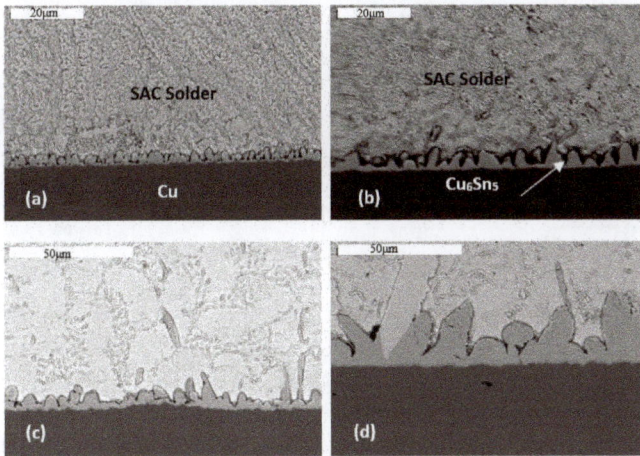

Figure 3. Interfacial IMC of SnAgCu/Cu interface after reflow for (**a**) 90 s, (**b**) 150 s, (**c**) 210 s and (**d**) 270 s.

For comparison, the IMC thickness between the SAC solder and the FeNi substrate did not show any obvious increase, as shown in Figure 4a–d. The thickness still kept blow 0.5 μm even after reflowing for almost 270 s. Moreover, the IMC type at the FeNi-Cu/SAC interface remained the single $FeSn_2$ phase.

It is worth noting that the $FeSn_2$ phase remains unchanged between the FeNi-Cu/SAC interface. Only the microstrucutre around the interface is slightly different. In Figure 4b, for example, the element Ni from UBM diffused into the solder rather than producing Ni-Sn IMC phase near the interface. The EDS result in Table 1 showed that some of the Cu_6Sn_5 phase within solder near the interface turned out to be Cu_6Sn_5 containing Ni, *i.e.*, $(Cu,Ni)_6Sn_5$, where some places of Cu atoms in Cu_6Sn_5 crystalline lattice were occupied by the Ni atoms [20].

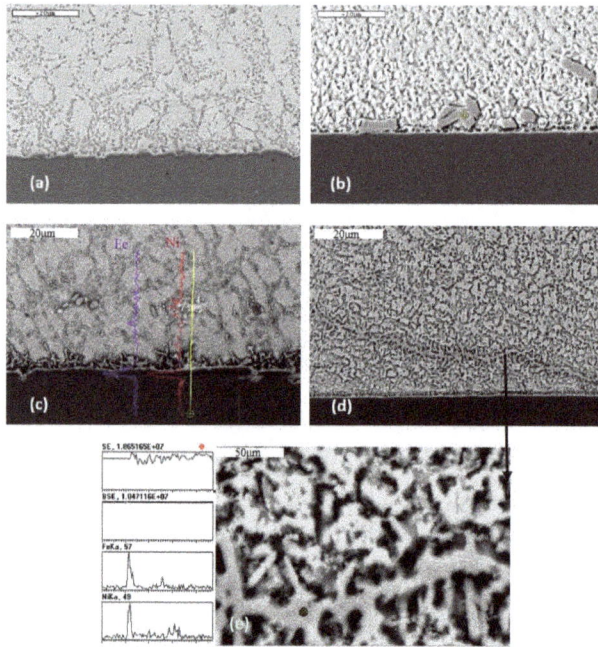

Figure 4. Interfacial IMCs of Sn-Ag-Cu/FeNi/Cu interface after reflow for (**a**) 90 s, (**b**) 150 s, (**c**) 210 s, (**d**) 270 s and (**e**) magnified image of (**d**).

Table 1. Compositions of marked spots in Figure 4e and the possible phases.

Markers	Elements Composition (at. %)					Possible Phases
	Fe	Ni	Cu	Ag	Sn	
1	0.40 *	15.87	39.98	0.13 *	43.63	$(Cu,Ni)_6Sn_5$
2	0.00 *	11.28	23.89	2.52	62.31	$(Cu,Ni)_6Sn_5$

* $\leqslant 2\sigma$.

With increasing reflow time, the Ni diffused in the solder and formed an Ni-containing Cu_6Sn_5 compound, which made the area near the interface getting coarser. We even observed spalling away from the interface like a band when reflowing time was prolonged to 270 s, see Figure 4c,d. The further line scanning and EDS result showed that this IMC could also be the Ni-containing Cu_6Sn_5.

For SAC/Cu joints, the amount of Cu_6Sn_5 phase was much larger with increasing reflow time. This means that the excessive Cu came from the Cu substrate during the reflowing process. In comparison, for SAC/FeNi-Cu joints, the amount of Cu_6Sn_5 within the solder did not have a noticeable increase after a long time period of reflowing, indicating that the electro-deposited FeNi layer blocked the Cu atoms effectively into the solder during a long time stay in liquid conditions.

3.3. IMCs Growth Depending on Reflow Time

The dependence of the IMCs thickness on reflow time for the SAC/Cu and SAC/FeNi-Cu joints are shown in Figure 5. It can be seen that the IMCs thickness increases almost linearly with prolonged reflow time, which is quite different from the results reported in [17], a parabola relationship. This difference might be caused by different internal mechanisms of IMC growth during the liquid/solid process. Generally, the growth of the IMCs between the solder and the UBM are regarded as a reaction and diffusion controlled process. During this process, a new IMC phase may form resulting from the chemical reaction between the liquid solder and the solid UBM. Thus, the elements required to create the new phase may diffuse from the UBM and solder respectively across the IMC to the other side. Under a certain temperatures, the reaction layers grow, depending on the evolution of IMCs in the interface. Therefore, the whole process can be co-accomplished by the surface chemical reaction and the internal atoms diffusion mechanisms. Here, d can be used to show the growth of the IMC phase, then its relationship with time can be expressed as: $d = (kt)^n$. Among them, k is the growth rate constant of the IMC layer, t is the IMC growth time, and n is the time index. When the compound growth is controlled by surface chemical reaction, the increase of the compound thickness with time prolonging should be a linear relationship ($n = 1$). When the reaction required elemental atoms diffusion dominates the compound growth, a parabola relationship between d and t begin to play a role ($n = 0.5$). From the results in Figure 5, the k values of growth rate could be calculated as 3.5×10^{-9} and 3.4×10^{-1} m·s^{-1} for the SAC/Cu and SAC/FeNi-Cu joints respectively at 260 °C. For the SAC/FeNi-Cu joint, because the element Fe acted as the dominant reaction element forming the interfacial $FeSn_2$ IMC. It can be deduced that the reaction rate of $FeSn_2$ phase formation are ten times slower than the Cu-Sn compound. Also, the FeNi layer prevented the Cu from forming Cu_6Sn_5 IMC along the interface.

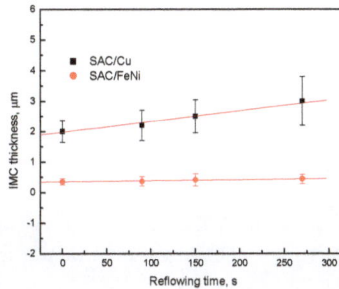

Figure 5. Dependence of the thicknesses for IMC layers on the reflow time.

3.4. Shear Properties and the Fracture Behavior

As shown in Figure 6, the maximum shear strength occurred at the joints reflowed for 90 s for the SAC/Cu and SAC/FeNi-Cu joints. It can be seen that both joints had a comparable strength during the reflow time range from 90 to 270 s. After reflowing for 90 s, the strength of SAC/Cu dropped as the reflow time increased, especially at the initial stage. Comparatively, the strength of the SAC/FeNi-Cu joint had a slight decrease after a short reflow time, and then remained almost constant as reflow time increased. This result showed that the SAC/FeNi-Cu joints were less sensitive to the reflow time.

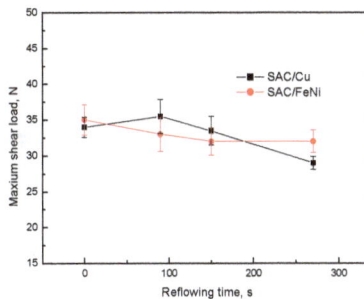

Figure 6. Dependence of the maximum shear force on the reflowing time for the SAC/Cu and SAC/FeNi-Cu connections.

The relationship between the shear strength and the shear displacement for the SAC/Cu and SAC/FeNi-Cu joints reflowed at 260 °C for 270 s and the corresponding fracture surface observations are displayed in Figure 7. The fracture surfaces of both joints showed a mixture feature of ductile dimples and shear bands within the solders. In general, the fracture mode of the SAC/FeNi-Cu joint presented a more ductile process and represented a much higher mechanical reliability than the fractures within or between the different IMC layers or along the IMC/substrate interface. This result indicated that the SAC/FeNi-Cu joint had a better mechanical reliability after a long time period of reflowing at 260 °C.

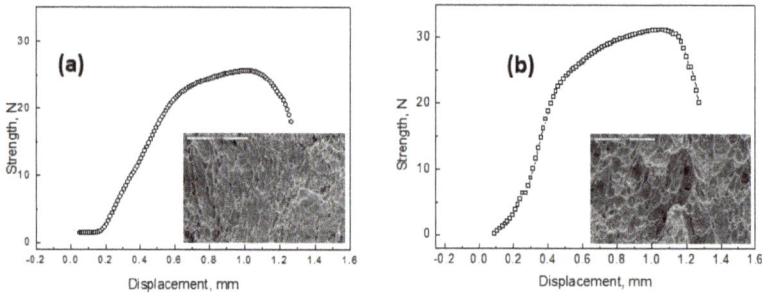

Figure 7. The shear displacement-shear strength curves and the corresponding fracture surface morphology of (**a**) SAC/ Cu and (**b**) SAC/FeNi-Cu joints after reflowing for 270 s.

4. Conclusions

During reflowing, a very thin $FeSn_2$ IMC layer was formed at the SAC/FeNi-Cu interface. The growth rate of the $FeSn_2$ layer was much lower than that of the Cu-Sn IMC at the SAC/Cu interface with prolonged reflow time. This indicated that the electro-deposited FeNi layer could effectively block the Cu atom diffusion into the solder during a much longer time period under liquid conditions. The solder lap shear test showed that the SAC/FeNi-Cu connection had an acceptable mechanical reliability after a long time period of reflowing at 260 °C.

Acknowledgments: This work was supported by special foundation of Shanghai Economic and Information Commission, No. JJ-YJCX-01-15-5718, Zhongche Nanjing Puzhen Vehicle Co. Ltd., No. C022.

Author Contributions: Yunxia Chen conceived and designed the experiments, analyzed the data and wrote the paper; Xulei Wu and Xiaojing Wang performed the experiments; Hai Huang contributed analysis tools.

Conflicts of Interest: The authors declare no conflict of interest.

References

1. Wong, C.P.; Luo, S.J.; Zhang, Z.Q. Microelectronics: Flip the Chip. *Science* **2000**, *290*, 2269–2270. [CrossRef] [PubMed]
2. Ho, C.E.; Yang, S.C.; Kao, C.R. Interfacial reaction issues for lead-free electronic solders. *J. Mater. Sci.* **2007**, *18*, 155–174.
3. Laurila, T.; Vuorinen, V.; Kivilahti, J.K. Interfacial reactions between lead-free solders and common base materials. *Mater. Sci. Eng. R* **2005**, *49*, 1–60. [CrossRef]
4. Ghosh, G. Interfacial microstructure and the kinetics of interfacial reaction in diffusion couples between Sn-Pb solder and Cu/Ni/Pd metallization. *Acta Mater.* **2014**, *53*, 205–211. [CrossRef]
5. Chen, C.; Zhang, L.; Lai, Q.; Li, C.; Shang, J.K. Evolution of solder wettability with growth of interfacial compounds on tinned FeNi plating. *J. Mater. Sci. Mater. Electron.* **2011**, *22*, 1234–1238. [CrossRef]
6. He, M.; Chen, Z.; Qi, G.J. Solid state interfacial reaction of Sn-37Pb and Sn-3.5Ag solders with Ni-P under bump metallization. *Acta Mater.* **2004**, *52*, 2047–2056. [CrossRef]

7. Guo, J.J.; Zhang, L.; Xian, A.P.; Shang, J.K. Solderability of Electrodeposited Fe-Ni Alloys with Eutectic SnAgCu Solder. *J. Mater. Sci. Technol.* **2007**, *23*, 811–816.
8. Dariavach, N.; Callahan, P.; Liang, J.; Fournelle, R. Intermetallic growth kinetics for Sn-Ag, Sn-Cu, and Sn-Ag-Cu lead-free solders on Cu, Ni, and Fe-42Ni substrates. *J. Electron. Mater.* **2006**, *35*, 1581–1592. [CrossRef]
9. Zhu, Q.S.; Guo, J.J.; Shang, P.J.; Wang, J.G.; Shang, J.K. Effects of Aging on Interfacial Microstructure and Reliability Between SnAgCu Solder and FeNi/Cu UBM. *Adv. Eng. Mater.* **2010**, *12*, 497–503. [CrossRef]
10. Liu, Y.; Zhang, J.; Yan, J.; Du, J.; Gan, G.; Yi, J. Direct Electrodeposition of Fe-Ni Alloy Films on Silicon Substrate. *Rare Met. Mater. Eng.* **2014**, *43*, 2966–2968.
11. Hwang, C.W.; Suganuma, K.; Saiz, E.; Tomsia, A.P. Wetting and interface integrity of Sn-Ag-Bi solder/Fe-42% Ni alloy system. *Trans. JWRI* **2001**, *30*, 167–172.
12. Suganuma, K. Advances in lead-free electronics soldering. *Curr. Opin. Solid State Mater. Sci.* **2001**, *5*, 55–64. [CrossRef]
13. Hwang, C.W.; Kim, K.S.; Suganuma, K. Interfaces in lead-free solder. *J. Electron. Mater.* **2003**, *32*, 1249–1256. [CrossRef]
14. Hwang, C.W.; Suganuma, K.; Lee, J.G.; Mori, H. Interface microstructure between Fe-42Ni alloy and pure Sn. *J. Mater. Res.* **2003**, *18*, 1202–1210. [CrossRef]
15. Ozaki, H.; Yamamoto, T.; Sano, T.; Hirose, A.; Kobayashi, K.F.; Ishio, M.; Shiomi, K.; Hashimoto, A. Formation of micro joints with high-melting point through rapid reaction between Sn-3.5Ag solder and Ni-Fe alloy substrate. In Proceedings of Materials Science & Technology Conference and Exhibition (MS & T'06), Cincinnati, OH, USA, 15–19 October 2006; pp. 1415–1424.
16. Yen, Y.W.; Hsiao, H.M.; Lin, S.W.; Lin, C.K.; Lee, C. Interfacial reaction in Sn/Fe-xNi couples. *J. Electron. Mater.* **2011**, *41*, 144–152. [CrossRef]
17. Yen, Y.W.; Syu, R.S.; Chen, C.M.; Jao, C.C.; Chen, G.D. Interfacial reactions of Sn-58Bi and Sn-0.7Cu lead-free solders with Alloy 42 substrate. *Microelectron. Reliab.* **2014**, *54*, 233–238. [CrossRef]
18. Wang, X.J.; Li, T.Y.; Chen, Y.X.; Wang, J.X. Current induced interfacial microstructure and strength weakening of SAC/FeNi-Cu connection. *Appl. Mech. Mater.* **2014**, *651–653*, 11–15. [CrossRef]
19. Li, J.F.; Agyakwa, P.A.; Johnson, C.M. Interfacial reaction in Cu/Sn/Cu system during the transient liquid phase soldering process. *Acta Mater.* **2011**, *59*, 1198–1211. [CrossRef]
20. Vuorinen, V.; Yu, H.; Laurila, T.; Kivilahti, J.K. Formation of Intermetallic Compounds Between Liquid Sn and Various CuNi$_x$ Metallizations. *J. Electron. Mater.* **2008**, *37*, 792–805. [CrossRef]

![metals logo] *metals*

MDPI

Article

Microstructure and Mechanical Properties of Friction Welding Joints with Dissimilar Titanium Alloys

Yingping Ji [1,2], Sujun Wu [2,*] and Dalong Zhao [2]

[1] School of Mechanical Engineering, Ningbo University of Technology, Ningbo 315211, China; yingping04@163.com

[2] School of Materials Science and Engineering, Beihang University, Beijing 100191, China; zhaodalong_buaa@126.com

* Correspondence: wusj@buaa.edu.cn; Tel.: +86-10-8231-6326

Academic Editor: Giuseppe Casalino

Received: 5 April 2016; Accepted: 5 May 2016; Published: 10 May 2016

Abstract: Titanium alloys, which are important in aerospace application, offer different properties via changing alloys. As design complexity and service demands increase, dissimilar welding of the titanium alloys becomes a particular interest. Linear friction welding (LFW) is a relatively novel bond technique and has been successfully applied for joining titanium alloys. In this paper, dissimilar joints with Ti-6Al-4V and Ti-5Al-2Sn-2Zr-4Mo-4Cr alloys were produced by LFW process. Microstructure was studied via optical microscopy and scanning electron microscopy (SEM), while the chemical composition across the welded samples was identified by energy dispersive X-ray spectroscopy. Mechanical tests were performed on welded samples to study the joint mechanical properties and fracture characteristics. SEM was carried out on the fracture surface to reveal their fracture modes. A significant microstructural change with fine re-crystallization grains in the weld zone (WZ) and small recrystallized grains in the thermo-mechanically affected zone on the Ti-6Al-4V side was discovered in the dissimilar joint. A characteristic asymmetrical microhardness profile with a maximum in the WZ was observed. Tensile properties of the dissimilar joint were comparable to the base metals, but the impact toughness exhibited a lower value.

Keywords: dissimilar friction weld; titanium alloy; evolution of microstructure; mechanical property

1. Introduction

Dissimilar weld is attracting increasing attention because it can take advantage of specific attributes of each material to enhance the performance of a product or introduce new functionalities. They are applied in various fields such as thermal power station, nuclear industries, automobile, aerospace, *etc.* A number of dissimilar joints with aluminum, titanium, ferrous and many kinds of materials have been successfully formed by various methods from fusion welding to friction welding process [1–5].

With high strength to weight ratio, corrosion resistance, and good strength sustainability at high temperatures, titanium alloys are important in aerospace applications. As design complexity and service demands increase, dissimilar welds with titanium alloys become a particular interest in the field of aerospace industry [3–5]. There have been a number of studies reporting the welding of dissimilar titanium alloys using various different welding processes, including friction stir welding [3], ultrasonic spot welding [4], linear friction welding (LFW) [5,6], tungsten inert gas welding [7] and electron beam welding [8].

LFW is a relatively novel solid-state joining process where two metals are welded together under reciprocating motion and apply force against each other [9]. Compared with traditional fusion welding technologies, LFW has many advantages such as less defect formation and the ability to join dissimilar materials and complex geometrical components, and it often negates the need for protective gas [10]. To date, LFW has been successfully used to join titanium alloys [9–12], nickel-base alloys [13,14] as well as other materials [15–17]. More importantly, the process can be viable for the production of dissimilar welds. For example, Bhamji *et al.* applied LFW process successfully to join aluminum and copper. The welds had good electrical and mechanical properties [18]. They also joined an aluminum alloy to a magnesium alloy by LFW and found that these welds had a reasonable strength [19]. Ma *et al.* produced a dissimilar Ti-6Al-4V and Ti-6.5Al-3.5Mo-1.5Zr-0.3Si joint by LFW and identified the microstructural evolution of the joint. They found different microstructure zones such as the TMAZs and weld zones in both sides of base metals. In addition, they investigated the mechanical properties and found the tensile strength of the joint was comparable to that of the parent [20]. Wen *et al.* made LFW dissimilar joints of Ti-6Al-4V and Ti-6.5Al-3.5Mo-1.5Zr-0.3Si and evaluated the microstructure, microhardness, and fatigue properties of the joint, which had essentially symmetrical hysteresis loops and an equivalent fatigue life to the base metals [21]. Zhao *et al.* investigated the influence of strain rate on the tensile properties of LFW dissimilar joints between Ti-6.5Al-3.5Mo-1.5Zr-0.3Si and Ti-4Mo-4Cr-5Al-2Sn-2Zr titanium alloys [22]. Frankel *et al.* compared the residual stresses between Ti-6Al-4V and Ti-6Al-2Sn-4Zr-2Mo [23].

Although LFW is a promising process for joining dissimilar titanium alloys, there are only a few public papers on this subject [20–23]. There are a great many publications about the LFW process for same titanium alloy. Karadge *et al.* detailed the texture of LFW Ti-6Al-4V joints by experiment [24]. Microstructural evolution of LFW Ti-6.5Al-3.5Mo-1.5Zr-0.3Si joint [25], the relationship between forging pressure and the microstructure of LFW Ti-6Al-4V joint were also revealed [26]. Interrelationship of microstructure and mechanical properties of LFW titanium joint was also investigated [27]. In addition, to predicting various weld responses, such as thermal fields and microstructural evolution, a great number of finite element models were established and the predictions of the models were found to be in good agreement with the experimental results [28–31].

These investigations on LFW joints with dissimilar and similar materials show that the microstructure and property can be established for given LFW joints through extensive experiments. However, due to the complicated nature of interaction between the LFW thermomechanical environment and the material microstructure, these findings cannot be easily extended to other LFW joints. Consequently, to expand the application of LFW process in dissimilar titanium alloy, it is necessary to investigate LFW joint with Ti-6Al-4V (Ti64) and Ti-5Al-2Sn-2Zr-4Mo-4Cr (Ti17). In this paper, two different titanium alloys consisting of one $\alpha + \beta$ alloy Ti64, and one near-β alloy Ti17, were welded by LFW process. The present study was focused on revealing the micro-structural characterization, mechanical properties, as well as the fracture mode of the dissimilar joints.

2. Materials and Methods

The base metals for the LFW process are Ti64 and Ti17 alloys, whose nominal chemical compositions are listed in Table 1.

Table 1. Chemical compositions of base metals (wt. %).

Alloy	Al	V	Sn	Zr	Mo	Cr	Fe	Si	C	N	H	O	Ti
Ti64	6.06	3.93	-	-	-	-	0.103	0.15	0.106	0.033	0.015	0.13	Balance
Ti17	5.05	-	2.13	2.07	4.12	4.13	0.30	-	0.05	0.05	0.013	0.08	Balance

The typical microstructures of the as-received materials revealed by scanning electron microscopy (SEM) are shown in Figure 1. SEM analysis was performed on an Apollo 300 (Camscan, Cambridge, UK). As shown in Figure 1a, Ti64 alloy is characterized by typical bimodal microstructure with globular primary α (α_p) distributed in the matrix of transformed β (β_t). The prior β grain size is 10–15 μm and the α_p size is about 20 μm. Ti17 alloy has a typical lamellar structure with lath α_p in 10–30 μm length and fine secondary α (α_s) embedded in β phases (Figure 1b).

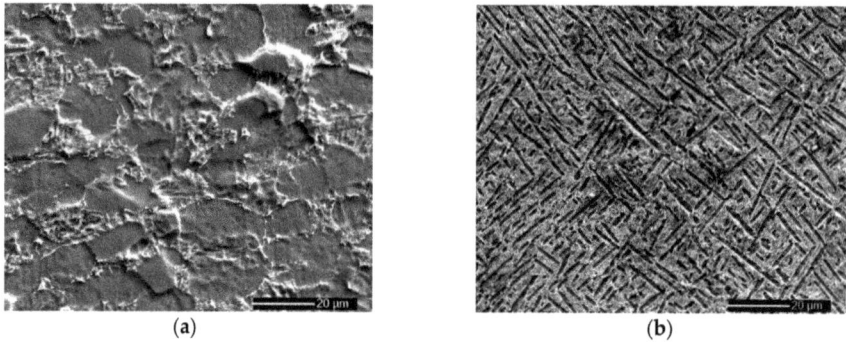

Figure 1. Microstructure of base metals: (**a**) Ti64 and (**b**) Ti17.

Attempts were made to weld samples of geometry 130 × 75 × 20 mm with a weld interface of 75 × 20 mm. Ti64 and Ti17 titanium alloys were used for the dissimilar LFW trials. Welds were produced on a homemade linear friction welding machine of LFW-20T. The Ti64 sample was reciprocated, while the Ti17 sample was held stationary. Prior to welding, the welding surfaces of the samples were ground and cleaned in an acetone bath. The welding parameters were selected as follows: amplitude of oscillation of 3 mm, frequency of oscillation of 50 Hz, friction force of 4.8 kN and friction time of 3 s. Post weld heat treatment was carried out at 630 °C for 3 h in vacuum to relieve residual stress. The welded specimens for investigation were free from surface defects and internal defects.

The LFW specimens for micro-structural observation were cut perpendicularly to the reciprocating motion direction and prepared by standard procedures followed being etched with Kroll's reagent (2 mL concentrated nitric acid, 1 mL hydrofluoric acid and 5 mL distilled water). Microstructure was investigated by light optical microscopy (OM, Olympus, Tokyo, Japan) and SEM. Energy dispersive X-rays (EDS, Kratos Analytical Ltd, Manchester, UK) was applied to analyze the compositional change across the welds.

Mechanical property studies included Vicker's micro-hardness tests, tensile tests and U notch impact toughness tests. Vickers micro-hardness tests were performed with a load of 500 g and a dwell time of 15 s. Tensile tests were carried out at room temperature in accordance with GB/6397-86 (China), using a fully computerized tensile testing machine at constant strain rates of 10^{-4} m·s^{-1}. A drop hammer impact testing machine was used to measure the impact toughness of specimens, which was performed according to GB/T229-1994 (China). For comparison, base material specimens were tested and had the same overall dimensions as those for the welded. For a given variant, at least three specimens were tested.

To ensure the center of U-notch located in the center of weld zone, the weldments for impact tests were polished on one side and etched by a Kroll's reagent before machining the U-notch. The configuration and size of specimens for tensile and impact tests are shown in Figure 2a,b, respectively.

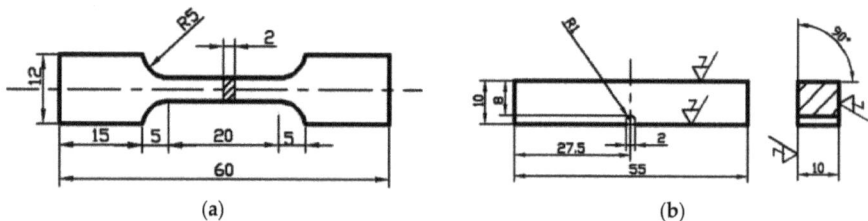

Figure 2. Configurations of specimens for (**a**) tensile and (**b**) impact toughness tests.

3. Results and Discussion

3.1. Macro and Microstructure

A low magnification overview of a dissimilar LFW joint using OM is presented in Figure 3. The weld interface appearing wavy is obvious between two base metals. On both sides, the microstructure shows a gradual change from the weld interface towards the base metals, but a less gradual micro-structural transition and smaller region with elongated grains is observed in the side of Ti17 than Ti64. According to the micro-structural characteristics, welded joints could be divided into four zones: weld zone (WZ), thermo-mechanically affected zone (TMAZ), heat affected zone (HAZ) and base metals (BM). Faint indications of grain boundaries are seen, but, overall, the grains along the weld center are not effectively revealed. To reveal the details of the weld microstructure, the corresponding SEM images of different zones at high magnification are provided in Figure 4.

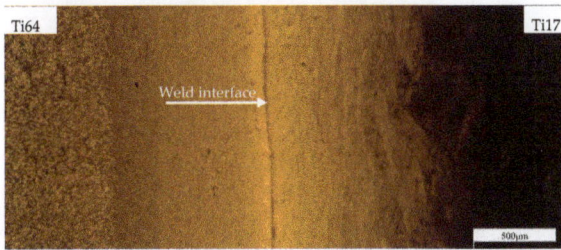

Figure 3. Cross section of the linear friction welding (LFW) joints with Ti64 and Ti17 alloys.

As shown in Figure 4a, in the HAZ of Ti64 side (Ti64-HAZ), an un-deformed bimodal microstructure that is characteristic of the base metal Ti64 (Figure 1a) can be observed clearly. However, the image of α_s in β_t becomes less clear while the shape of α_p remains unchanged, suggesting that the dissolution of the α_s occurred in this region during welding. With the decreasing distance to the weld center, the thermo-mechanically affected zone of Ti64 (Ti64-TMAZ) with severe plastic deformation is observed, where the α_p are elongated and the residual β are reoriented along the oscillation direction. In addition, some necklace-shaped microstructures distributed along the grain boundaries of deformed grains are also observed in this region (Figure 4b). This indicates that partially re-crystallization occurred to Ti64-TMAZ adjacent to weld center during the LFW process. Unlike HAZ and TMAZ, WZ has completely different microstructures, which consists of fully re-crystallized grains with fine size around 20 μm (Figure 4c). In the case of microstructures in the Ti17 side, they are similar to those in the Ti64 side. As shown in Figure 4d,e, grains are elongated and reoriented with their long dimension perpendicular to the applied force in the thermo-mechanically affected zone of Ti17 (Ti17-TMAZ) while some retention and annihilation of the Ti17 microstructures are observed in heat affected zone of Ti17 (Ti17-HAZ).

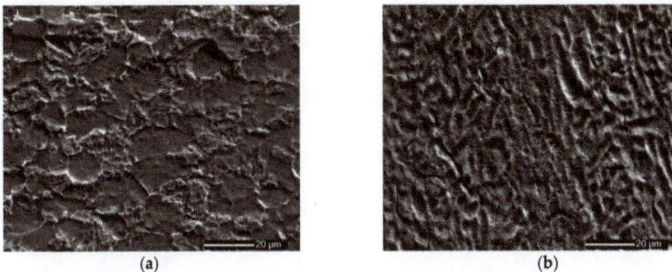

(a)   (b)

Figure 4. *Cont.*

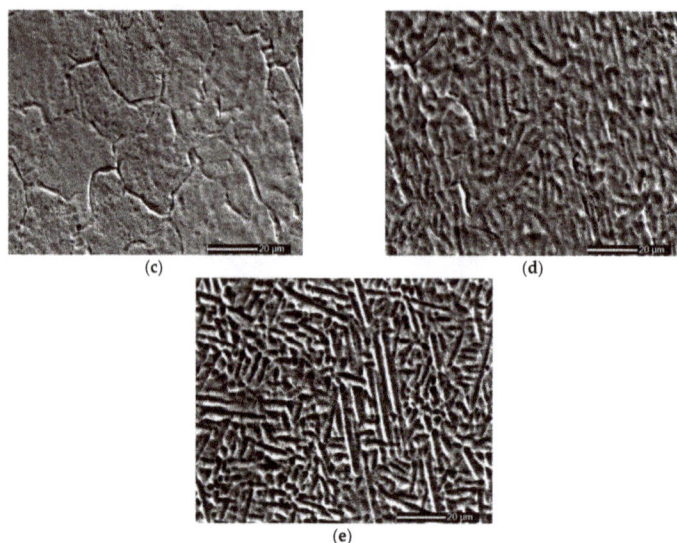

Figure 4. Typical microscopic structures across a friction welded joint with Ti64 and Ti17 alloys:
(a) Ti64-heat affected zone (HAZ); (b) Ti64-thermo-mechanically affected zone (TMAZ); (c) weld zone
(WZ); (d) Ti17-TMAZ and (e) Ti17-HAZ.

As described in the above section, a significant micro-structural change is observed across
the dissimilar joints. The gradient of temperatures and strains along the joints attribute to the
difference. During LFW process, WZ simultaneously underwent serious plastic deformation and
sufficient frictional heat due to the linear movement influenced by friction and upset pressures, where
temperatures above the β-transus ($β_t$) point [32,33]. This plastic deformation introduced a large
number of dislocations in the materials of welding interface. As the density of these dislocations
increased, they tended to form sub-grain cell structures. These low-angle grains rotated to form
high-angle strains free grains, resulting in very fine equal-axed grains in WZ. In the case of materials
in the TMAZ, they experienced thermo-mechanical deformation in sub $β_t$ temperature [33,34] and
the deformation was in a smaller extent, so there was no sufficient deformed energy to fully activate
re-crystallization and the part re-crystallization occurred to the deformed grain boundaries. In the case
of HAZ, a little heat was conducted from the WZ to the zone, leading to the dissolution of $α_s$.

3.2. Compositional Analysis

A line scan EDS profile across the weld interface (scan in Figure 5a) is shown in Figure 5b. It clearly
shows the existence of a mixed layer at the interface with a composition between that of Ti64 and Ti17.
The composition abruptly changes at the Ti64 interface while the transition is more gradual on the
Ti17 side of the joint. It indicates that the dissimilar welds had a compositional heterogeneity and
element diffusion occurred to Ti64 and Ti17 during LFW process. For example, a significant diffusion
of V (The curve with red colure) into Ti17 side is obvious. However, no long range inter-diffusion
crossing the interface could be observed, which is in agreement with the study on LFW of Ti-6Al-4V
and Ti-6Al-2Sn-4Zr-6Mo [10]. This may be related to the limited atomic diffusion resulting from the
short time above the βt temperature during the LFW process.

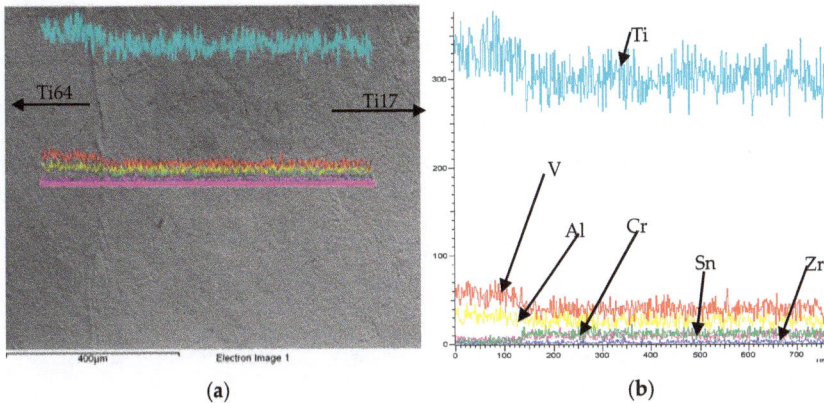

Figure 5. (**a**) SEM micrograph of LFW joints with Ti64 and Ti17 alloys and (**b**) the corresponding SEM-EDS line scans across the interface.

3.3. Microhardness

The microhardness across the weld interface of dissimilar LFW joints is presented in Figure 6. An asymmetrical microhardness profile across the weld is obvious with an average value of 332 HV for Ti64 and approximately 350 HV for Ti17, respectively. In addition, a distinct increase from each base metal up to the weld interface is also observed and a maximum of 420 HV occurs to WZ. The micro-hardness development in the LFW joints with dissimilar Ti64 and Ti17 alloys is similar to what has been reported in the literature studied on LFW joints with titanium alloys [33]. The increase in hardness of WZ could be ascribed to the grain refinement resulted from dynamic re-crystallization. It is expected that the high hardness within the WZ is going to influence the fracture properties of the welds.

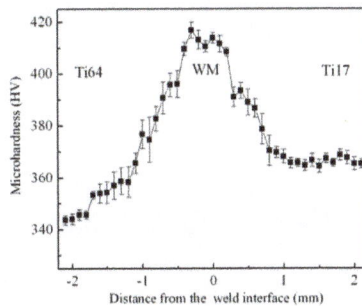

Figure 6. Transverse hardness of LFW joints with Ti64 and Ti17 alloys.

3.4. Tensile Properties

The average tensile properties of four LFW joints are given in Table 2. For a comparison, the tensile properties of base metals Ti64 and Ti17 are also included, where the values are the mean of three specimens for each base metal. For base metals, Ti64 has yield strength (YS) of 838 MPa, ultimate tensile strength (UTS) of 904 MPa and ductility of 14.6%, while Ti17 has higher strength and lower ductility. The higher strength of Ti17 than Ti64 is responsible for the smaller width of Ti17-TMAZ than Ti64-TMAZ as presented in Figure 3. In the case of LFW joints, they show a superior strength and lower ductility than base metal Ti64, but exhibit a contrary trend compared with base metal Ti17.

It should be stated that the failure of LFW specimens in the tensile tests located in the Ti64 side, approximately 1.2 mm away from the welding interface. It suggests that a highly durable and sound dissimilar joint of Ti64 and Ti17 was achieved via LFW process. The phenomenon was also observed in the tensile and fatigue tests on dissimilar LFW joints of Ti-6Al-4V to Ti-6.5Al-3.5Mo-1.5Zr-0.3Si [20,21]. This may be resulting from the lower strength of Ti64 and the reaction zone formed at Ti64 side. Firstly, compared with the regions of Ti17 side close to the interface, the regions in the Ti64 side had lower hardness and strength, where cracks were apt to initiate. Secondly, a reaction zone with equi-axil grains formed in Ti64 side, this would be the weakest zone for the crack to initiate. However, further work is still needed to clearly reveal this phenomenon.

Table 2. Tensile properties of base metals and joints.

Specimen	UTS/MPa	YS/MPa	Elongation δ%
Ti64	904	838	14.6
Ti64/Ti17 Joints	950	888	11.9
Ti17	1134	1044	10.6

The tensile fracture surfaces exhibited in Figure 7 indicate that all specimens crack in a ductile mode. It is evident that the dimples on the surface of dissimilar welds are in globular shaped (Figure 7b), which are similar to that on the fracture surface of base metal Ti64 (Figure 7a). This resulted from the fact that the failure of welding joints occurred in the zone of Ti64 side. In the case of Ti17, the dimples are strip-shaped (Figure 7c), which are smaller and shallower than those on the surface of Ti64 and joints, corresponding to the smaller elongation.

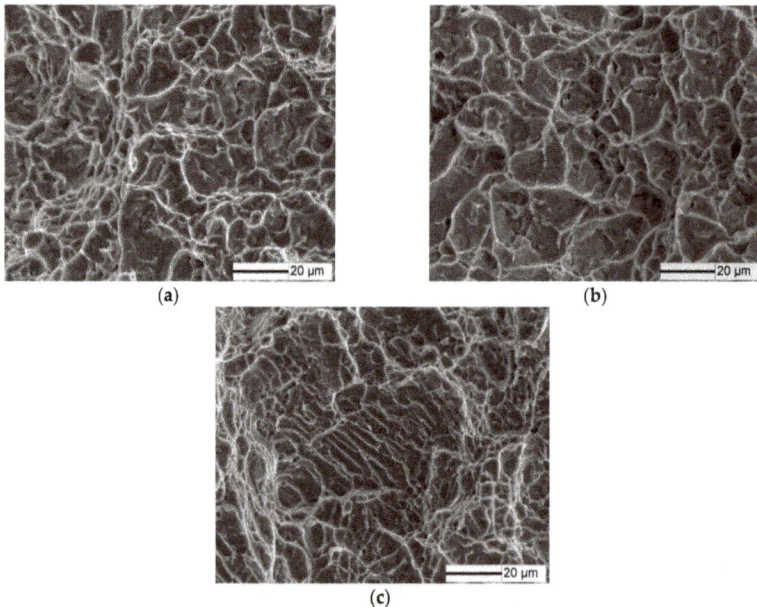

(a)

(b)

(c)

Figure 7. Fracture surface of (a) Ti64; (b) joints and (c) Ti17.

3.5. Impact Toughness

The impact toughness is characterized by a_k, a value defined through dividing the impact energy by the minimum cross-sectional area of the starting sample [35]. The impact toughness of joints and base metals are shown in Table 3, which are the average values of three specimens for the given variant. It is found that Ti17 has superior impact toughness than Ti64. In addition, WZ presents a lower toughness than base metals with an average value about 38.7 J/cm². This is in disagreement with the previous research on impact toughness of linear friction welded Ti-6Al-4V alloy joints, where the impact toughness of the weld was higher than that of the base metal Ti-6Al-4V [35].

Table 3. Impact toughness of WZ and base metals.

Sample	Ti64	WZ	Ti17
a_k (J/cm²)	44.5	38.7	53.6

The corresponding fracture surfaces of impact toughness specimens are shown in Figure 8. It is obvious that the fracture surfaces of base metals both show trans-granular fracture mode (Figure 8a,c), which are similar to that of tensile tests. In contrast to base metals, some small dimples and facts corresponding with the fine grains are shown in WZ (Figure 8b), suggesting that a mixture failure mode of trans-granular coupled with inter-granular occurred in WZ. Failure in the inter-granular mode that occurred in the WZ of LFW joint was also observed in fracture toughness tests [36].

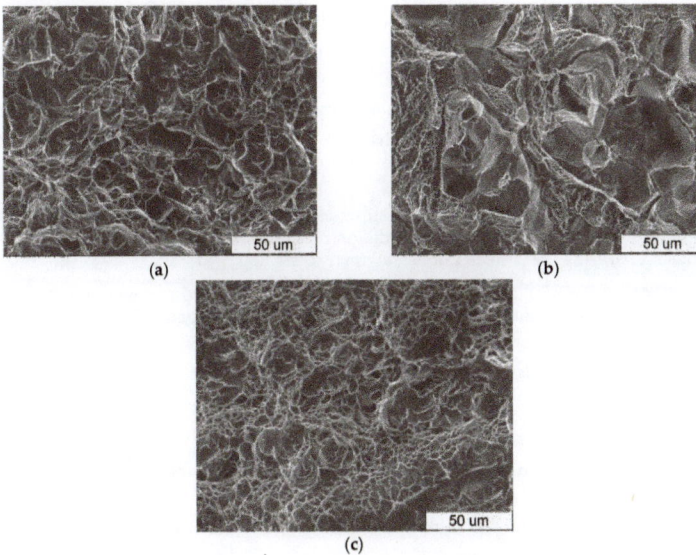

Figure 8. The typical fracture surface of (**a**) Ti64; (**b**) WZ and (**c**) Ti17.

For base metals, Ti17 alloy shows a higher toughness than Ti64 alloy, which results from the microstructures. On the one hand, the boundaries of α_p are the preferred sites for micro-crack nucleation and provide a relatively easy path for fracture propagation. Therefore, with a decreased fraction of α_p, the nucleation sites of micro-cracks in Ti17 alloy at the α_p phase decreased, leading to superior impact toughness than Ti64. On the other hand, the crack path is apt to deflect at grain boundaries, colony boundaries, or arrested and deviates at α/β interface in titanium alloys, which consumes more of the plasticity energy path and results in improved toughness. In the present test, the lamellar microstructure with more colony boundaries displayed a more tortuous and deflected crack path than the bimodal microstructure, leading to the superior toughness of Ti17 alloy.

It is unexpected that a marginally lower toughness occurred for WZ with superfine microstructures in the present research. The significant decrease of impact toughness in WZ could be related to a combination of factors. Firstly, the soft phase of continuous α layer that lined the prior β grain boundary was effectively less constrained by the harder surrounding intra-granular structure, leading to the inter-granular failure and degradation of toughness. Secondly, with high oxidation tendency, it is inevitable that the titanium alloy oxidation takes place during the friction weld process, even though most of the oxide layer was expelled from the friction surface during the welding process. Some of the nano-scale oxides, however, could be left in the weld joints, and segregated in the fine grain boundaries during re-crystallization, resulting in the weakened grain boundaries in the WZ. This attributes to a further degradation of the toughness of WZ.

4. Conclusions

The micro-structural evolution, micro-hardness, tensile properties and impact toughness of LFW dissimilar welds with Ti64 and Ti17 alloys were investigated. The following conclusions were drawn.

(1) The microstructure across the linear friction welding dissimilar joints with titanium alloys displayed marked change, mainly consisting of a re-crystallized grain zone in the weld center, deformed grains and partial re-crystallization in the thermo-mechanical affected zones, and dissolved secondary α in the heat affected zones.

(2) The maximum hardness is located in the weld metal, which may result from the fine grains arising from the rapid cooling during the welding process.

(3) The linear friction welding dissimilar joints obtained higher tensile strength than base metal Ti64 with lower strength. The failure located in the Ti64 side approximately 1.2 mm away from the welding interface.

(4) Base metals had superior impact toughness and fractured in a trans-granular mode, but weld zone exhibited decreased toughness and failed in a mixture of trans-granular and inter-granular fracture modes.

Acknowledgments: This work has been financially supported by "Fracture Mechanism of Dissimilar Titanium Alloy Welded Joints" Ningbo Natural Science Foundation program (No. 2015A610071). As part of these grants, we received funds for covering the costs to publish in open access.

Author Contributions: Yingping Ji and Sujun Wu conceived and designed the experiments; Yingping Ji and Dalong Zhao performed the experiments; Yingping Ji analyzed the data; Yingping Ji and Sujun Wu contributed to writing and editing the manuscript.

Conflicts of Interest: The authors declare no conflict of interest.

References

1. Mvola, B.; Kah, P.; Martikainen, J. Dissimilar ferrous metal welding using advanced gas metal arc welding processes. *Rev. Adv. Mater. Sci.* **2014**, *38*, 125–137.
2. Martinsen, K.; Hu, S.J.; Carlson, B.E. Joining of dissimilar materials. *CIRP Ann.-Manuf. Technol.* **2015**, *64*, 526–533. [CrossRef]
3. Chen, Y.C.; Nakata, K. Microstructural characterization and mechanical properties in friction stir welding of aluminum and titanium dissimilar alloys. *Mater. Des.* **2009**, *30*, 469–474. [CrossRef]
4. Zhang, C.Q.; Robson, J.D.; Prangnell, P.B. Dissimilar ultrasonic spot welding of aerospace aluminum alloy AA2139 to titanium alloy TiAl6V4. *J. Mater. Process. Technol.* **2016**, *231*, 382–388. [CrossRef]
5. Tao, B.H.; Li, Q.; Zhang, Y.H.; Zhang, T.C.; Liu, Y. Effects of post-weld heat treatment on fracture toughness of linear friction welded joint for dissimilar titanium alloys. *Mater. Sci. Eng. A* **2015**, *634*, 141–146. [CrossRef]
6. Zhao, P.; Fu, L.; Chen, H. Low cycle fatigue properties of linear friction welded joint of TC11 and TC17 titanium alloys. *J. Alloy. Compd.* **2016**, *675*, 248–256. [CrossRef]
7. Xu, C.; Sheng, G.; Wang, H.; Feng, K.; Yuan, X. Tungsten Inert Gas Welding-Brazing of AZ31B Magnesium Alloy to TC4 Titanium alloy. *J. Mater. Process. Technol.* **2016**, *32*, 167–171. [CrossRef]
8. Wang, S.Q.; Li, W.Y.; Zhou, Y.; Li, X.; Chen, D.L. Tensile and fatigue behavior of electron beam welded dissimilar joints of Ti-6Al-4V and IMI834 titanium alloys. *Mater. Sci. Eng. A* **2016**, *649*, 146–152. [CrossRef]

9. Vairis, A.; Frost, M. High frequency linear friction welding of a titanium alloy. *Wear* **1998**, *217*, 117–131. [CrossRef]
10. Guo, Y.; Chiu, Y.; Attallah, M.M.; Li, H.; Bray, S.; Bowen, P. Characterization of Dissimilar Linear Friction Welds of α-β Titanium Alloys. *J. Mater. Eng. Perform.* **2012**, *21*, 770–776. [CrossRef]
11. Wanjara, P.; Jahazi, M. Linear friction welding of Ti-6Al-4V: Processing, microstructure, and mechanical-property inter-relationships. *Metall. Mater. Trans. A* **2005**, *36*, 2148–2164. [CrossRef]
12. Li, W.Y.; Ma, T.J.; Yang, S.Q. Microstructure evolution and mechanical properties of linear friction welded Ti-5Al-2Sn-2Zr-4Mo-4Cr (Ti17) titanium alloy joints. *Adv. Eng. Mater.* **2010**, *12*, 35–43. [CrossRef]
13. Ma, T.; Yan, M.; Yang, X.; Li, W.; Chao, Y.J. Microstructure evolution in a single crystal nickel-based superalloy joint by linear friction welding. *Mater. Des.* **2015**, *85*, 613–617. [CrossRef]
14. Chamanfar, A.; Jahazi, M.; Gholipour, J.; Wanjara, P.; Yue, S. Analysis of integrity and microstructure of linear friction welded Waspaloy. *Mater. Charact.* **2015**, *104*, 148–161. [CrossRef]
15. Astarita, A.; Curioni, M.; Squillace, A.; Zhou, X.; Bellucci, F.; Thompson, G.E.; Beamish, K.A. Corrosion behaviour of stainless steel-titanium alloy linear friction welded joints: Galvanic coupling. *Mater. Corros.* **2015**, *66*, 111–117. [CrossRef]
16. Grujicic, M.; Yavari, R.; Snipes, J.S.; Ramaswami, S. A linear friction welding process model for Carpenter Custom 465 precipitation-hardened martensitic stainless steel: A weld microstructure-evolution analysis. *J. Eng. Manuf.* **2015**, *228*, 1887–2020. [CrossRef]
17. Rotundo, F.; Marconi, A.; Morri, A.; Ceschini, A. Dissimilar linear friction welding between a SiC particle reinforced aluminum composite and a monolithic aluminum alloy: Microstructural, tensile and fatigue properties. *Mater. Sci. Eng. A* **2013**, *558*, 852–860. [CrossRef]
18. Bhamji, I.; Moat, R.J.; Preuss, M.; Threadgill, P.L.; Addison, A.C. Linear friction welding of aluminium to copper. *Sci. Technol. Weld. Join.* **2012**, *17*, 314–320. [CrossRef]
19. Bhamji, I.; Preuss, M.; Moat, R.J.; Threadgill, P.L.; Addison, A.C. Linear friction welding of aluminium to magnesium. *Sci. Technol. Weld. Join.* **2012**, *17*, 368–374. [CrossRef]
20. Ma, T.J.; Zhong, B.; Li, W.-Y.; Zhang, Y.; Yang, S.Q.; Yang, C.L. On microstructure and mechanical properties of linear friction welded dissimilar Ti-6Al-4V and Ti-6.5Al-3.5Mo-1.5Zr-0.3Si joint. *Sci. Technol. Weld. Join.* **2012**, *17*, 9–12. [CrossRef]
21. Wen, G.D.; Ma, T.J.; Li, W.Y.; Wang, S.Q.; Guo, H.Z.; Chen, D.L. Strain-controlled fatigue properties of linear friction welded dissimilar joints between Ti-6Al-4V and Ti-6.5Al-3.5Mo-1.5Zr-0.3Si alloys. *Mater. Sci. Eng. A* **2014**, *612*, 80–88. [CrossRef]
22. Zhao, P.; Fu, L. Strain hardening behavior of linear friction welded joints between TC11 and TC17 dissimilar titanium alloys. *Mater. Sci. Eng. A* **2015**, *621*, 149–156. [CrossRef]
23. Frankel, P.; Preuss, M.; Steuwer, A.; Withers, P.J.; Bray, S. Comparison of residual stresses in Ti-6Al-4V and Ti-6Al-2Sn-4Zr-2Mo linear friction welds. *Mater. Sci. Technol.* **2009**, *25*, 640–650. [CrossRef]
24. Karadge, M.; Preuss, M.; Lovell, C.; Withers, P.J.; Bray, S. Texture development in Ti-6Al-4V linear friction welds. *Mater. Sci. Eng. A* **2007**, *458*, 182–181. [CrossRef]
25. Lang, B.; Zhang, T.C.; Li, X.H.; Guo, D.L. Microstructural evolution of a TC11 titanium alloy during linear friction welding. *J. Mater. Sci.* **2010**, *45*, 6218–6224. [CrossRef]
26. Romero, J.; Attallah, M.M.; Preuss, M.; Karadge, M.; Bray, S.E. Effect of the forging pressure on the microstructure and residual stress development in Ti-6Al-4V linear friction welds. *Acta Mater.* **2008**, *57*, 5582–5592. [CrossRef]
27. Corzo, V.; Casals, O.; Alcalá, J.; Mateo, A.; Anglada, M. Mechanical evaluation of linear friction welds in titanium alloys through indentation experiments. *Weld. Int.* **2007**, *21*, 125–128. [CrossRef]
28. Schroeder, F.; Ward, R.M.; Turner, R.P.; Walpole, A.R.; Attallah, M.M.; Gebelin, J.-C.; Reed, R.C. Validation of a Model of Linear Friction Welding of Ti6Al4V by Considering Welds of Different Sizes. *Metall. Mater. Trans. B* **2015**, *46*, 2326–2331. [CrossRef]
29. Turner, R.; Gebelin, J.-C.; Ward, R.M.; Reed, R.C. Linear friction welding of Ti-6Al-4V: Modelling and validation. *Acta Mater.* **2011**, *59*, 3792–3803. [CrossRef]
30. McAndrew, A.R.; Colegrove, P.A.; Addison, A.C.; Flipo, B.C.D.; Russell, M.J. Modelling the influence of the process inputs on the removal of surface contaminants from Ti-6Al-4V linear friction welds. *Mater. Des.* **2015**, *66*, 183–195. [CrossRef]

31. Grujicic, M.; Arakere, G.; Pandurangan, B.; Yen, C.-F.; Cheeseman, B.A. Process Modeling of Ti-6Al-4V Linear Friction Welding (LFW). *J. Mater. Eng. Perform.* **2012**, *21*, 2011–2023. [CrossRef]
32. Ma, T.; Chen, T.; Li, W.-Y.; Wang, S.; Yang, S. Formation mechanism of linear friction welded Ti-6Al-4V alloy joint based on microstructure observation. *Mater. Charact.* **2011**, *62*, 130–135. [CrossRef]
33. Ji, Y.; Chai, Z.; Zhao, D.; Wu, S. Linear friction welding of Ti-5Al-2Sn-2Zr-4Mo-4Cr alloy with dissimilar microstructure. *J. Mater. Process. Technol.* **2014**, *214*, 878–887. [CrossRef]
34. Zhao, P.K.; Fu, L.; Zhong, D.C. Numerical simulation of transient temperature and axial deformation during linear friction welding between TC11 and TC17 titanium alloys. *Comput. Mater. Sci.* **2014**, *82*, 325–333. [CrossRef]
35. Ma, T.J.; Li, W.-Y.; Yang, S.Y. Impact toughness and fracture analysis of linear friction welded Ti-6Al-4V alloy joints. *Mater. Des.* **2008**, *30*, 2128–2132. [CrossRef]
36. Ji, Y.; Wu, S. Study on microstructure and mechanical behavior of dissimilar Ti17 friction welds. *Mater. Sci. Eng. A* **2014**, *586*, 32–40. [CrossRef]

![metals logo] *metals*

MDPI

Article

Low Cycle Fatigue Behaviors of Alloy 617 (INCONEL 617) Weldments for High Temperature Applications

Rando Tungga Dewa [1], Seon Jin Kim [1,*], Woo Gon Kim [2] and Eung Seon Kim [2]

[1] Department of Mechanical Design Engineering, Pukyong National University, Busan 608-739, Korea; rando.td@gmail.com
[2] Korea Atomic Energy Research Institute (KAERI), Daejeon 305-353, Korea; wgkim@kaeri.re.kr (W.G.K.); kimes@kaeri.re.kr (E.S.K.)
* Correspondence: sjkim@pknu.ac.kr; Tel.: +82-51-629-6163; Fax: +82-51-629-6250

Academic Editor: Giuseppe Casalino
Received: 15 March 2016; Accepted: 25 April 2016; Published: 28 April 2016

Abstract: In this study, we comparatively investigate the low cycle fatigue behavior of Alloy 617 (INCONEL 617) weldments by gas tungsten arc welding process at room temperature and 800 °C in the air to support the qualification in high temperature applications of the Next Generation-IV Nuclear Plant. Axial total-strain controlled tests have been performed with the magnitude of strain ranges with a constant strain ratio ($R_\varepsilon = -1$). The results of fatigue tests consistently show lower fatigue life with an increase in total strain range and temperature at all testing conditions. The reduction in fatigue life may result from the higher cyclic plastic strain accumulation and the material ductility at high temperature conditions. A constitutive behavior of high temperature by some cyclic hardening was observed. The occurrence of serrated yielding in the cyclic stress response was also observed, suggesting the influence of dynamic strain aging during high temperature. We evaluated a well-known life prediction model through the Coffin-Manson relationship. The results are well matched with the experimental data. In addition, low cycle fatigue cracking occurred in the weld metal region and initiated transgranularly at the free surface.

Keywords: alloy 617; low cycle fatigue (LCF); gas tungsten arc welding (GTAW); weldment; high temperature; dynamic strain aging (DSA); life prediction; fracture behavior

---

1. Introduction

The very high temperature reactor (VHTR) is one of the most promising nuclear systems among the Generation-IV reactors to economically produce electricity and hydrogen. The VHTR major components include the reactor internals, reactor pressure vessel, piping, hot gas ducts (HGD), and intermediate heat exchangers (IHX). These components are required to have good mechanical properties, creep-fatigue resistance, and phase stability at high temperatures. There have been no materials approved by ASME Section III Subsection NH which is a nuclear code for temperatures reaching 1000 °C. Currently, Alloy 617 is the leading candidate material to prolong the design life of IHX and HGD of helium-cooled VHTR systems due to its creep-fatigue resistance at high temperatures [1–3].

Nowadays, the Korea Atomic Energy Research Institute (KAERI) is developing a nuclear hydrogen development and demonstration plant with a capacity of 200 MW$_{th}$ with thermal and core outlet temperature of 950 °C [1,2,4,5]. The components have a projected plant design service life of 40–60 years operation and 3–8 MPa in He impurities. The most important consideration is the creep-fatigue and fatigue behavior of the materials, especially in welded structure [3–5]. The Alloy 617 for the IHX and other components is expected to operate at room temperature, which is assumed as a start-up condition, and at temperatures between 800–950 °C.

Low cycle fatigue (LCF) loading is expected to be an important damage mode for the IHX as a result of operating conditions that generate power transients and a temperature-gradient induced from thermal strain during operation as well as startups, shutdown work and load fluctuating, each of which produces cyclic loading [3]. The consideration could be used to determine material resistance against the cyclic loading. For mechanical structures, welded components are necessary and some of the components are joined using various welding techniques. Therefore, weldments are unavoidable and may have several inherent defects. It is necessary to understand fatigue damage of weldments, because it is more brittle than base metal (BM), and it is micro-structurally and mechanically heterogeneous, which could form a source of fatigue failures [4–6].

Many attempts have been made in the past two decades to evaluate the LCF and/or creep-fatigue behavior in Alloy 617 BM and weldments at ambient and elevated temperatures [3,7–14]. Based on their results, it is very difficult to separate the effects of temperature, strain rate and environment as they have complex interrelationships. However, they did not investigate the LCF behavior of the weldments as a comparative study at room temperature and 800 °C thermal conditions over different total strain ranges as investigated in this study, and also at a specific rate of change in strain deformation. It is important to provide a baseline data of the LCF properties of weldments to ensure the reliability of welded structures.

To support the modeling and understanding of the LCF behavior of Alloy 617, weldments was made from gas tungsten arc welding (GTAW) butt-welded plates. In this paper, KAERI and Pukyong National University (PKNU) investigate the LCF behavior for Alloy 617 weldments, in order to provide the commencement of high temperature LCF behavior. We performed an initial strain-controlled LCF test in air at room temperature and 800 °C under different total strain ranges of 0.6, 0.9%, 1.2% and 1.5%, and with a constant strain ratio ($R_\varepsilon = -1$), and the results were consistent with those listed from ASTM Standard E606 [15]. The tensile test measurements were first done as a reference data for Alloy 617 weldments for better understanding and explanation of fracture behavior [4]. The microscopic investigations were also examined with selected specimens to compare LCF fracture behavior of Alloy 617 weldment specimens.

2. Materials and Methods

2.1. As-Received Alloy 617 Weldments

Figure 1 shows the cross-view of etched Alloy 617 weldments microstructure. The microstructure consisted of BM with well-uniformed equiaxed grains, following the heat affected zone (HAZ) structure with a fusion line and coarser grain boundaries. Lastly, the dendritic structure was formed in the weld metal (WM) region because of the solidification during the welding process. The chemical composition of the as-received Alloy 617 used in the present study is shown in Table 1. The shape of the weldments has a single V-groove with an angle of 80° and 10 mm root gap from a 25 mm thick rolled plate. A filler metal was used for KW-T617. It was prepared according to AWS specifications, AWS A 5.14-05 ERNiCrMo-1 (UNS N06617), and the diameter was 2.4 mm. In order to prevent a bending deformation of the weldments, some passes to the back side were added for each specimen. A post heat treatment was not conducted because a Ni-based superalloy is not normally applied. After the welding process, the soundness of the weldments was qualified through an ultrasonic test (UT), a tensile test, and a bending test. The bending testing results coincide well with ASME specifications, which means the micro-crack is within 3.2 mm. It was also observed that the weldments exhibited acceptable ductility. Nevertheless, the soundness of the weldments has no problems.

LCF and tensile weldment specimens were machined from a weld pad in the transverse direction against the welding direction as shown in Figure 2a. The cross section of the weld specimen involves a gage length, consisting of WM and HAZ materials only: the WM is in the center of the section and two HAZs are next to it. For further details, Figure 2b shows the shape and dimension of the LCF specimen with 6.0 mm diameter in the reduced section with a parallel length of 18 mm and a gauge length of

12.5 mm. However, the flat tensile test specimens were machined into a rectangular cross section with a gage length of 28.5 mm, width of 6.25 mm, and thickness of 1.5 mm.

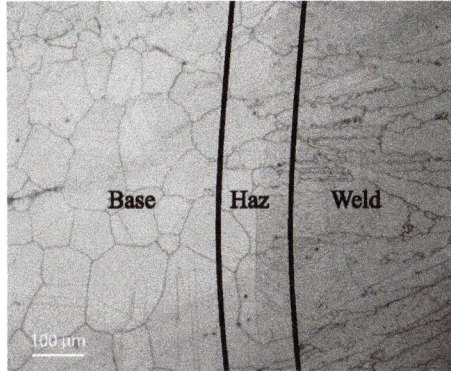

Figure 1. Microstructure of the cross-view for Alloy 617 weldments.

Table 1. The chemical composition of Alloy 617 plate (wt. %).

		C	Ni	Fe	Si	Mn	Co	Cr	Ti	P	S	Mo	Al	B	Cu
ASTM	Min	0.05	Bal.	-	-	-	10	20	-	-	-	8	0.8	-	-
Spec	Max	0.15	Bal.	3.0	1.0	1.0	15	24	0.6	0.015	0.015	10	1.5	0.006	0.5
Alloy 617	-	0.08	53.11	0.95	0.08	0.03	12.3	22	0.41	0.003	<0.002	9.5	1.06	<0.002	0.027

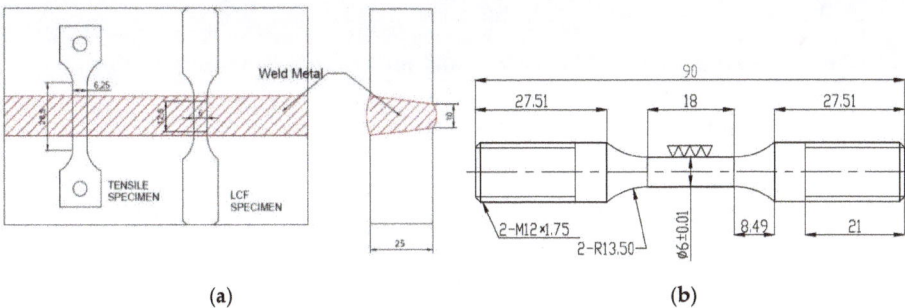

(a)   (b)

Figure 2. (**a**) The weld pad configuration of weldment specimens; (**b**) The cylindrical low cycle fatigue (LCF) specimen's shape and geometry.

2.2. Experimental Methods

We conducted fully reversed (R_ε = −1) total axial strain-controlled LCF tests of Alloy 617 weldments in air environment by using a servo hydraulic machine (Instron 8516 capacity 100 kN, Instron Korea, Seoul, Korea) under 0.25 Hz constant frequency regarding four total strain ranges of 0.6%, 0.9%, 1.2%, and 1.5% at room temperature and 800 °C. The strain rate was varied from 3×10^{-3} s^{-1} to 7.5 × 10^{-3} s^{-1} depending on total strain range. We use a closed loop servo hydraulic system equipped with a tube furnace and three temperature controllers on the top, center, and bottom region. Therefore, the temperature was remained within 800 °C ± 2 °C of the nominal temperature throughout the test. The specimen was held at a target temperature with zero load for about 30 min to allow temperatures to stabilize before the commencement of the test. We applied triangular waveforms with equal push-pull mode to specimens. As in the literature review [3,9,10,15], we also defined the failure criteria as the number of cycles, which means a 20% level drop off from the initial level of the peak tensile stress. We used a personal computer through a high precision extensometer with 12.5 mm gage length that was attached to the gauge section of the specimen. The gauge section mainly covers the weld and HAZ materials only, to produce a strain/time records as well as load/time records and load/strain loops of each test condition.

In order to examine the LCF fracture morphologies, we investigated the post fracture analysis of the selected specimens which interrupted into two pieces in preliminary observation by using high magnification scanning electron microscopy (SEM Hitach JEOL JSM 5610, JEOL Korea Ltd., Seoul, Korea) and optical microscopy (OM Olympus GX-51, Olympus Korea Co. Ltd., Seoul, Korea). We prepared the specimens for OM observations by cutting them into two parts. The cutting surfaces were polished after mounting and they were sequentially etched in solutions of hydrochloric acid, ethanol, and copper II chloride. The tensile tests have been first done as a reference data for better understanding and explanation of fracture behavior, with a strain rate of 5.85×10^{-4} s^{-1}. Table 2 shows the results of the tensile tests for Alloy 617 weldments. However, it is notable that the increase in temperature resulted in a lower yield stress (YS) and the ultimate tensile strength (UTS) values, although the tensile elongation (or material ductility) slightly increases with increasing temperature.

Table 2. The results of tensile tests on Alloy 617 weldments.

Temperature (°C)	YS (MPa)	UTS (MPa)	EL (%)
Room temperature	462	764.9	24.6
800 °C	314	382.7	27.3

3. Results and Discussion

3.1. Low Cycle Fatigue Behaviors of Alloy 617 Weldments

The fatigue resistance can be considered as: the microstructural or material property, the fatigue rate, and the temperature influence. We listed a summary of the results in Table 3. Most of the test results show that the increasing of strain range and temperature condition resulted in a reduction of fatigue life. At room temperature condition shows a comparatively higher cyclic stress response with increasing strain ranges compared to the 800 °C testing conditions, although the plastic strain accumulation of the 800 °C testing conditions is comparatively higher. From the results, we can deduce that the higher plastic strain deformation and the material ductility of the 800 °C testing conditions to have a major role in the shorter fatigue life.

Table 3. Summary of LCF tests results of Alloy 617 weldments.

Specimen Reference		Cycles to Failure	Elastic Strain Amp (mm/mm)	Plastic Strain Amp (mm/mm)	Stress Amp (MPa)
Temperature	Total Strain Amp (mm/mm)				
Room temperature	0.0075	416	0.0037	0.0038	665.9
	0.0060	1485	0.0033	0.0027	656.2
	0.0045	2924	0.0029	0.0016	621.1
	0.0030	28422	0.0027	0.0003	545.6
800 °C	0.0075	230	0.0034	0.0041	566.3
	0.0060	334	0.0034	0.0026	565.3
	0.0045	655	0.0029	0.0016	523.9
	0.0030	3282	0.0026	0.0004	417.2

In engineering, materials usually exhibit cyclic hardening or softening to some level in the cyclic plastic deformation process. Figure 3 shows the cyclic stress response curves with respect to the number of cycles regarding four total strain ranges under continuous cyclic loading. The cyclic stress response at room temperature conditions showed a cyclic softening region for the major portion of the life after a short period of initial hardening. The short period of cyclic initial hardening was observed for about 2–200 cycles, and remained during the softening phase until failure. At the lowest strain range (0.6%), the saturation region was also observed. Dissimilar cyclic stress response behavior was found in both temperature conditions. However, the results at 800 °C distinctly showed a cyclic hardening for the major portion of the fatigue life. At the end of the test, the cyclic stress response decreased rapidly, which indicates the initiation of a macro-crack.

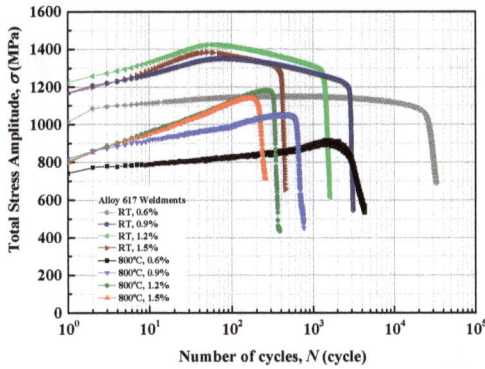

Figure 3. The cyclic stress response curves regarding to four total strain ranges under continuous cyclic loading for Alloy 617 weldments.

In order to clearly understand the relationship between cyclic stress response and crack behavior, Figure 4 shows the normalized stress amplitude on the fraction of fatigue life, which was used to deduce the number of cycles regarding macro-crack initiation, N_i. The fraction of fatigue life is the ratio between the number of cycles and the total cycles to failure. We defined the crack initiation as the point at which the stress amplitude rapidly decreased from an initial trend. From the figure, the profile significantly shows that the crack initiation occurs earlier at approximately 90% of the total cycles to failure. Meanwhile, at 0.6% total strain range, it was occurred in approximately 60% of the total cycles to failure, and remained during propagation life. It is noted that, under low total strain range, the fatigue crack propagation occurred earlier and it is particularly important to overall fatigue life when it is utilized in the high temperature condition.

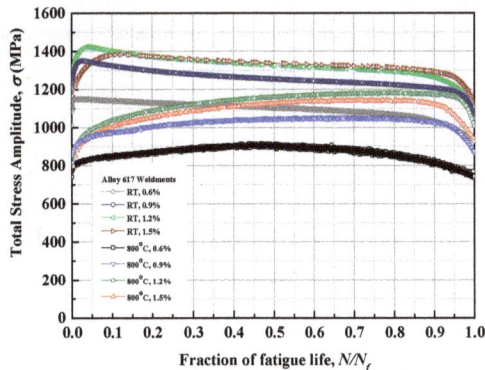

Figure 4. The normalizing fatigue life on cyclic stress response regarding to four total strain ranges of Alloy 617 weldments at room temperature and 800 °C.

In this study, we found that the work hardening behavior is an effect of initial cyclic hardening. Figure 5 shows the summary of cyclic stress-plastic strain curves of Alloy 617 weldments. From this graph, we can find that the stress increases with increasing plastic strain. In order to quantitatively determine the work hardening behavior, the Ramberg-Osgood equation is adopted:

$$\frac{\Delta \epsilon_T}{2} = \frac{\Delta \epsilon_p}{2} + \frac{\Delta \epsilon_e}{2} = \frac{\Delta \sigma}{E} + \left(\frac{\Delta \varepsilon_p}{2K'} \right)^{n'} \tag{1}$$

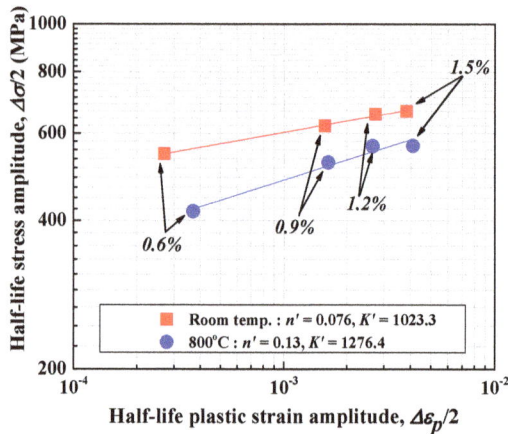

Figure 5. The cyclic stress-plastic strain curves of Alloy 617 weldments at room temperature and 800 °C.

The results are also given in Figure 5, where, K' and n' are cyclic strength coefficient and cyclic strain hardening exponent, respectively. From this figure, K' and n' were determined from the log-log relationships between half-life value (stable condition) of stress amplitude and plastic strain amplitude. The work hardening tendency of the weldments linearly increased through the whole total strain range conditions. However, the degree of work hardening of Alloy 617 weldments depends on the stress amplitude and the plastic strain value.

In this manifestation, we recognized the occurrence of dynamic strain aging (DSA) at 800 °C regarding four total strain ranges under continuous cyclic loading. The results showed a transformation of work hardening at 800 °C. DSA (sometimes termed as Portevin Le-Chatelier effect) is the hardening mechanism, which manifests itself by fluctuating or serration plastic flow. In the manifestation of DSA, we synchronously noticed the serrated flows on the tension-compression stresses at all testing conditions. Figure 6 shows the tensile peak stresses tested at 800 °C that exhibit serrations as a result of the locking-unlocking interactions of moving atoms. Meanwhile, J.-D. Hong *et al.* [16] revealed that the cyclic hardening might be increased with the occurrence of the DSA, and the ratio of the cyclic hardening could be used to evaluate the degree of DSA. Therefore, Figure 7 shows the ratio of cyclic hardening measured by the ratio of the maximum stress amplitude (σ_{peak}) to the stress amplitude of the first cycle ($\sigma_{initial}$). From the figure, it is obvious that the ratio of cyclic hardening increases along with the transformation of LCF test environment from previous work [4,5] at room temperature to the present investigation at 800 °C. As in the literature survey [11], they also acknowledged that the DSA occurs in the temperature range of about 650–900 °C for this Alloy 617.

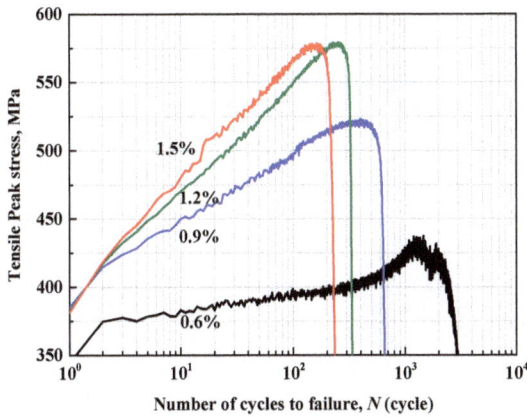

Figure 6. The serrated flows in the tensile peak stress curves of Alloy 617 weldments at 800 °C.

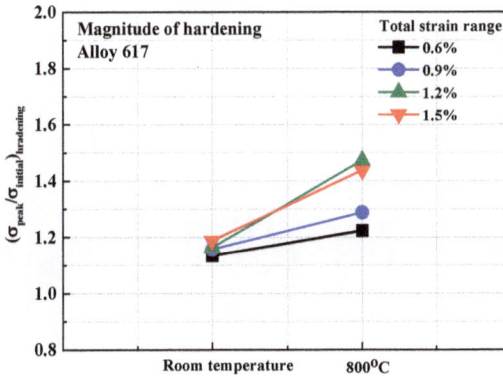

Figure 7. The comparison of ratio of cyclic stress amplitude as a magnitude of hardening of Alloy 617 weldments at room temperature and 800 °C.

3.2. Low Cycle Fatigue Life Prediction

Many scientists have made an enormous effort in order to predict the lifetime of metallic materials at elevated temperatures. The state of the art suggests that an issue has been addressed mainly in terms of strain. Plastic deformation is expected to play a major role in this process. Several researchers, e.g., Lazzarin, *et al.* and Susmel, *et al.* [17,18], have primarily considered a very high-temperature fatigue problem with the high cycle fatigue regime. This was initiated from the pioneering Coffin-Manson relationship as a definition of strain elements at LCF regime. In this issue, an appropriate life assessment using a conventional Coffin-Manson relationship is proposed for designing against the LCF of Alloy 617 weldments. The combining Coffin-Manson equation, better known as the strain-life relationship, is a function of the total strain range, $\Delta\varepsilon_T$, and the number of cycles to failure, N_f, by a similar power law function. The total strain as a dependence parameter can be separated into the plastic and elastic strain ranges and expressed by:

$$\frac{\Delta\varepsilon_T}{2} = \frac{\Delta\varepsilon_P}{2} + \frac{\Delta\varepsilon_e}{2} = \frac{\sigma'_f}{E}\left(2N_f\right)^b + \varepsilon'_f\left(2N_f\right)^c \tag{2}$$

where, $\Delta\varepsilon_T/2$ is the total strain amplitude, $\Delta\varepsilon_e/2$ is the elastic strain amplitude, $\Delta\varepsilon_p/2$ is the plastic strain amplitude, $2N_f$ is the number of reversals to failure. σ_f' is the fatigue strength coefficient, b is the fatigue strength exponent, ε_f' is the fatigue ductility coefficient, c is the fatigue ductility exponent and E is the elastic modulus.

Figure 8a,b show the variety of LCF lives plotted with total strain ranges in the continuous cycling of the Alloy 617 weldment specimens at room temperature and 800 °C, respectively. These data were analyzed by means of the least squares fit method. In both logarithmic coordinates, the strain-life curves were simply fitted and they are related to Equation (2). The coefficients and exponents in the Coffin-Manson equation were determined, and are listed in Table 4. As previous works reported [3,15], when it was plotted on a log-log scale, the c slope typically may vary from $-0.5\sim-1.0$. The authors also reported the exponent c of -1.12 of Alloy 617 weldments at 800 °C, and it can be concluded that the value is comparable in the present study. From the figure, the transition of fatigue life in reversals $(2N_t)$ represents the intersection of the elastic and plastic straight lines which means the stabilized hysteresis loop has equal plastic and elastic strain components. From this point to the left, where the fatigue life is lower than the transition, the fatigue life of the material is mainly dominated by ductility. Otherwise, where the fatigue life is higher than the transition, is controlled by the strength. Table 5 lists the transition of fatigue life at room temperature and 800 °C. From the results could be assumed that the strain amplitude at the transition life decreases with increasing temperature.

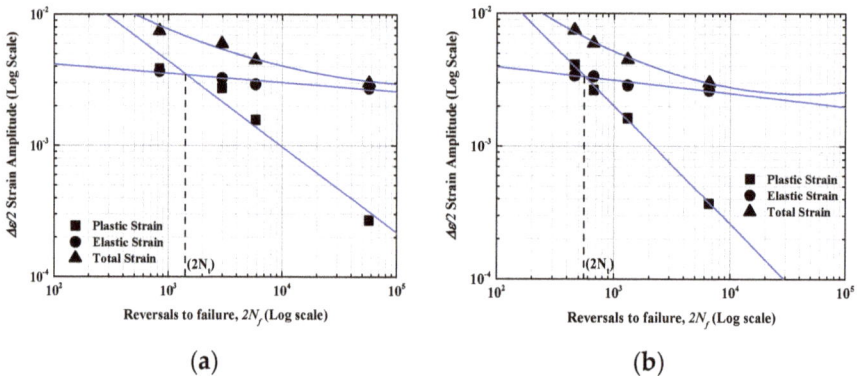

(a)

(b)

Figure 8. The Coffin-Manson curves of Alloy 617 weldments at (a) Room temperature; (b) 800 °C condition.

Table 4. Coefficients and exponents of Alloy 617 weldments in the Coffin-Manson equation.

Temperature	ε_f'	c	σ_f' (MPa)	b	E (GPa)	n'	K' (MPa)
Room temperature	0.398	−0.654	1213	−0.104	214.4	0.118	1086.43
800°C	0.924	−0.889	973	−0.100	155.9	0.14	1276.4

Table 5. The transition of fatigue life of Alloy 617 Weldments at room temperature and 800 °C.

Temperature	$2N_t$	$\Delta\varepsilon/2$
Room temperature	1410	0.0035
800°C	554	0.0033

In the process of the validation of the predicted life, Figure 9 shows the experimental data which are compared to the predicted lives by the Coffin-Manson relationship. Figure 9 shows a good correlation between the predicted and experimental fatigue lives. The validations are in good agreement within a factor of 2.0. An excellent correlation can be observed for the Coffin-Manson relationship with 86.37% of accuracy.

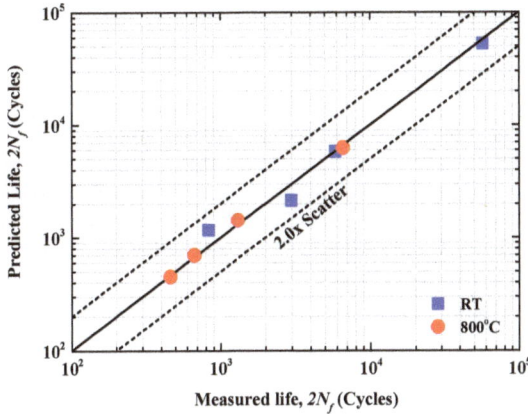

Figure 9. Validation of the predicted lives by Coffin-Manson relationship for Alloy 617 weldments.

3.3. Fracture Morphologies of Alloy 617 Weldments

Regardless of the effect of temperature, the fracture surface morphologies of the weldments at room temperature and 800 °C are very similar, as shown in Figure 10. From previous work [4], the weldments showed a relatively wedge-type crack with some deviations oriented at 45° to the loading direction. Figure 10 also shows the fractured specimens, which indicates the failure occurring inside of the gauge section in the WM region. Therefore, we denote that the possible weak spot is in the WM area of the weldments. However, the possible weakness of the WM as well as the presence of microstructural heterogeneity, and also the WM zone was comprised of large columnar grains with dendritic structure. A large plastic strain thereby accumulates, depending upon the material strength of individual microstructure, which could become a source of fatigue failures. In addition, the results of hardness measurement on the WM showed a lower value than that of the HAZ, but still higher than the BM.

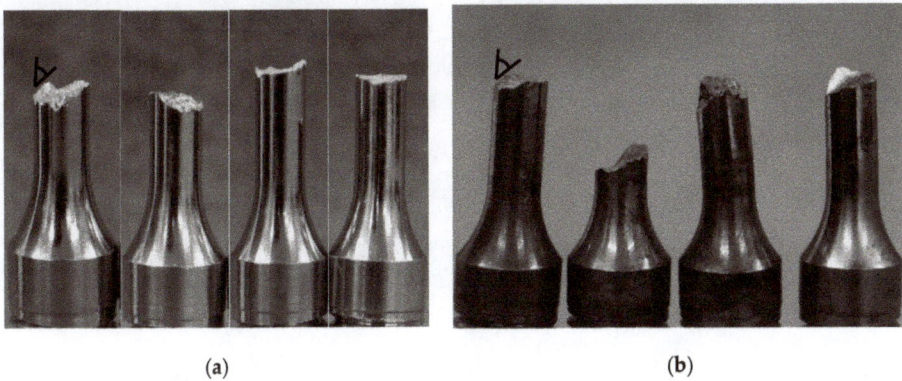

(a)   (b)

Figure 10. Comparison of the fracture surface morphologies of Alloy 617 weldments at: (**a**) Room temperature; (**b**) 800 °C condition. From left to right *i.e.*, total strain range of 0.6%, 0.9%, 1.2%, 1.5%.

In nickel based Alloy 617 [6–12], a reduction of fatigue life was attributed to creep, oxidation, and DSA at elevated temperature. In this study, however, the difference in fatigue resistance cannot be fully interpreted by an effect of fracture morphology. In the regime of DSA, there were no evidences of creep and less effect of oxidation to account for the difference in the fatigue life with temperature in the present investigation. Figure 11a–d show a typical micrograph of the Alloy 617 weldment specimens at room temperature and 800 °C, respectively, which also show a transgranular crack initiation and propagation along with the interdendritic paths. Hong *et al.* [19] stated that the fatigue crack initiation is related to the plastic strain localization on the slip bands. The slip bands that contain extrusions, and intrusions, if connected to the surface, will lead to the production of crack initiation. Crack initiation occurred along slip planes, and this was evidenced by the cleavage facets in Figure 11c (stage I). This mechanism resulted in multiple points of the crack initiation site on their surfaces, and reduced the crack initiation life. This type continued through only a few grains, and transformed into stage II transgranular cracking. Figure 11d shows the fatigue striations that occurred on the fracture surface during crack propagation. Since DSA-induced work hardening process may cause the rapid crack propagation due to atoms interactions, the fatigue life also will decrease in the regime of DSA.

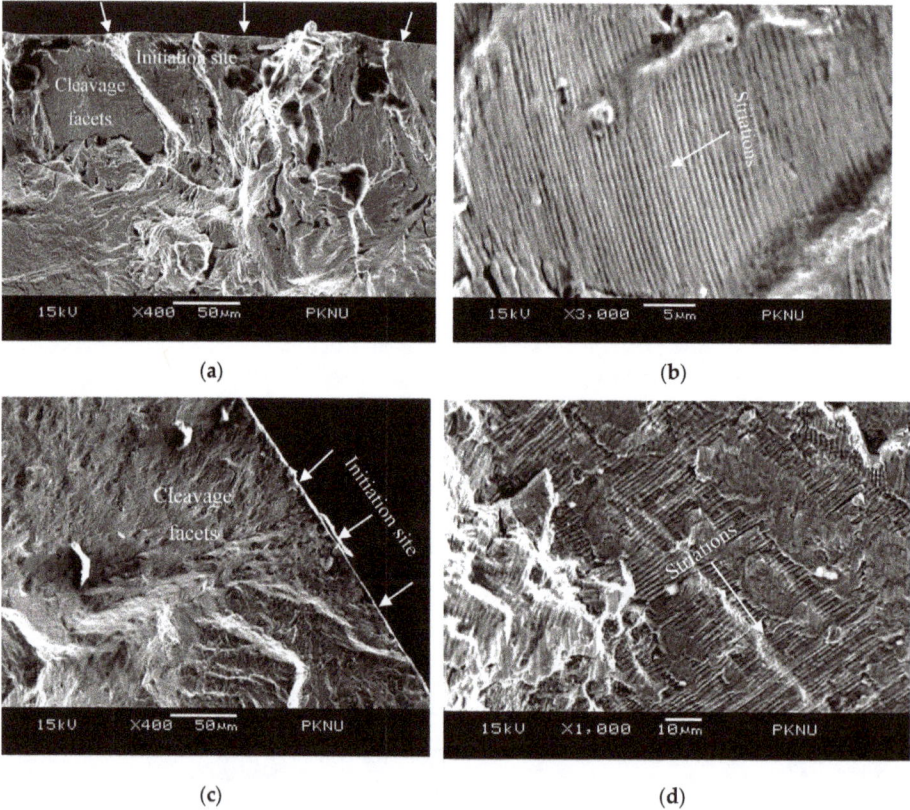

Figure 11. Typical SEM micrographs of the fractured surface for Alloy 617 weldments at 0.6% total strain range: (**a**,**b**) Room temperature; (**c**,**d**) 800 °C condition.

4. Conclusions

A fully reversed ($R_\varepsilon = -1$) total axial strain-controlled LCF tests of the Alloy 617 (INCONEL 617) weldments fabricated by GTAW process were conducted at room temperature and 800 °C in the air, with a total strain range of 0.6, 0.9, 1.2, and 1.5%. The following key conclusions are drawn:

1. The increasing of strain ranges and temperature condition resulted in a reduction of fatigue life. From the results, we can deduce that the higher plastic strain deformation and the material ductility of the 800 °C testing conditions have a major role in the shorter fatigue life.
2. For room temperature conditions, initial hardening was observed below 200 cycles, after that softening was observed until failure. Meanwhile, the 800 °C condition distinctly showed a cyclic hardening for the major portion of the fatigue life. In the 800 °C condition, we synchronously noticed the serrated flows on the tension-compression stresses at all testing conditions as dynamic strain aging (DSA) phenomena. Since the DSA-induced work hardening process due to atoms interactions may cause the rapid crack propagation, the fatigue life will also decrease in the regime of DSA.
3. The observed LCF life was well characterized by the Coffin-Manson relationship, and they are well compared with previous works. The parameter of c slope typically varied from $-0.5 \sim -1.0$, and, in this study, the c slope was obtained by -0.654 and -0.889 at room temperature and 800 °C conditions, respectively.
4. The LCF cracking in weldments occurred inside of the gauge section in the WM region, and showed a wedge-type crack with some deviations oriented at 45° to the loading direction. A transgranular crack initiation with some cleavage facets (stage I) and propagation with some striations (stage II) along with the interdendritic paths were observed.

Acknowledgments: The authors would like to recognize KAERI, and acknowledge that this research was supported by Nuclear Research & Development Program through the National Research Foundation of Korea (NRF) funded by the Ministry of Science, ICT & Future Planning (NRF-2015M2A8A2021963).

Author Contributions: Seon Jin Kim formulated this research with cooperation from Woo Gon Kim and Eung Seon Kim (KAERI). Rando Tungga Dewa performed the experiment works, with the help of Seon Jin Kim, interpreted the results, prepared and revised the manuscript. All co-authors contributed to manuscript proof and submissions.

Conflicts of Interest: The authors declare no conflict of interest.

References

1. Kim, W.G.; Park, J.Y.; Ekaputra, I.M.W.; Kim, S.J.; Kim, M.H.; Kim, Y.W. Creep deformation and rupture behavior of Alloy 617. *Eng. Fail. Anal.* **2015**, *58*, 441–451. [CrossRef]
2. Kim, W.G.; Yin, S.N.; Kim, Y.W.; Ryu, W.S. Creep behaviour and long-term creep life extrapolation of alloy 617 for a very high temperature gas-cooled reactor. *Trans. Indian Inst. Met.* **2010**, *63*, 145–150. [CrossRef]
3. Wright, J.K.; Carroll, L.J.; Wright, R.N. *Creep and Creep-Fatigue of Alloy 617 Weldments*; Annual Report Idaho National Laboratory: Fremont, CA, USA; August; 2014.
4. Kim, S.J.; Dewa, R.T.; Kim, W.G.; Kim, M.H. Cyclic Stress Response and fracture behaviors of Alloy 617 base metal and weldments under LCF loading. *Adv. Mater. Sci. Eng.* **2015**, *2015*. [CrossRef]
5. Kim, S.J.; Choi, P.H.; Dewa, R.T.; Kim, W.G.; Kim, M.H. Low cycle fatigue properties of Alloy 617 base metal and weldment at room temperature. *Proc. Mater. Sci. Eur. Conf. Fract.* **2014**, *3*, 2201–2206.
6. Prasad Reddy, G.V.; Sandhya, R.; Valsan, M.; Bhanu Sankara Rao, K. High temperature low cycle fatigue properties of 316(N) weld metal and 316L(N)/316(N) weldments. *Int. J. Fatigue* **2008**, *30*, 538–546. [CrossRef]
7. Bhanu Sankara Rao, K.; Schiffers, H.; Schuster, H.; Nickel, H. Influence of time and temperature dependent processes on strain controlled LCF behavior of Alloy 617. *Metall. Trans. A* **1988**, *19A*, 359–371.
8. Meurer, H.P.; Gnirss, G.K.H.; Mergler, W.; Raule, G.; Schuster, H.; Ullrich, G. Investigations on the fatigue behavior of high temperature gas cooled reactor components. *Nucl. Technol.* **1984**, *66*, 315–323.

9. Totemeier, T.C.; Tian, H. Creep-fatigue interactions in INCONEL 617. *Mater. Sci. Eng. A* **2007**, *468–470*, 81–87. [CrossRef]

10. Carroll, L.J.; Cabet, C.; Wright, R.N. The role of environment on high temperature creep-fatigue behaviour of Alloy 617. In Proceedings of the ASME 2010 Pressure Vessel and Piping Division Conference, Bellevue, WA, USA, 18–22 July 2010; pp. 907–916.

11. Wright, J.K.; Carroll, L.J.; Simpson, J.A.; Wright, R.N. Low cycle fatigue of Alloy 617 at 850 °C and 950°C. *J. Eng. Mater. Technol. ASME* **2013**, *135*, 1–8. [CrossRef]

12. Chen, X.; Sokolov, M.A.; Sham, S.; Erdman, D.L.; Busby, J.T.; Mo, K.; Stubbins, J.F. Experimental and modeling results of creep-fatigue life of Inconel 617 and Haynes 230 at 850 °C. *J. Nucl. Mater.* **2013**, *432*, 94–101. [CrossRef]

13. Kewther Ali, M.; Hashmi, M.S.J.; Yilbas, B.S. Fatigue properties of the refurbished INCO-617 Alloy. *J. Mater. Process. Technol.* **2001**, *118*, 45–49. [CrossRef]

14. Totemeier, T.C. High-temperature creep-fatigue of Alloy 617 base metal and weldments. In Proceedings of ASME 2007 Pressure Vessels and Piping Conference, San Antonio, TX, USA, 22–26 July 2007; pp. 255–260.

15. ASTM E 606-92. Standard Practice for Strain-Controlled Fatigue Testing. In *Annual Book of ASTM Standards*; ASTM International: Baltimore, MD, USA, 2002; p. 569.

16. Hong, J.D.; Lee, J.; Jang, C.; Kim, T.S. Low cycle fatigue behavior of alloy 690 in simulated PWR water—Effects of dynamic strain aging and hydrogen. *Mater. Sci. Eng. A* **2014**, *611*, 37–44. [CrossRef]

17. Gallo, P.; Berto, F.; Lazzarin, P. High temperature fatigue tests of notched specimens made of titanium Grade 2. *Theor. Appl. Fract. Mech.* **2015**, *76*, 27–34. [CrossRef]

18. Louks, R.; Susmel, L. The linear-elastic Theory of Critical Distances to estimate high-cycle fatigue strength of notched metallic materials at elevated temperatures. *Fatigue Fract. Eng. Mater. Struct.* **2015**, *38*, 629–940. [CrossRef]

19. Hong, S.G.; Lee, S.B. Mechanism of dynamic strain aging and characterization of its effect on the low-cycle fatigue behavior in type 316L stainless steel. *J. Nucl. Mater.* **2005**, *340*, 307–314. [CrossRef]

MDPI

St. Alban-Anlage 66

4052 Basel

Switzerland

Tel. +41 61 683 77 34

Fax +41 61 302 89 18

www.mdpi.com

Metals Editorial Office

E-mail: metals@mdpi.com

www.mdpi.com/journal/metals

www.ingramcontent.com/pod-product-compliance
Lightning Source LLC
Chambersburg PA
CBHW051841210326

41597CB00033B/5735